Small-Angle Scattering from Confined and Interfacial Fluids

Yuri B. Melnichenko

Small-Angle Scattering from Confined and Interfacial Fluids

Applications to Energy Storage and Environmental Science

Springer

Yuri B. Melnichenko
Biology and Soft Matter Sciences Division
Neutron Scattering Directorate
Oak Ridge National Laboratory
Oak Ridge, TN, USA

This manuscript has been authored by UT-Battelle, LLC under Contract No. DE-AC05-00OR22725 with the U.S. Department of Energy. The United States Government retains and the publisher, by accepting the article for publication, acknowledges that the United States Government retains a non-exclusive, paidup, irrevocable, world-wide license to publish or reproduce the published form of this manuscript, or allow others to do so, for United States Government purposes. The Department of Energy will provide public access to these results of federally sponsored research in accordance with the DOE Public Access Plan (http://energy.gov/downloads/doe-public-access-plan).

ISBN 978-3-319-34646-5 ISBN 978-3-319-01104-2 (eBook)
DOI 10.1007/978-3-319-01104-2

Springer Cham Heidelberg New York Dordrecht London

Softcover re-print of the Hardcover 1st edition 2016

Printed on acid-free paper

Springer International Publishing AG Switzerland is part of Springer Science+Business Media (www.springer.com)

To Gala, Oleg and Yuri

Preface

The phase behavior of bulk fluids is now well understood and their properties can be predicted accurately using equations of state over a wide range of pressures and temperatures. The behavior of bulk fluids changes dramatically when they are injected into small pores, due to increasing importance of the boundary conditions and molecule-surface interactions. Thus, confinement leads to the emergence of a new set of variables that impact the phase behavior in tight pores but may be neglected in the thermodynamic limit. Examples of such variables are the pore size, shape, and interconnectivity as well as the chemical composition of the pore walls and fluid-surface interaction potential. Due to involvement of numerous system-specific parameters, a comprehensive understanding of the influence of confinement on the fluid behavior is only beginning to emerge.

In addition to their fundamental interest, the ability to understand and predict the phase behavior and dynamics of fluids in natural and engineered porous solids is crucial for a variety of the environment- and energy storage-related technologies. These include the capture and sequestration of anthropogenic greenhouse gases, hydrogen storage, membrane separation of gases, environmental remediation, and catalysis. Until recently, the adsorption of fluids and structure of pores in various porous materials have been routinely explored using volumetric and gravimetric methods, mercury porosimetry, and sorption isotherms. These traditional tools, however, have their limitations. First, they provide data averaged over the entire sample volume and thus fail to elucidate how pores of different sizes contribute to the integral parameters as a function of pressure and temperature. Second, they are invasive, which eliminates contribution from the closed-off regions of the pore space and may affect the integrity of the solid matrix. In contrast, noninvasive small-angle scattering (SAS) techniques offer the unique opportunity to "look but not touch" inside pores and monitor changes in the adsorption behavior of fluid molecules confined in pores of different sizes and topology, as well as to detect the pores inaccessible to the invading fluid. For this reason, a few years ago researchers began to develop and refine scattering techniques and their interpretations as a reliable tool for probing properties of confined and interfacial fluids in natural and engineered porous materials with different structural properties. These efforts resulted in evolving new methods of the SAS data analysis and interpretation as well as developing the new generation of the state-of-the-art high pressure cells that are being used by researchers interested in studying structural and adsorption properties of confined supercritical fluids and gases under pressure.

This book examines the macro-, meso- and microscopic aspects of the fluid behavior in porous solids using noninvasive methods of small-angle neutron and x-ray scattering (SANS and SAXS) as well as ultra small-angle neutron and x-ray scattering (USANS and USAXS). There is a great deal of similarity between the x-ray and neutron scattering and therefore both methods are presented and discussed together so that readers may become familiar with both and appreciate the advantages and disadvantages of each type of radiation for the specific system or type of experiment. Both neutrons and x-rays penetrate porous solids and are scattered on the solid/void interface. At ambient conditions, the resulting scattering patterns are governed by the geometry and topology of the pore space on the scale from about 1 nm to about 10 μm and provide quantitative data about the total porosity, pore size distribution, and the specific area of the scattering interface. SAS experiments performed on fluid saturated samples maintained in cells with controlled pressure and temperature conditions, combined with isotopic substitution of invading fluids, facilitate contrast variation experiments. The simplest application—direct contrast matching between the solid matrix and the invading fluid—discriminates between the open (accessible) and closed (inaccessible) porosity. Both SANS/USANS and SAXS/USAXS enable the observation of pore-size-specific invasion of the pore space by fluids and help to access important information on the volume fraction of the adsorbed phase and its average physical density. For geological samples, this can be done *in situ* at a subsurface-like temperature and pressure conditions.

The book is meant as a reference for active researchers in the field, but also may serve as a comprehensive guide for university faculty members and students, who may be insufficiently aware of the range of opportunities provided by the small-angle scattering techniques. The book commences with introductory chapters, which describe major principles of SAS techniques and are sufficiently comprehensive to be useful to researchers interested in structural characterization of various types of materials in different fields of science. Chapters 1–4 introduce the basic properties of neutrons and x-rays, provide brief description of the available neutron and x-ray sources, and give illustrative examples of SAS instrumentation and sample environment. This is followed by discussion of the practical aspects of SAS experiments, sample preparation methods, optimal instrument configurations, and basic principles of the data reduction and analysis presented in Chaps. 5 and 6. Chapter 7 deals with the SAS structural characterization of various porous solids, and Chap. 8 describes studies of confined vapors below saturating pressure with the emphasis on vapor adsorption and capillary condensation. Chapter 9 is concerned with studies of confined and interfacial liquids. Experimental SAS investigations of high-pressure adsorption of supercritical fluids and gases in various engineered and natural porous materials are discussed in Chap. 10.

The author has enjoyed and benefitted from longstanding collaboration with T.P. Blach, N.C. Gallego, C.I. Contescu, M. Mastalerz, J.R. Morris, A.P. Radlinski, J.A. Rupp, L.F. Ruppert, R. Sakurovs, and G.D. Wignall. Special thanks are due to my younger colleagues J. Bahadur, S.M. Chathoth, G. Cheng, and L. He who contributed their enthusiasm and talent in many studies of confined fluids described in this book. It is a great pleasure to acknowledge M.M. Agamalian, J.M. Carpenter, A.P. Radlinski, and G.D. Wignall for reading select chapters of the manuscript and offering valuable comments. The assistance of Reneé Manning and Genevieve Martin in preparing high quality artwork is greatly appreciated.

Oak Ridge, TN, USA Yuri B. Melnichenko

Contents

1 **Basic Definitions and Essential Concepts of Small-Angle Scattering** . . . 1
1.1 Interaction of X-Rays and Neutrons with Matter; Scattering Length . . . 1
1.2 Scattering Vector and Scattering Cross Section . . . 4
1.3 Scattering Length Density and Contrast . . . 7
1.4 Absorption and Transmission of X-Ray and Neutron Beams . . . 11
1.5 Form and Structure Factors . . . 12
1.6 Complementarity of Neutron and X-Ray Scattering Techniques . . . 14
References . . . 17

2 **Radiation Sources** . . . 19
2.1 Steady State Reactors . . . 19
2.2 Spallation Neutron Sources . . . 23
2.3 Photon Sources . . . 29
References . . . 32

3 **Constant Flux and Time-of-Flight Instrumentation** . . . 35
3.1 General Purpose SANS Instrument at HFIR, ORNL . . . 35
3.2 Extended Q SANS Instrument at SNS, ORNL . . . 43
3.3 Perfect Crystal USANS Instrument at NIST . . . 46
3.4 Time-of-Flight USANS at SNS . . . 50
3.5 SAXS and USAXS Instruments at APS, ANL . . . 51
References . . . 55

4 Sample Environment 57
4.1 Sample Cells for Ambient Conditions 57
4.2 SANS High-Pressure Cells 59
4.3 SAXS High-Pressure Cells 63
References 67

5 Practical Aspects of Planning and Conducting SAS Experiments 69
5.1 Applying for Beam Time 69
5.2 Choice of the Instrument Configuration 71
5.3 Detector Sensitivity and Instrument Backgrounds 76
5.4 Optimal Sample Thickness, Transmission, and Multiple Scattering 77
5.5 Subtraction of the Sample Background 79
5.6 Data Acquisition Time, Masking, and Radial Averaging 81
5.7 Absolute Calibration 83
5.8 Instrument Resolution 85
5.9 Effective Thickness of Powder Samples 89
5.10 Contrast Variation with Liquids and Gases 92
5.11 Average Scattering Length Density of Multicomponent Samples 100
References 104

6 Fundamentals of Data Analysis 109
6.1 Correlation Functions: Mathematical Form and Geometrical Meaning 109
6.2 Scattering from Two-Phase Random Systems: The Porod Invariant 113
6.3 Asymptotic Behavior: The Porod Law 115
6.4 Radius of Gyration 117
6.5 Asymptotic Behavior: The Guinier Approximation 117
6.6 Structural Parameters of the Two-Phase Porous Medium 119
6.7 Bridging the Asymptotic Behavior: The Unified Scattering Function 120
6.8 Scattering from Fractal Systems and the Polydisperse Spherical Model 122
6.8.1 Scattering from Mass, Surface, and Pore Fractals 122
6.8.2 Polydisperse Spherical Model 126
6.9 Beyond the Two-Phase Model 127
6.9.1 Partial Scattering Functions of Multiphase Systems 127
6.9.2 Scattering Contrast and the Invariant of a Three-Phase System 130
6.9.3 Oscillatory Deviations from the Porod Law 131
6.10 Interrelation Between the Reciprocal and Real Space 134
References 136

7 **Structural Characterization of Porous Materials Using SAS** 139
7.1 Porous Media for Energy, Environmental, and Biomedical Applications 139
7.2 Porous Silica 147
7.2.1 Porous Vycor Glass 147
7.2.2 Silica Aerogels 148
7.2.3 Porous Fractal Silica 149
7.2.4 Ordered Mesoporous Silica 150
7.3 Porous Carbons 155
7.3.1 Activated Carbons 155
7.3.2 Glassy Carbon 158
7.3.3 Carbon Aerogel 162
7.4 Alumina Membranes 165
7.5 Porous Polymer Monoliths 165
7.6 Ceramics, Alloys, and Composite Materials 166
7.7 Structure of Sedimentary Rocks 167
References 168

8 **Neutron and X-Ray Porosimetry** 173
8.1 Principles of the Scattering-Based Porosimetry 173
8.2 Structure of Nanoporous Low-Dielectric-Constant Films 176
8.3 Vapor Adsorption in Porous Silica 180
8.3.1 Contrast Matching SANS 180
8.3.2 Synchrotron SAXS 187
8.4 Carbonaceous Materials 189
8.5 Kinetics of Sorption and Desorption 194
8.5.1 Dynamic Micromapping of CO_2 Sorption in Coal 195
8.5.2 Vapor Adsorption in MCM-41 198
8.5.3 Vapor and Water Uptake in Nafion Membranes 199
References 201

9 **Individual Liquids and Liquid Solutions Under Confinement** 205
9.1 Confined Electrolytes 205
9.1.1 Ion Adsorption in Electrolyte Saturated Porous Carbons 206
9.1.2 Ionic Liquids Under Confinement 210
9.2 Detection of the Oil Generation in Hydrocarbon Source Rocks 213
9.3 Cavitation on Hydrophobic Nanostructured Surfaces 217
9.4 Liquid–Liquid Demixing in Mesopores 232
9.5 Supercooled Water in Confined Geometries 238
9.6 Order-Disorder Transitions in Liquid Crystals 240
References 246

10 Supercritical Fluids in Confined Geometries 251
10.1 Specifics of the Supercritical Fluid Adsorption 251
10.2 Density Fluctuations Near the Liquid–Gas Critical Point of Confined Fluids 254
10.3 Adsorption of Supercritical CO_2 in Porous Silica 258
10.3.1 Silica Aerogels 258
10.3.2 Porous Fractal Silica 267
10.4 Methane in Porous Carbons 271
10.5 Hydrogen Storage in Activated Carbons 276
10.6 CO_2 Sequestration in Coal 282
10.7 Pore Interconnectivity and Accessibility to Fluids in Coal and Shales 289
10.8 Structural Stability of Porous Materials Under Pressure 296
Appendix: Derivation of the Eq. (10.23) for Accessible Porosity 300
References 305

Index 311

Author Biography

Yuri B. Melnichenko was educated as a physicist in the USSR, receiving his Ph.D. from Kiev State University (1984) and later a prestigious Doctor of Physics and Mathematics degree from the Academy of Sciences of USSR (1992). He is a Humboldt Foundation Fellow (1993, Germany) and a recipient of Max Planck Society award (1994, Germany). Visiting researcher, the Max Planck Institut für Polymerforschung (Mainz, Germany), in 1993–1995. Since 1995—a research staff member, Oak Ridge National Laboratory. He conducts research in the field of soft matter materials and confined fluids using small-angle and quasi-elastic neutron scattering techniques and is an author and coauthor of more than 160 peer-reviewed scientific articles. Co-editor of the book "Computational Studies, Nanotechnology, and Solution Thermodynamics of Polymer Systems" (Kluwer Academic/Plenum Publishers, New York, 2001). In 2005 elected a Fellow of American Physical Society for "Significant contribution to the fundamental science underlying universal aspects of macromolecules in polymer solutions, supercritical mixtures and polymer blends." Most recent interests are in the area of high pressure adsorption and dynamics of fluids confined in pores of engineered and natural porous materials.

Abbreviations

1D	One dimensional
2D	Two dimensional
ACF	Activated carbon fiber
AFM	Atomic force microscopy
ANL	Argonne National Laboratory
ANSTO	Australian Nuclear Science and Technology Organisation
APS	Advanced Photon Source
ASAXS	Anomalous SAXS
BENSC	Berlin Neutron Scattering Center
BET	Brunauer-Emmett-Teller
Bio-SANS	Biological SANS instrument
BJH	Barrett, Joyner, and Halenda
CCD	Charge-coupled device
CDC	Carbide-derived carbon
CG2	Cold guide 2
CG3	Cold guide 3
CHESS	Cornell High Energy Synchrotron Source
CHRNS	Center for High Resolution Neutron Scattering
CLD	Chord length distribution
CMS	Contrast-matched solution
CM-SANS	Contrast matching SANS
CO_2	Carbon dioxide
COF	Covalent organic frameworks
CPG	Controlled porous glass
CR	Count rate
DCD	Double-crystal diffractometer
DFT	Density functional theory
EC	Empty cell
ECBM	Enhanced coalbed methane
EDLC	Electrochemical double layer capacitor
EP	Ellipsometric porosimetry

EQ-SANS	Extended Q-range small-angle neutron scattering
Erf	Error function
ESRF	European Synchrotron Radiation Facility
ESS	European Spallation Source
FTIR	Fourier transform infrared
FWHM	Full width at half maximum
Gi-SANS	Grazing incidence SANS
GP-SANS	General-Purpose Small-Angle Neutron Scattering
GUI	Graphical user interface
HANARO	High-Flux Advanced Neutron Application Reactor
HB	Horizontal beam tube
HFIR	High Flux Isotope Reactor
HFR	High Flux Reactor
HSQ	Hydrogen silsesquioxane
iBA	Isobutyric acid
ID	Inner diameter
ILL	Institut Laue-Langevin
J-PARC	Japan Proton Accelerator Research Complex
KEK	High Energy Accelerator Research Organization
KP	Kirste-Porod
LC	Liquid crystal
$^{6}Li_2CO_3$	Lithium-6 carbonate
linac	Linear accelerator
MM	Mineral matter
MOF	Metal organic frameworks
MOP	Microporous organic polymers
MPCVD	Microwave plasma-assisted chemical vapor deposition
MSQ	Methylsilsesquioxane
NCNR	National Center for Neutron Research
NG	Neutron guide
NIST	National Institute of Standards and Technology
NMR	Nuclear magnetic resonance
NR	Neutron reflectometry
OD	Outside diameter
ORNL	Oak Ridge National Laboratory
PCP	Porous coordination polymer
PDSP	Polydisperse spherical pore (model)
PFAC	Polyfurfuryl alcohol-derived activated carbon
PFS	Porous fractal silica
PFSA	Perfluorinated sulfonic acid
PMMA	Poly(methyl methacrylate)
PSI	Paul Scherrer Institute
QENS	Quasi-elastic neutron scattering
RF	Resorcinol formaldehyde

RFIM	Random-field Ising model
RTD	Resistance temperature detector
SANS	Small-angle neutron scattering
SANSADSO	SANS adsorption
SAS	Small-angle scattering
SAX	Small-angle x-ray scattering
SC	Supercritical
SCF	Supercritical fluid
SDD	Sample-to-detector distance
SEM	Scanning electron microscopy
SILP	Supported ionic liquid phase
SLD	Scattering length density
SNS	Spallation Neutron Source
SRS	Short-range smectic
SSD	Sample-source distance
STEM	Scanning transmission electron microscopy
SXR	Specular x-ray reflectometry
TEATFB	Tetraethylammonium tetrafluoroborate
TEM	Transmission electron microscopy
TiC-CDC	Titanium carbide-derived carbon
TOC	Total organic carbon
TOF	Time of flight
UMC	Ultramicroporous carbon
USANS	Ultrasmall-angle neutron scattering
USAS	Ultrasmall-angle scattering
USAXS	Ultrasmall-angle x-ray scattering
VSANS	Very small-angle SANS
ZAC	Zero average contrast

Chapter 1
Basic Definitions and Essential Concepts of Small-Angle Scattering

This chapter provides introductory information about the specifics of x-ray and neutron interactions with matter, as well as the basics of small-angle scattering (SAS). The scattering principles of x-rays and neutrons are discussed based on a single approach. Although the physics of small-angle light scattering is similar to that of x-rays and neutrons, it is not discussed here because it cannot be used to study confined fluids in generally nontransparent porous solids. In most cases formula derivations have been intentionally omitted as they are already found in the literature. The text presented is not intended as a substitute for general books and chapters on scattering and diffraction theory, and an extensive list of advanced texts and World Wide Web resources for further reading is given at the end of the chapter.

1.1 Interaction of X-Rays and Neutrons with Matter; Scattering Length

X-rays are electromagnetic waves that are scattered by electrons. The x-ray scattering process can be thought of as involving absorption of the x-ray photon, accompanied by excitation of the electronic system, immediately followed by de-excitation and reemission of an x-ray photon. Because every electron has the same x-ray scattering amplitude (also known as "scattering length" b_X), each element scatters proportionally to its number of electrons:

$$b_X = b_0 Z, \tag{1.1}$$

where $b_0 = 0.282\ 10^{-12}$ cm is the Thompson scattering factor of one electron and Z is the atomic number (the number of electrons in the electronic shell of an atom). X-ray scattering lengths increase rapidly with atomic number (Table 1.1, Fig. 1.1).

Y.B. Melnichenko, *Small-Angle Scattering from Confined and Interfacial Fluids*, DOI 10.1007/978-3-319-01104-2_1

Table 1.1 Coherent (bound) neutron (b) and x-ray (b_X) scattering lengths, coherent (σ_{coh}) and incoherent (σ_{inc}) neutron cross sections, neutron (σ_{abs}) and x-ray ($\sigma_{abs,\,X}$) absorption coefficients of some common elements and isotopes

Atom	Neutrons, b, (10^{-12} cm)	X-rays, b_X, (10^{-12} cm)	Z	$\sigma_{coh}=4\pi b^2$ (barn)	σ_{inc} (barn)	σ_{abs} (barn)	$\sigma_{abs,\,X}$ (barn)
H	−0.374	0.282	1	1.76	79.7	0.33	0.655
D	0.667	0.282	1	5.59	2	0.0005	0.655
C	0.665	1.692	6	5.6	0	0.0035	89.9
N	0.94	1.974	7	11.1	0.3	1.9	173
O	0.58	2.256	8	4.23	0	0.0002	304
Al	0.345	3.666	13	1.49	0	0.23	2220
Si	0.415	3.948	14	2.16	0	0.17	2970

Z is atomic number. The absorption values for neutrons and x-rays are for $\lambda = 1.8$ Å and $\lambda = 1.54$ Å, respectively [1, 4]. 1 barn $= 10^{-24}$ cm^2

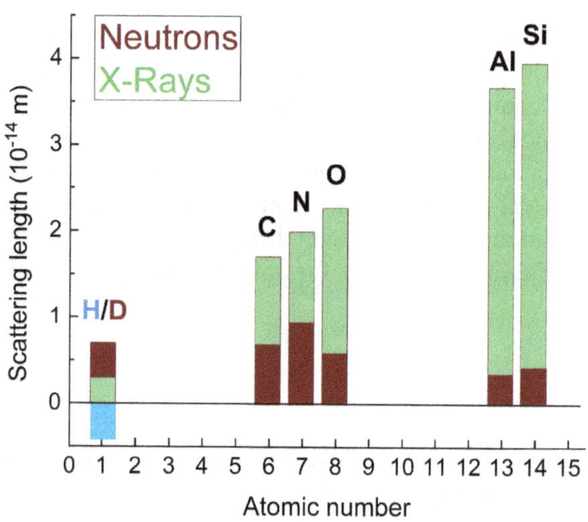

Fig. 1.1 X-ray and neutron scattering lengths of atoms listed in Table 1.1. For x-rays, the scattering length is proportional to the atomic mass/number of electrons. Neutron scattering lengths vary irregularly with the atomic number and can be different for different isotopes of each element. Scattering length of hydrogen (H, marked in *blue*) and deuterium (D) differ by both sign and amplitude

This means that the scattering pattern measured during an x-ray experiment will be mostly defined by the scattering from heavy atoms. Neutrons represent uncharged elementary particles, and their scattering centers are the nuclei of atoms. Each atom has a characteristic neutron scattering length (b). The magnitude of b determines the strength of neutron scattering, and its sign indicates whether the incident and scattered waves are in or out (180°) of phase.

Neutron scattering lengths are determined by quantum mechanics of the neutron–nucleus interaction. They are isotope-dependent and vary irregularly across the periodic table. To a good approximation, neutron scattering lengths are independent of their wavelength (λ) for low-energy (cold and thermal) neutrons. The problem of ab initio calculation of b has not yet been resolved, and the values of b are determined experimentally and then tabulated (e.g., see [1]).

The scattering length of a complex scattering object is the sum of the scattering lengths of its constituent atoms for x-rays and neutrons:

$$b = \sum_{i=1}^{n} b_i. \tag{1.2}$$

For example, the x-ray scattering length of SiO_2 can be calculated as $b_X = (3.948 + 2 \times 2.256) \times 10^{-12}$ cm $= 8.46 \times 10^{-12}$ cm. At the same time, the neutron scattering length of SiO_2 is $b = (0.415 + 2 \times 0.58) \times 10^{-12}$ cm $= 1.575\ 10^{-12}$ cm. As can be seen in Table 1.1, the x-ray scattering lengths of hydrogen and deuterium are identical, whereas their neutron scattering lengths differ significantly and the value for b of hydrogen is actually negative, while that of deuterium is positive. The negative scattering length of hydrogen, as well as some other atoms such as lithium, titanium, and vanadium, indicates that neutrons scattered from these elements are subject to a 180° phase shift compared with other atoms. The resulting large difference in scattering length between hydrogen and deuterium forms the basis of deuterium labeling techniques that represents a unique feature of neutron scattering, giving it considerable advantages over corresponding x-ray methods.

Another important difference between x-ray photons and neutrons is their energy. The energy of an x-ray photon E_{ph} with the wavelength $\lambda = 1.54$ Å is $E_{ph} = 12.398/1.54 \cong 8$ keV. The kinetic energy of a thermal neutron is almost six orders of magnitude lower and close to the energy associated with atomic motions: $E_{am} = 0.0867\,T \cong 25$ meV for temperatures T around 300 K. Therefore, the energy exchange between x-rays and atoms might be neglected, and small-angle x-ray scattering (SAXS) experiments are always conducted in "elastic" mode under conditions when there is no energy exchange between radiation and molecules. The same assumption (i.e., elasticity of scattering) also applies to small-angle neutron scattering (SANS) experiments to a good approximation. However, because of their much lower energy, in principle neutrons can exchange their energy with atoms to a degree that can be measured experimentally. Such experiments are performed using specially designed quasi-elastic and inelastic neutron scattering spectrometers that deliver information on diffusion, as well as on the internal vibrational and rotational modes of atoms and molecules [2]. Finally, because neutron has a magnetic moment, it can interact with the magnetic moments of unpaired electrons, making it an excellent probe for determination of the static and dynamic magnetic properties of matter such as magnetic ordering phenomena, magnetic excitations, spin fluctuation, etc. [3]. Discussion of this type of scattering, however, is beyond the scope of this book.

1.2 Scattering Vector and Scattering Cross Section

Both x-rays and neutrons exhibit wave-particle duality. Whether neutrons and x-rays appear to behave as particles or as waves depends on the type of experiment being conducted. X-rays are scattered elastically by electrons, and neutrons are scattered elastically or inelastically, coherently or incoherently, by nuclei. In SAS experiments, the only concern is with the elastic component of scattering. The observed scattering profiles depend on the phase difference between coherent neutron or x-ray waves scattered at different points of the sample.

The principal aim of an SAS experiment is to determine the probability that a neutron or a photon, which is incident on the sample with wave vector $\boldsymbol{k_0}$, is scattered into the state with wave vector $\boldsymbol{k}$. The intensity of scattered radiation is thus measured as a function of the momentum transfer (Fig. 1.2):

$$\hbar \boldsymbol{Q} = \hbar(\boldsymbol{k} - \boldsymbol{k}_0), \tag{1.3}$$

where $\boldsymbol{Q}$ is known as the scattering vector. The magnitude of a wave vector is defined as its wave number: $|\boldsymbol{k}| = k = 2\pi/\lambda$, where λ is the wavelength. For the elastic scattering process, $|\boldsymbol{k}| = |\boldsymbol{k}_0| = k$ and the magnitude of scattering vector $\boldsymbol{Q}$ (sometimes called momentum transfer) is given by

$$|\boldsymbol{Q}| = Q = \frac{4\pi}{\pi} \sin \Theta, \tag{1.4}$$

where 2Θ is the scattering angle. It completely characterizes the scattering geometry—the incident and scattered beam directions as well as the wavelength (Fig. 1.2). The majority of literature on neutron scattering gives the scattering vector in terms of $\boldsymbol{Q}$, whereas researchers working on x-ray diffraction prefer a different definition of the scattering vector, designated as $\boldsymbol{s}$:

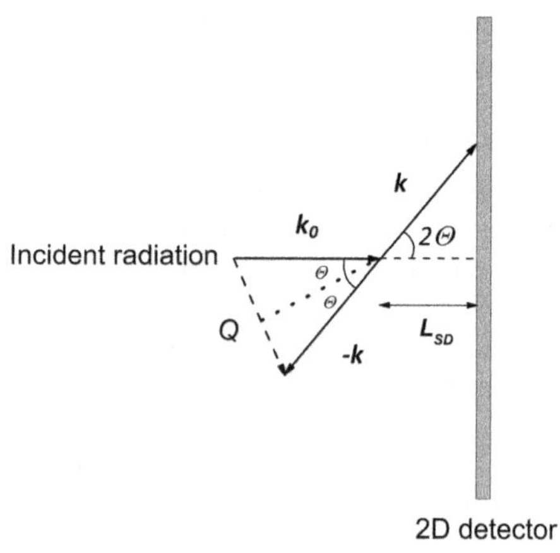

Fig. 1.2 The vector diagram for elastic SAS experiment, when $|\boldsymbol{k}_0| = |\boldsymbol{k}| = 2\pi/\lambda$. A fraction of x-rays or neutrons scattered through angle 2Θ are recorded on a 2D detector at distance L_{SD}

$$|\boldsymbol{s}| = s = \frac{2}{\lambda}\sin\Theta, \tag{1.5}$$

From which follows that $Q = 2\pi s$.

The previously introduced scattering length of each element is related to another important variable—cross section (σ). Assuming that flux of neutrons or photons incident on the sample is Φ_0 [particles s^{-1} cm^{-2}], a number of scattering events occurring each second (scattering intensity I_S) may be defined through the relation

$$I_S = \Phi_0\sigma, \tag{1.6}$$

where σ has the dimension of an area and a commonly used unit for such areas is called "barn": 1 barn $= 10^{-24}$ cm^2.

As discussed in Sect. 1.1, practically any sample is composed of several atomic species, i, each of which represents a mixture of several isotopes that possess a nuclear spin. Because the neutron scattering power depends not only on the nature of the nucleus but also on the total spin of the nucleus–neutron system, neutron scattering lengths are different for different isotopes. The scattering lengths of isotopes are characterized by an average value and a standard deviation:

$$b = \langle b \rangle \pm \Delta b. \tag{1.7}$$

The average scattering length $\langle b \rangle$ over all isotopes and spin state populations is called the coherent scattering length. At the same time, the incoherent scattering length is defined as the root mean square deviation of b from $\langle b \rangle$. The coherent and incoherent scattering cross sections are related to the corresponding scattering lengths as follows:

$$\sigma_{coh} = 4\pi\langle b \rangle^2. \tag{1.8}$$

$$\sigma_{inc} = 4\pi\langle \Delta b \rangle^2. \tag{1.9}$$

The coherent and incoherent cross sections are calculated from the bound (i.e., not free) scattering lengths for nuclei. The actual cross sections depend on the incoming neutron energy (wavelength), especially for light elements. For low-energy neutrons (i.e., cold neutrons), the tabulated cross sections are generally a lower limit. If the isotope has no spin (e.g., ^{12}C or ^{16}O), then $b^2 = \langle b \rangle^2 = \langle \Delta b^2 \rangle$ and there is no incoherent scattering.

Only the coherent scattering cross section contains information on the interference effects arising from spatial correlation of the scattering elements in the system, in other words, the structure of the sample. The incoherent cross section contains no information on interference effects and forms an isotropic background that must be subtracted from the raw data. The incoherent component of the scattering does, however, contain information on the motion of single atoms and molecules, which can be investigated by studying the changes in energy of the incident beam [2].

Compilations of the scattering lengths and cross sections can be readily found in the literature (e.g., [1, 4]) and on the World Wide Web (e.g., www.ncnr.nist.gov/resources/n-lengths/). Values of b and σ for some selected nuclei are given in Table 1.1.

Unlike neutrons, for photons there is no strict analog of incoherent scattering. X-ray Compton scattering is similar in that it contains no information on interference effects (i.e., the structure of the sample) and contributes a background to the coherent signal. However, in the SAS limit $Q \Rightarrow 0$, this background goes to zero and can be neglected to a good approximation.

The aim of the SAS experiment is to measure the flux of the scattered beam Φ_S normalized to the flux of the incident neutrons Φ_0 as a function of scattering angle (or scattering vector) and, based on this variation, make some conjectures about the structure of the sample. Φ_0 [particles s^{-1} cm^{-2}] refers to a plane incident wave and Φ_S [particles s^{-1}sterad^{-1}] to a spherical scattered wave. Hence, the ratio Φ_S/Φ_0 has dimension of area per solid angle and is called *the differential scattering cross section*:

$$\frac{d\sigma}{d\Omega} = \frac{\Phi_S}{\Phi_0}\left[\frac{cm^2}{sterad}\right]. \tag{1.10}$$

It represents the probability that a photon or neutron is scattered into a unit solid angle in the given direction (Fig. 1.3). Integrating the differential scattering cross section throughout the solid angle Ω gives an equation that describes the total scattering cross section that has dimension of area:

$$\begin{aligned}\sigma_{tot} &= \int_0^{2\pi}\int_0^{\pi}\left(\frac{d\sigma}{d\Omega}\right)\sin\Theta\, d\Theta\, d\Phi \\ &= \frac{\text{total number of particles scattered in all directions per second}}{\text{flux of the incident beam}}.\end{aligned} \tag{1.11}$$

Even after normalizing to the incident flux, the same sample will show different differential scattering cross sections depending on its size. It is therefore useful to introduce yet another normalization factor $1/V$, where V is the sample volume illuminated by the beam:

$$\frac{d\Sigma}{d\Omega}(Q) = \frac{1}{V}\frac{d\sigma}{d\Omega}(Q). \tag{1.12}$$

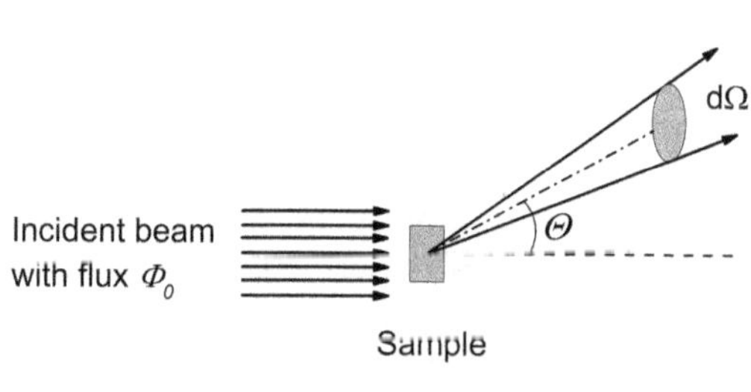

Fig. 1.3 Schematic of the SAS experiment

As defined, the differential scattering cross section per unit volume has the dimension of cm^{-1} $sterad^{-1}$. In the scattering community it is common to omit the dimensionless unit of the solid angle and present scattering curves in so-called absolute units of cm^{-1}. Note that the differential cross section per unit volume is often, and incorrectly, referred to in the literature as the intensity of scattering, and the scattering data are represented by the symbol $I(Q)$. Use of this symbol is justified only if it is clearly stated that it denotes the differential neutron scattering cross section per unit volume (in units of cm^{-1}), in which case $I(Q) \equiv d\Sigma/d\Omega(Q)$. With this reservation, in this book the terms intensity and differential cross section are used interchangeably.

1.3 Scattering Length Density and Contrast

Neutron and x-ray scattering techniques have been widely employed to provide information on the spatial arrangements of building blocks in random materials and to determine crystal structures using Bragg's law:

$$\lambda = 2D \sin \Theta, \tag{1.13}$$

where D is the distance between crystallographic planes. Combining Eqs. (1.4) and (1.13) gives

$$D = 2\pi/Q. \tag{1.14}$$

Hence, to study structural inhomogeneities on length scales of ~1,000 Å, one needs to measure scattering at Q-values as low as of $\sim 10^{-3}$ Å^{-1} and collect data at small angles ($\Theta < 10°$) using long-wavelength ($5 < \lambda < 20$ Å), that is "cold" neutrons. At the same time, investigation of structural properties on a short length scale of ~10 Å requires taking data at much wider scattering angles ($\Theta \sim 40°$).

In many engineered and natural materials such as porous solids, bones, cements, rocks, and clays, structural inhomogeneities can span length scales from nanometers to microns, and a complete characterization of such multi-hierarchical structures at all levels requires the application of complementary neutron and x-ray scattering methods with adequate resolution. Modern SANS and SAXS instruments can probe length scales from few Angstroms to about 1000 Å. Moreover, ultrasmall-angle neutron scattering (USANS) and ultrasmall-angle x-ray scattering (USAXS) instruments can extend the characterization capabilities of length scales up to several microns [5]. This range can be further extended up to a macroscopic scale (millimeters and centimeters) by the use of non-diffractive neutron and x-ray imaging methods (radiography and tomography). Because of their high penetration power, and high interaction probability with light elements in general, and hydrogen in particular—neutrons are especially suitable for nondestructive imaging and resolution of small amounts of organic substances inside various inorganic materials [6].

As mentioned previously, the main objective of an SAS experiment is to determine the differential scattering cross section per unit volume $d\Sigma/d\Omega(Q)$ since this quantity contains information on the shape, size, and correlations between the scattering centers in the sample. It should be emphasized that the non-zero SAS signal can be detected only if the scattering power fluctuates from point to point in the sample, as, in general, radiation incident on a medium whose scattering power is independent of position is scattered only in the forward direction ($\theta = 0$) and all scattering at $\theta \neq 0$ cancels out.

The scattering lengths of several neighboring atoms can be averaged over volume with dimensions on the order of the wavelength of the radiation and in the limit of low and intermediate Q SAS can be considered as scattering from continuum. On the other hand, in the limit of high $Q \gg 1$ Å^{-1}, scattering occurs from individual atoms (neutron or x-ray crystallography). The crossover between the continuum and single atom scattering regimes is not well defined. It should correspond to situations where the volume of the building block in the system is small enough that a constant average value of the scattering length cannot be defined, which makes the SAS theory inapplicable.

The scattering power of the volume υ is proportional to the *scattering length density* (SLD), which is defined for neutrons as

$$\rho^* = \frac{\sum_{i=1}^{n} b_i}{\upsilon}, \tag{1.15}$$

where b is a bound coherent scattering length and the summation is over every atom (i) in a molecule. In turn, for x-rays the SLD (which is proportional to the electron density) is simply

$$\rho_X^* = \frac{\sum_{i=1}^{n} b_0\, Z}{\upsilon}. \tag{1.16}$$

The reference volume υ can be chosen as the volume of a molecule (e.g., water or SiO_2 or CO_2) and be calculated as:

$$\upsilon = \frac{\sum_{i=1}^{n} M_i}{N_A \rho}, \tag{1.17}$$

where M_i is atomic mass of the constituent atoms, ρ the physical density, and N_A Avogadro's number. Neutron SLDs of several solids and liquids are given in Table 1.2.

The amplitude of the differential scattering cross section $d\Sigma/d\Omega(Q)$ is proportional to the (neutron or x-ray) contrast factor that is defined as the square of the difference of the SLD of a scattering object and the surrounding medium:

Table 1.2 Neutron scattering lengths (b), physical densities (ρ), molecular volumes (υ), and scattering length densities $(\rho^*)_n$ of some solids and liquids

Atom or molecule	Formula	Scattering length, b, (10^{-12} cm)	Physical density, ρ, (g/cm^3)	Molecular volume, υ, (10^{-24} cm^3)	$(\rho^*)_n$ (10^{10} cm^{-2})	$(\rho^*)_X$ (10^{10} cm^{-2})
Silicon	Si	0.415	2.33	20	2.075	20.07
Silicon dioxide:	SiO_2	1.575				
Fused			2.2	45.5	3.47	18.87
Quartz			2.65	37.7	4.18	29.41
Carbon:	C	0.665				
Amorphous			1.8	11.1	6	15.31
Graphite			2.267	8.82	7.55	19.28
Diamond			3.515	5.69	11.7	29.89
Aluminum oxide	Al_2O_3	2.43	3.95	43	5.67	33.25
Water	H_2O	−0.168	1.0	30	−0.56	9.47
Heavy water	D_2O	1.914	1.11	30	6.39	9.45
Toluene	C_7H_8	1.663	0.87	177	0.94	8.1
Toluene-d8	C_7D_8	9.991	0.943	177	5.64	8.0
Methanol	CH_4O	−0.251	0.792	67.3	−0.373	7.58
Methanol-d4	CD_4O	3.91	0.888	67.6	5.79	7.55
Acetone	C_3H_6O	0.331	0.791	122	0.27	7.42
Acetone-d6	C_3D_6O	6.577	0.872	122	5.39	7.41
Hexane	C_6H_{14}	−1.246	0.655	217	−0.571	6.46
Hexane-d14	C_6D_{14}	13.328	0.767	217	6.39	6.51

For deuterated materials, it is assumed that the number of molecules per unit volume is independent of deuteration. X-ray scattering length densities $(\rho^*)_X$ are calculated for λ = 1.54 Å corresponding to photon energy 16 keV. Online calculators are available at http://www.ncnr.nist.gov/resources/sldcalc.html and http://sld-calculator.appspot.com/

$$\frac{d\Sigma}{d\Omega}(Q) \sim (\Delta\rho^*)^2 = \left(\rho_1^* - \rho_2^*\right)^2. \tag{1.18}$$

Consider porous glass that consists of an SiO_2 matrix with air-saturated pores. In this case there exists a significant contrast between the SLD of air (which is virtually zero as the physical density of air is ~0) and the SLD of a silica matrix (the neutron SLD = 3.47×10^{10} cm^{-2}, and the x-ray SLD = 18.61×10^{10} cm^{-2}). The presence of pores generates strong fluctuations between the SLD of the regions consisting of SiO_2 and air, which results in a strong SAS signal, with the scattering power of x-rays exceeding that of neutrons by a factor of ~5.

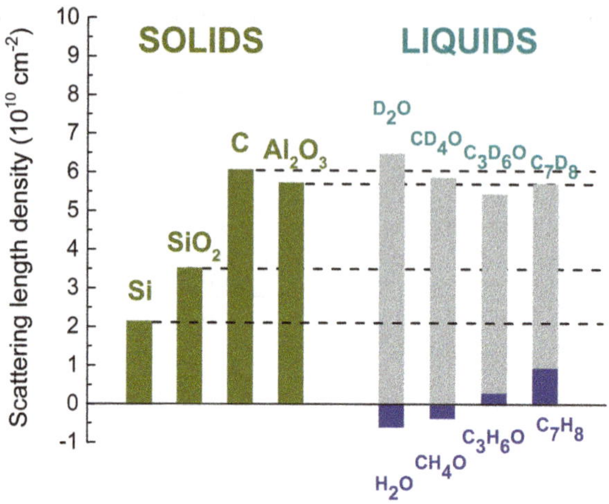

Fig. 1.4 Neutron SLDs of a few solids and organic solvents (protonated and deuterium substituted analogues). As shown in the figure, the SLD of silicon and SiO_2 can be matched by (H + D) mixing any solvents at a certain contrast-matching volume fraction of the components (ϕ_{CM}), whereas the SLD of carbon and aluminum oxide can be matched only by (H + D) mixtures of water, methanol, and toluene at corresponding ϕ_{CM}. Such contrast matching cannot be conducted using x-rays as the electron densities of h- and d-solvents are practically the same and a factor of 3–5 lower than that of SiO_2 or Al_2O_3 (Table 1.2)

Neutron scattering from porous glass can be completely eliminated if all pores are saturated with a fluid (liquid or gas) that has the same SLD as that of SiO_2, for example, a mixture of toluene and toluene-d8 with an average SLD equal to that of Al_2O_3 (i.e., 3.47×10^{10} cm^{-2}) (Table 1.2 and Fig. 1.4). From an experimental point of view this situation corresponds to the *contrast matching condition* at which $\rho_1^* = \rho_2^*$ and therefore $d\Sigma/d\Omega(Q) \Rightarrow 0$ in Eq. (1.18). Changing contrast (contrast variation) between the constituting system elements (or phases) is a powerful tool that can help deliver information about the sample structure. In both SANS and SAXS it provides a mean of dramatically simplifying interpretation of the scattering patterns from multicomponent systems.

As an example, imagine that one needs to determine the shape and size of a small fraction of nanoparticles (e.g., Al_2O_3) dispersed in a silica matrix of porous glass. If the size of the pores is much larger than that of the nanoparticles, the SAS signal from this three-component system will be dominated by scattering from pores as the SAS signal increases with the linear size of the scattering objects as a power of 6 (see Sect. 1.5). However, if the pores are filled with a contrast matching solution, neutron contrast between the SiO_2 matrix and pores disappears and the measured SAS signal will contain structural information about nanoparticles only. Equally, contrast variation can be used to determine the SLD (and thus physical density) of one of the components in a binary (or two-phase) system if the density of the other component is known.

Contrast variation using deuterium substitution is only possible using SANS because of the very large difference in the scattering length of deuterium and

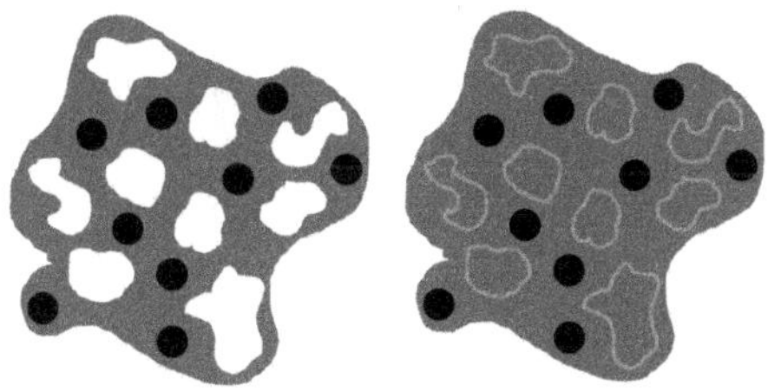

Fig. 1.5 SiO_2 matrix (*grey*) with few spherical nanoparticles (*black*) and pores (*white*). Filling pores with a contrast matching mixture of protonated and deuterated solvent eliminates scattering from pores and therefore allows for analyzing shape and spatial distribution of nanoparticles directly from the measured SAS signal

hydrogen. These isotopes are indistinguishable to x-rays (Table 1.1 and Fig. 1.1). Neutron SLDs of common solids, as well as solvents typically used for contrast matching experiments of porous materials, are shown in Table 1.2 and are illustrated in Fig. 1.5. Examples of experimental use of the contrast variation method are given in Sect. 5.10.

1.4 Absorption and Transmission of X-Ray and Neutron Beams

When radiation penetrates a sample, a fraction of the incident beam is scattered, and the x-ray photons or neutrons change their direction of propagation. Another fraction of the incident particles might be absorbed and converted into other forms of energy. As a result, the transmitted beam that propagates in the original incident direction is attenuated because of the scattering and absorption. When a neutron is absorbed, it becomes trapped by the nucleus. A compound nucleus forms an excited state and it decays into the ground state, in most cases, by emitting γ rays or α particles. The compound nucleus can decay by fission, like uranium in a reactor core. Absorption of x-rays occurs mostly through photoionization of atoms, followed by emission of electrons from their inner atomic shells. The energy of the released electron is close to that of the absorbed photon, and the difference is transformed into thermal energy or fluorescent radiation. Detectors used in SAS instruments are designed in such a way that they do not detect secondary radiation emitted from excited atoms or nuclei.

The efficiency of the absorption process can be quantified in terms of the absorption cross section σ_{abs}, which is defined as the number of photons (or neutrons) absorbed per second by an atom (or a nucleus), divided by the flux of the incident radiation. Hence, the total cross section becomes a sum of three terms:

$$\sigma_{tot} = \sigma_{coh} + \sigma_{inc} + \sigma_{abs}. \tag{1.19}$$

The transmission factor T of a sample of thickness L is defined as the ratio of the transmitted Φ_t and incident fluxes in the low Q limit ($Q \to 0$):

$$T = \frac{\Phi_t}{\Phi_0} = \exp(-\mu L), \tag{1.20}$$

where μ is the linear attenuation coefficient (cm^{-1}), which for a sample of single isotopic composition is related to the number density (N) of atoms or nuclei:

$$\mu = \sigma_{tot} N, \quad \text{and} \tag{1.21}$$

$$N = \left(\frac{N_A}{M}\right)\rho, \tag{1.22}$$

where $N_A = 6.022\ 10^{23}$ atom/mole is the Avogadro number and M is the atomic mass. The transmission factor (or, simply, transmission) of a particular sample is usually determined experimentally, although it can also be calculated with high accuracy if the chemical composition and density of the sample are known. At low neutron energies (on the order of a few millielectronvolts), σ_{abs} is a smooth function of neutron energy, proportional to the wavelength of the neutron $\sigma_{abs} \sim \lambda$. Values of σ_{abs} for thermal neutrons with $\lambda = 1.8$ Å are tabulated in Table 1.1 and can be rescaled for other wavelengths. In general, absorption of x-rays increases with wavelength approximately as λ^3; however, monotonic variation of μ vs. λ frequently shows an abrupt drop, which is the effect known as *absorption edge* [7].

The total cross section of hydrogen exceeds σ_{tot} of most other atoms by more than an order of magnitude, which makes neutron transmission extremely sensitive to concentrations of organic compounds in inorganic matrices. Both neutron and x-ray transmission values also depend strongly on the sample density (see Eqs. (1.20)–(1.22)). Thus, transmission measurements could provide important information on absolute adsorption, as well as on the average density of the adsorbed fluids in pores [8].

1.5 Form and Structure Factors

Equation (1.18) can be rewritten in a general form:

$$\frac{d\Sigma}{d\Omega}(Q) = NV_p^2(\Delta\rho^*)^2 P(Q)S(Q) + B, \tag{1.23}$$

where N is the number density of scattering particles N_p per unit volume, V_p is the particle volume, $(\Delta\rho^*)^2$ is the contrast factor, and B is the background signal (Sect. 5.5). The term $P(Q)$ in Eq. (1.23) is a dimensionless function known as *the form factor* that describes scattering from a single particle and is dependent on, and sensitive to, both its shape and size. To obtain sufficient scattering intensity, SAS measurements are conducted on samples containing many (ideally identical) scattering objects. If the objects are sufficiently separated from each other, they scatter independently to a good approximation and the total scattering intensity is the sum of the intensities from the individual scatterers. Analytical expressions for $P(Q)$ have been worked out for most common shapes [9]. Examples of the form factors of the simplest shapes are given here:

Sphere of radius R:

$$P(Q) = 9\left[\frac{\sin(QR) - (QR)\cos(QR)}{(QR)^3}\right]^2. \tag{1.24}$$

Thin disk of radius R:

$$P(Q) = \frac{2}{(QR)^2}\left[1 - \frac{J_1(2QR)}{QR}\right]. \tag{1.25}$$

Thin rod of length L:

$$P(Q) = \frac{2S_i(QL)}{QL} - \frac{\sin^2(QL/2)}{(QL/2)}. \tag{1.26}$$

In Eq. (1.25), J_1 is a first-order Bessel function, and in Eq. (1.26), S_i is the sine integral function. Examples of the calculated $P(Q)$ of a sphere, disk, and rod are shown in Fig. 1.6. Note that in practice, well-defined sharp modulations of $P(Q)$ are smeared because of a limited Q-resolution of SAS instruments, as well as finite polydispersity (i.e., variation of the particle sizes) that is always present in a physical sample. Two limiting scattering regimes, Guinier region ($Q \Rightarrow 0$) and Porod law ($Q \to \infty$) are discussed in Chap. 2.

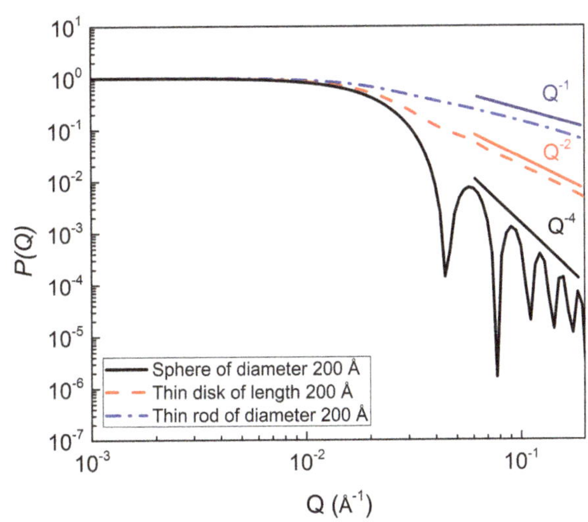

Fig. 1.6 Single particle form factor $P(Q)$ of a sphere, thin disk and a thin rod. *Straight lines* show the variation of scattering for the intermediate Q range (Sect. 4.3)

Form factor can be directly measured from a system in which the number density of scattering objects is small and, therefore, the distance between the objects is much larger than their sizes. Under such conditions, interference between scattering from different objects might be neglected and the measured signal contains information about the shape and size of individual objects. If the particle concentration is increased, at some point the scattering interference effects between different scattering objects begin to be important. The interference effects are described by another dimensionless function, $S(Q)$ in Eq. (1.23), and is called the *structure factor*. This parameter contains information about spatial arrangements of the scattering objects, that is, the correlations in their position. It can be represented as:

$$S(Q) = 1 + 4\pi N \int_0^\infty \left[(g(r) - 1) r^2 \frac{\sin Qr}{Qr} \right] dr, \tag{1.27}$$

where $g(r)$ is the particle pair distribution function that describes the spatial distribution of the particles (pores) as a function of the mean separation distance, r. In solid porous materials, positions of pores are "frozen" in space and time and $S(Q)$ is entirely defined by the position component of the structure factor. In other systems, such as soft matter systems (e.g., interacting colloids), the structure factor is also dependent on the interaction potential between scattering objects and can provide additional information about thermodynamic properties of the system, such as osmotic pressure and compressibility.

Finally, note that form factors and structure factors are normalized such that $\lim_{Q \Rightarrow 0} P(Q) = 1$ and $\lim_{N \Rightarrow 0} S(Q) = 1$. Thus, for dilute systems in the small—Q limit, the zero-angle scattering $I(Q \Rightarrow 0)$ is simply

$$\frac{d\Sigma}{d\Omega}(Q = 0) = NV_p^2 (\Delta\rho^*)^2 + B. \tag{1.28}$$

It follows from Eq. (1.28) that SAS depends linearly on the number (or concentration) of scattering objects, whereas its variation as a function of the particle volume is quadratic and thus $d\Sigma/d\Omega(Q)$ increases with the particle size R as $d\Sigma/d\Omega(Q) \sim R^6$. Therefore, the magnitude of SAS is defined predominantly by the linear particle sizes rather than their concentration in the system. Several measurements of SAS from a system of dilute scattering objects of different (small) concentrations of particles N_P can be used for determining particle volume and molecular weight $M_W = \rho V_P N_A$, if its physical density, ρ, is known.

1.6 Complementarity of Neutron and X-Ray Scattering Techniques

SAS of x-rays and neutrons has been widely used for the microscale and nanoscale structural characterization of hard and soft matter systems in terms of parameters such as averaged particle sizes, shapes, distribution, and surface-to-volume ratio, correlations, and interactions between the scattering objects. The materials can be solid or liquid and can contain solid, liquid, or gaseous domains (e.g., fluid-filled pores) of the same or another material in any combination. The structure of both disordered and ordered systems (like lamellae and fractal-like materials) can be investigated. Applications are broad and include colloids of all types, porous media, cement, oil, polymers, plastics, proteins, foods, and pharmaceuticals. Non-diffraction methods of neutron and x-ray tomography and radiography can deliver information on the internal structure and composition of macroscopic objects with micron-scale resolution and are often applied in research, as well as in quality control.

Neutron scattering and x-ray scattering techniques are complementary rather than competitive. However, differences in the charge and energy, and hence the way they interact with matter, define certain advantages and disadvantages of each type of radiation for conducting a specific type of experiment or dealing with a specific type of sample. One of the most relevant advantages of neutrons in the context of this book is their high penetration capability, which makes it easier to use metal containers for non-ambient studies at high pressure, as well as at low and high temperatures. Because some metals have positive and some have negative values of b_{coh}, they can be blended to produce alloys with $< b_{coh} > = 0$. For instance, machinable alloys of zirconium ($b_{coh} = 7.16 \times 10^{-12}$ cm) and titanium ($b_{coh} = -3.44 \times 10^{-12}$ cm) are used to make high-pressure vessels with $< b_{coh} > \sim 0$ that contribute no coherent parasitic scattering to SANS. There is no equivalent combination of materials available for x-ray experiments.

The low energy of thermal and cold neutrons facilitates in situ studies of the unperturbed fluid phase in bulk and in the pores without a danger of local heating and or damaging the sample. Neutron scattering originates from the bulk of the sample rather than from its surface. A large cross section of the neutron beam neutron beam (on the order of 1 cm^2) allows sampling of the whole sample volume under investigation, and the data collected can be considered to be representative of the whole object. In addition, neutrons are particularly well suited for investigations that take advantage of isotope substitution since hydrogen and deuterium have very different scattering powers. Isotope (or deuterium) substitution allows studies of closed porosity in porous matrices as a function of the pore size by filling pores with contrast matching mixtures of deuterated and hydrogenous solvents (e.g., $H_2O + D_2O$) or gases (e.g., deuterated methane, CD_4) at the contrast matching pressure.

The major weakness of the neutron scattering technique stems from the limited flux that can be delivered to the sample at conventional reactor sources. Most powerful research reactors, such as the 58 MW High Flux Reactor (HFR) at the Institut Laue-Langevin (ILL, Grenoble, France), the FRM-II reactor (Garching, Germany), and the 85 MW High Flux Isotope Reactor (HFIR) at Oak Ridge National Laboratory (ORNL, Oak Ridge, Tennessee, USA) deliver neutron fluxes on the order of 10^{15} (s^{-1} cm^{-2}), and further increase can be expected from newer pulsed spallation sources [10]. Third-generation x-ray sources are capable of generating photon fluxes approximately 15 orders of magnitude larger than existing reactors. Use of such high-flux beams in SAS research requires significant beam attenuation to avoid sample damage. Even after attenuation and focusing of the beam down to a few tens of micrometers, the resulting flux on the sample is high enough to ensure counting time on the order of seconds, which allows one to conduct position-sensitive scans and investigate fast kinetic phenomena such as penetration of supercritical CO_2 into coal. An increasing number of researchers benefit from using both types of radiation as it can be done conveniently at research centers with both x-ray storage rings and nuclear reactors on the same campus, such as the European Synchrotron Radiation Facility (ESRF/ILL) at Grenoble, France, and Diamond/ISIS at Didcot, UK. A comparison of x-ray and neutron properties and applications is summarized in Table 1.3.

Table 1.3 Comparative properties of neutrons and x-rays for SAS studies

Parameter/property	Neutrons	X-rays
Radiation is scattered by	Nuclei	Electrons
Availability	Large facilities only	Tabletop or large facilities
Flux ($s^{-1}cm^{-2}$)	Low, ~ 10^{15}	High, ~10^{24}
Typical wavelengths (Å)	1.5–20	Characteristic line spectra (CuK_α $\lambda = 1.54$ Å, MoK_α $\lambda = 0.71$ Å) or wavelength band at synchrotron sources
Energy range (meV)	Cold: 0.1–25 Thermal: 25–100	100–10^7
SAS: Typical length scales (Å)	5–1000	5–1000
USAS: Typical length scales	1000 Å – 3 μm	1000 Å – 1 μm[a]
Typical sample volumes (cm^3)	0.05–1.5	0.0001–0.5
Typical counting time per sample	Minutes to hours	Seconds to minutes
Use metallic containers	Easy	Hard/expensive (e.g., diamond, beryllium)
Effect of isotopic substitution	Significant	Negligible
Absolute intensity calibration	Straightforward	Difficult. Interpretation depends on precise knowledge of atomic scattering factor curves
Radiation/heat damage to sample	Negligible	Very likely
Magnetic scattering	Routine	Possible
Absorption coefficient	Very small, much less than scattering	Large, greater than scattering
Incoherent background	Comparable to weakly scattering samples	Negligible
Inelastic scattering	Appreciable change in λ	Small change in λ

Adapted with modifications from [11, 12]

[a]Q-resolution of USAXS is by a factor of 3–4 lower than that of USANS (~ 2×10^{-5} Å^{-1}) due to a much higher coherent scattering length of Si for x-rays compared to neutrons (Table 1.1). The resolution may be increased and made comparable to that of USANS by using higher-order reflection optics Si(440) instead of Si(220) and x-ray energies greater than 20 keV [13]

References

1. Lovesey, S.W.: Theory of Neutron Scattering from Condensed Matter. Clarendon, Oxford (1984)
2. Bee, M.: Quasielastic Neutron Scattering. Principles and Applications in Solid State Chemistry. Biology and Materials Science. Adam Hilger, Bristol, PA (1988)
3. Chattereji, T. (ed.): Neutron Scattering from Magnetic Materials. Elsevier, Amsterdam, Boston, (2006)
4. International Tables for Crystallography. Kluwer Academic, Dordrecht (1992)
5. Triolo, R., Agamalian, M.: The combined Ultra-small and Small-angle Neutron Scattering (USANS/SANS) technique for earth sciences. In: Liang, L., Rinaldi, R., Schober, H. (eds.) Neutron Applications in Earth, Energy and Environmental Sciences. Springer, New York (2009)
6. Lehmann, E.H.: Neutron imaging methods and applications. In: Liang, L., Rinaldi, R., Schober, H. (eds.) Neutron Applications in Earth, Energy and Environmental Sciences. Springer, New York (2009)
7. Warren, E.: X-Ray Diffraction. Dover, Mineola, NY (1990)
8. Melnichenko, Y.B., Wignall, G.D.: Small-angle neutron scattering in materials science: recent practical applications. J. Appl. Phys. Rev. **102**, 0211101 (2007)
9. http://www.ncnr.nist.gov/resources/
10. Mezei, F.: New perspectives from new generations of neutron sources. C. R. Physique **8**, 909 (2007)
11. King, S.M.: Small-angle neutron scattering. In: Pethrick, R.A., Dawkins, J.V. (eds.) Modern Techniques for Polymer Characterization. John Wiley, Chichester, NY (2003)
12. Parise, J.B.: Introduction to neutron properties and applications. Rev. Mineral. Geochem. **63**, 1 (2006)
13. Ilavsky, J., Allen, A.J., Levine, L.E., Zhang, F., Jemian, P.R., Long, G.G.: High-energy ultra-small-angle X-ray scattering instrument at the advanced photon source. J. Appl. Cryst. **45**, 1318 (2012)

General

- Lindner, P., Zemb, T. (eds.): Neutron, X-Ray and Light Scattering: Introduction to an Investigative Tool for Colloidal and Polymeric Systems. North-Holland, Amsterdam, Oxford, New York, Tokyo (1991)
- Feigin, L.A., Svergun, D.I.: Structure Analysis by Small-Angle X-Ray and Neutron Scattering. Plenum, New York, London (1987). PDF file with this book is available at:http://www.embl-hamburg.de/biosaxs/reprints/feigin_svergun_1987.pdf
- Cowley, J.M.: Diffraction Physics. Elsevier, Amsterdam (1995)
- Sivia, D.S.: Elementary Scattering Theory for X-Ray and Neutron Users. Oxford University Press, Oxford, New York (2011)
- Roe, R.J.: Methods of X-Ray and Neutron Scattering in Polymer Science. Oxford University Press, New York, Oxford (2000)

Neutron Scattering

- Egelstaff, P.A.: Thermal Neutron Scattering. Academic, London (1965)
- Scherm, R.: Fundamentals of neutron scattering by condensed matter. Ann. Phys. **7**, 349 (1972)
- Bacon, G.E.: Neutron Diffraction. Clarendon, Oxford (1975)
- Squires, G.L.: Introduction to the Theory of Thermal Neutron Scattering. Dover, New York (1996)

- Furrer, A., Mesot, J., Strassle, T.: Neutron Scattering in Condensed Matter Physics. World Scientific, New Jersey, London, Singapore (2009)
- Hammouda, B.: Probing Nanoscale Structures – the SANS Toolbox, http://www.ncnr.nist.gov/staff/hammouda/the_SANS_toolbox.pdf

X-ray Scattering

- Guinier, A., Fournet, G.: Small Angle Scattering of X-Rays. Wiley Interscience, New York (1955)
- Guinier, A.: X-Ray Diffraction in Crystals, Imperfect Crystals, and Amorphous Bodies. Freeman, San Francisco (1963)
- Glatter, O., Kratky, O. (eds.): Small Angle X-Ray Scattering. Academic, London, New York (1982). A PDF file with this book is available at:http://physchem.kfunigraz.ac.at/sm/Software.htm
- James, R.W.: The Optical Principles of the Diffraction of X-Rays. Ox Bow, Woodbridge, CN (1982)

There are also numerous websites with lectures given by experts in the theory and experimental applications of small angle scattering:

- Pynn, R.: An introduction to neutron and x-ray scattering. http://neutrons.ornl.gov/conf/nxs2013/lecture/pdf/PynnIntroductoryLectures_2013.pdf
- Schaefer, D.: Exploring the nanoworld with small-angle scattering. http://neutrons.ornl.gov/conf/nxs2013/lecture/pdf/NXSchool2013ANLORNL.pdf
- Sinha, S.K.: Introduction to neutron and x-ray scattering. http://physchem.kfunigraz.ac.at/sm/Software.htm
- Urban, V.: Small angle scattering of neutrons and x-rays. http://neutrons.ornl.gov/conf/nxs2010/pdf/lectures/Small-Angle-Scattering_SAS_V-Urban-ORNL-vu3.pdf
- Kohlbrecher, J.: Principles of small angle scattering. http://www.slri.or.th/en/index.php?option=com_attachments&task=download&id=319

Chapter 2
Radiation Sources

The main sources of neutrons are steady state reactors and spallation neutron sources. In the first case, neutrons are continuously produced by fission processes in the reactor core. In the second case, a pulsed neutron beam is generated by the collision of high-energy protons or electrons, which fragment heavy atoms in a solid heavy-metal target such as uranium or tungsten. A liquid target (liquid mercury) was designed and used for the first time at the Spallation Neutron Source (SNS) at ORNL. The main sources of x-rays are photon sources, such as synchrotrons. In a synchrotron, a beam of electrons or positrons is accelerated to nearly the speed of light and made to circulate on a closed orbit under high vacuum in a storage ring emitting high-flux electromagnetic radiation. In this chapter we discuss major components of large facilities used for neutron and x-ray production and compare key parameters of the currently operating and future neutron and photon sources.

2.1 Steady State Reactors

Nuclear reactors operate on the physical basis of the fission chain reaction. When slow neutrons are absorbed by ^{235}U, the excited nuclei decay into a cascade of fission products. On average, 2.5 neutrons are produced by the fission of one ^{235}U nucleus. These neutrons possess elevated energies of about 1 MeV and are not suitable for maintaining the fission process in ^{235}U. The fast, high-energy neutrons are slowed down by *moderators*, usually light (H_2O) or heavy (D_2O) water. Typical examples of continuous sources are HFIR at ORNL, the High Flux Reactor (HFR) at the Institut Laue-Langevin (ILL) in Grenoble, France, and the new reactor FRM-II in Garching, Germany. All of these reactors have similar performance (thermal flux) on the order of 10^{15} n/cm^2 s [1] (see Fig. 2.1).

The most common type of research reactor is the pool-type reactor. The core of such a reactor represents an assembly of fuel elements in a large pool of light or heavy water. HFIR is a beryllium-reflected, light-water-cooled and moderated,

Y.B. Melnichenko, *Small-Angle Scattering from Confined and Interfacial Fluids*, DOI 10.1007/978-3-319-01104-2_2

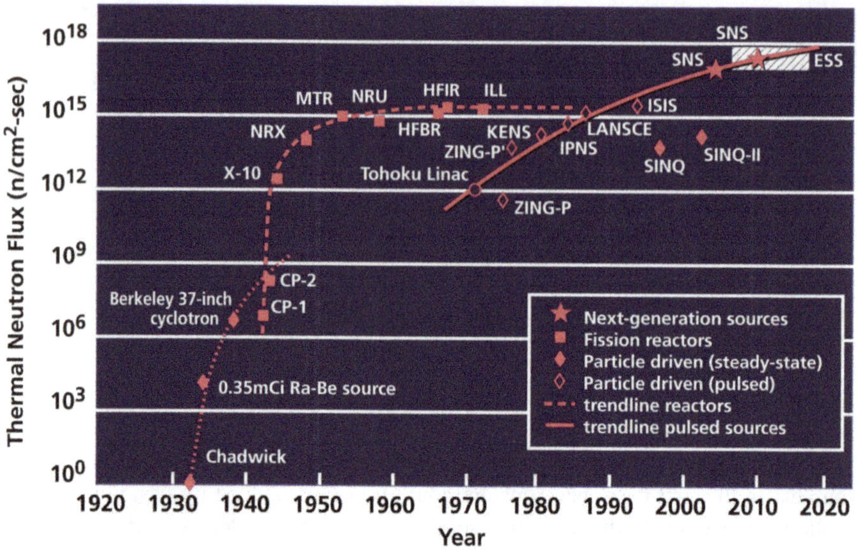

Fig. 2.1 Historical evolution of the performance of neutron sources over time. Data for fission reactors and spallation sources are time-averaged fluxes and peak fluxes, respectively [1]. Reproduced with permission

Fig. 2.2 Aerial view of the HFIR campus, Oak Ridge, Tennessee, USA

flux-trap-type reactor that uses highly enriched ^{235}U as the fuel. An aerial photograph of the HFIR campus is shown in Fig. 2.2. Figure 2.3 shows a cutaway of the reactor with the pressure vessel and some of the auxiliary facilities. The reactor core assembly is contained in a ~2.5 m diameter pressure vessel located in a pool of light water (Fig. 2.4). The reactor core is cylindrical, ~0.6 m high, and 0.4 m in diameter. The fuel region is composed of two concentric fuel elements (Fig. 2.5). The fuel

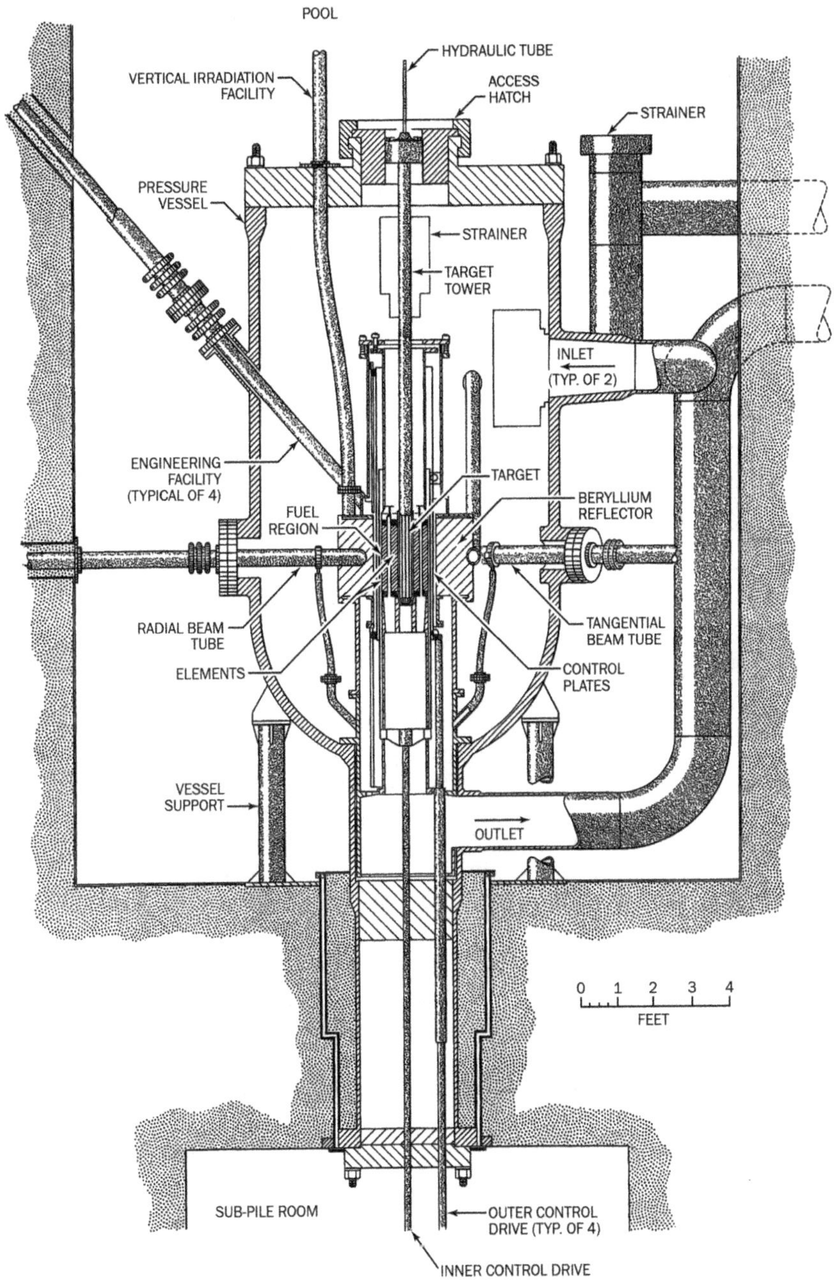

Fig. 2.3 Cutaway of the HFIR reactor

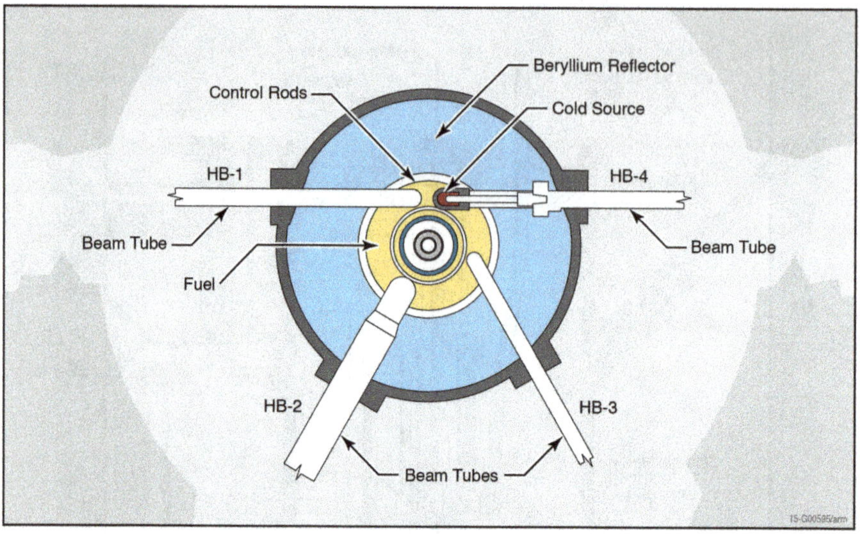

Fig. 2.4 Reactor core assembly and the schematic of horizontal beam (HB) tubes

Fig. 2.5 HFIR's two concentric fuel elements

plates are made of U_3O_8-Al cermet and are curved in the shape of an involute, thus providing a constant coolant channel width. The primary coolant (water) enters the pressure vessel through two pipes above the core, passes through the core, and exits through a pipe beneath the core. The inlet coolant temperature is 49 °C, and the corresponding exit temperature is 69 °C. From the reactor, the coolant flow is directed to heat exchangers. A secondary coolant system removes heat from the primary system and transfers it to the atmosphere by passing water over a cooling tower.

The reactor can be shut down using the control plates (in the form of two thin, neutron-absorbing concentric cylinders), which are located in an annular region between the outer fuel element and the beryllium reflector. These plates are driven in opposite directions to open and close a window at the core mid-plane. The fuel region is surrounded by a concentric ring of beryllium reflector ~0.30 m thick. The beryllium reflects neutrons back to the reactor core; this helps to reduce the number of neutrons escaping from the core and increase the neutron density in the fission reaction, thus increasing its efficiency. The whole reactor assembly is immersed into a pool of water, which also acts as a reflector of effectively infinite thickness. Neutron flux in the beryllium reflector is tapped by extending four empty "beam" tubes into the reflector, thus allowing neutrons to be delivered to one of two experiment halls (one for cold neutrons and one for thermal, or "warm," neutrons), where they are used in neutron scattering experiments. The reactor is currently operating at 85 MW. The average core lifetime that defines the length of the reactor cycle is approximately 23 days followed by an end-of-cycle outage for refueling. The length of these outages varies to allow for calibrations, maintenance, and inspections. The reactor is run on a six-cycle-per-year schedule and generates neutrons that are used by more than 500 researchers per year. More information about HFIR design is available in [2]. Table 2.1 compares parameters of the existing medium- and high-flux reactors available throughout the world for neutron scattering experiments.

2.2 Spallation Neutron Sources

Two known types of nuclear reactions can be used for neutron production. *Electron* accelerators can produce neutrons by accelerating electrons to energies of ~100 MeV and then directing the electron beam at a heavy metal target (e.g., depleted uranium). Although electron accelerators are relatively inexpensive to construct, this type of neutron production suffers from low efficiency (about 5 neutrons are generated for every 100 electrons), and its performance is limited by the large amount of heat generated in the target that must be removed by extensive cooling. The other method of neutron production uses high-energy *protons* (H^+) that are generated by an ion source and accelerated by a linear accelerator (linac). At SNS, the linac accelerates the proton beam from 2.5 MeV to ~1 GeV in vacuum, corresponding to about 90 % of the speed of light (see Fig. 2.6). The proton beam is passed into an accumulation ring, where it is accumulated in bunches and compressed by a factor of 1000 and after approximately 1200 circuits. Each bunch of protons is released from the ring as a short microsecond-long pulse that is delivered to the liquid mercury target at a frequency of 60 Hz. When the beam strikes the target, neutrons are chipped off the target nuclei in a process called *spallation*.

The efficiency of this process is high, with each proton yielding about 30 neutrons [5]. The high-energy neutrons coming out of the target are slowed down (moderated) by passing them through cells filled with water (to produce room-temperature neutrons) or through containers of liquid hydrogen at a temperature of 20 K (to produce cold neutrons). The latter method is analogous to the moderation process used in a cold source in a nuclear reactor (Sect. 2.1). Design of

Table 2.1 Existing medium- and high-flux steady-state reactor sources and their key parameters (adopted with modifications from [3]). Consult the facility web sites for updated information. Links are available at http://www.neutron.anl.gov/facilities.html and http://www.ncnr.nist.gov/nsources.html

Country	USA	USA	Canada	France	France	Germany	Germany	Russia	Australia	Korea	Japan
Neutron source	HFIR	NBSR	NRU	HFR	ORPHEE	BENSC	FRM-II	IBR-2	OPAL	HANARO	JRR-3M
Organization	Oak Ridge National Laboratory	National Institute for Standards and Technology	Rutherford Appleton Laboratory	Institut Laue-Langevin	Laboratoire Leon Brollouin	Helmholtz-Zentrum Berlin	Institute of High Energy Physics	Joint Institute for Nuclear Research	Australian Nuclear Science and Technology Organization	Korean Atomic Energy Research Institute	Japan Atomic Energy Agency
Power (MW)	85	20	125	58	14	10	20	2 (average thermal power) 1500 (peak power in pulse)	20	24	20
Flux (n/cm^2 s)	1.5×10^{15}	3×10^{14}	3×10^{14}	1.5×10^{15}	3×10^{14}	2×10^{14}	8×10^{14}	8×10^{12}(time averaged)	3×10^{14}	2×10^{14}	3×10^{14}
Number of cold/hot sources	1/0	1/0	0/0	2/1	1/1	1/0	1/1	1/0	1/0	1 (planned)/0	1/0
Number of SANS/USANS instruments	3/1	3/1	0/0	4/1	3/0	1/1	2/1 (VSANS)*	1/0	1/1	2/1	2/1
Since/planned in	1965	1970	1957	1972 (refurbished 1993)	1980	1973	2004	1982	2006	1997	1990

*VSANS is a very small-angle neutron scattering instrument, which uses the focusing mirror principle and is available at the Research Neutron Source Heinz-Maier Leibnitz (FRM-II reactor) in Garching, Germany. The instrument allows performing scattering experiments with a scattering vector values between $Q = 3 \times 10^{-5}$ Å^{-1} (with $\lambda = 12.8$ Å), and 3×10^{-3} Å^{-1}, bridging a gap between Bonse-Hart and pinhole instruments [4]. The achievable low Q limit of VSANS is close to that of Bonse-Hart USANS instruments ($Q \sim 4 \times 10^{-5}$ Å^{-1}) at $\lambda = 2.4$ Å

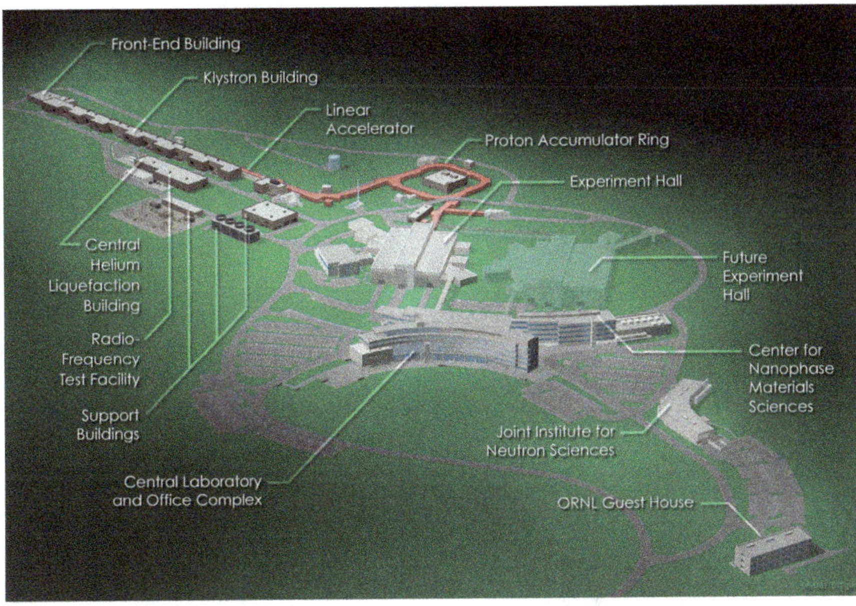

Fig. 2.6 Layout of SNS in Oak Ridge, Tennessee, USA

the moderators is crucial to forming the shape and sharpness of the pulse [6]. Apart from the pulse shape, the repetition rate of the pulse is an important design parameter as it defines the so-called duty cycle of the source (the ratio of the pulse length over repetition time). In the case of the SNS coupled moderator, the duty cycle is about 2 %. The pulsed, moderated neutrons are delivered through beam guides to specially designed instrument stations. The pulsed nature of accelerator-based sources offers significant advantages in experiments using the time-of-flight (TOF) method, in which the speed/wavelength of neutrons is measured by timing their flight from the moderator to the moment when the neutrons reach the detector [3].

Each instrument at SNS is unique and is designed for certain types of experiments with specific types of samples. An aerial image of the SNS campus is shown in Fig. 2.7, and the layout of the SNS experimental hall is shown in Fig. 2.8. At the time of writing, the nominally1.4 MW (upgradable to 2 MW) SNS and 1 MW Japan Proton Accelerator Proton Complex (J-PARC, Tokai, Japan) are the only spallation sources to achieve 1 MW power. The planned European Spallation Source (ESS), to be built in Lund, Sweden, will operate at 5 MW power and will deliver 14 long (2.86 ms) neutron pulses every second. These long pulses will have a direct impact on the number of neutrons generated in the target station, which will exceed by a factor of 6 per MW the neutron fluxes generated by the existing short (~1 μs) pulse sources, resulting in an unparalleled intensity. ESS is planning to deliver its first neutrons in 2019, and the whole project is scheduled for completion by 2024.

Fig. 2.7 Aerial view of the SNS campus

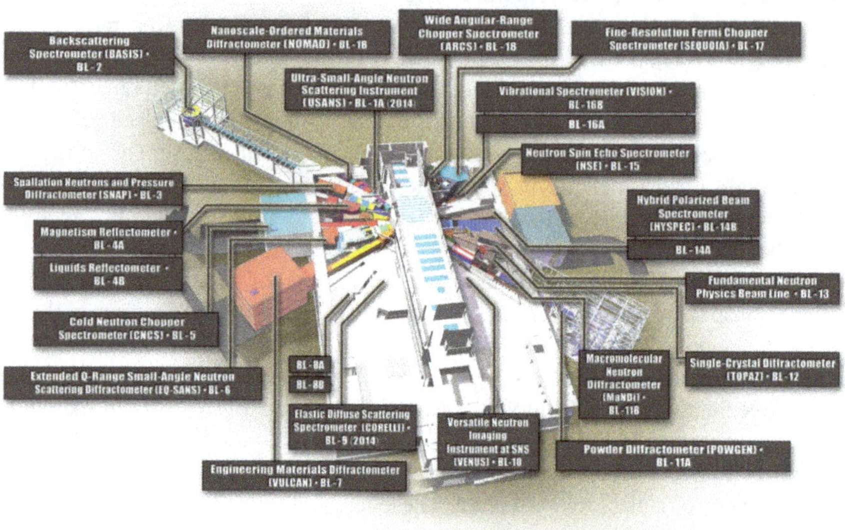

Fig. 2.8 Layout of the SNS experiment hall

Table 2.2 Operating and future spallation neutron sources and their key parameters (adopted with modifications from [3, 7, 8])

Country	USA	USA	UK	Switzerland	Japan	Europe	China
Neutron source	SNS	LANSCE	ISIS	SINQ	J-PARC	ESS	CSNS
Organization	Oak Ridge National Laboratory	Los Alamos National Laboratory	Rutherford Appleton Laboratory	Paul Scherrer Institute	Japan Atomic Energy Agency	Pan-European project	Institute of High Energy Physics
Proton energy (MeV)	1000	800	800	570	3000	1330	1600
Current (μA)	1400	70	200	1500	333	7500	70
Proton beam power	1.4 MW (nominal)	56 kW	160 kW	1 MW	1 MW	5 MW	100 kW
Repetition rate (Hz)	60	20	50/10 (2 targets)	Continuous	25	16 (long pulse)	25
Target material	Liquid mercury	Tungsten	Tantalum	Liquid PbBi	Liquid mercury	Liquid mercury	Tungsten
Moderator	L-H_2/H_2O	L-H_2/H_2O	L-H_2/L-CH_4/ H_2O	L-D_2/D_2O	L-H_2	L-H_2	H_2OL-CH4/ L-H_2
Time average thermal neutron flux (n/cm^2·s)	1.2×10^{14}	~10^{11}	2×10^{12}	1×10^{14} (continuous)	2.5×10^{14}	2.5×10^{14} (long pulse)	8×10^{12}
Number of SANS/ USANS instruments	1/1 (TOF USANS under construction)	2/0	4/0	2/1	2/1	TBD	TBD
Since/planned in	2006	1983	1985 (TS1) 2008 (TS2)	1996	2008	First neutrons 2019 Completion 2025	2014

Consult the facility websites for updated information. Links to the corresponding websites are available at http://www.neutron.anl.gov/facilities.html and http://www.ncnr.nist.gov/nsources.html

In Table 2.2, the key parameters of the operating and future proton spallation neutron sources are compared. The Table shows, as a key characteristic, time-averaged thermal neutron fluxes produced by different spallation sources compared with the fluxes generated by steady state reactors. Peak intensities (see Fig. 2.1), often cited in the literature, can be confusing because they do not represent the actual neutron fluxes that researchers have to deal with in their experiments. For instance, the peak thermal flux generated by SNS is 8×10^{16} neutron/cm^2 s, whereas the time-averaged flux is only 1.2×10^{14} neutron/cm^2 s [8]. The first number exceeds by ~1.5 orders of magnitude the thermal flux generated by the steady state HFIR (Table 2.1). At the same time, however, the time-averaged flux of SNS is about an order of magnitude *lower* than the HFIR flux. It should be noted, however that the TOF instruments at the pulsed sources make use of most of the neutrons generated, while the reactor-based instruments utilize only a small fraction of the spectrum, which reduces the difference between time-averaged fluxes of both types of facilities.

Table 2.3 presents a list of the available neutron scattering facilities (both reactor and accelerator sources). Direct links to the web sites of the listed facilities are available at:

- http://www.neutron.anl.gov/facilities.html.
- http://www.ncnr.nist.gov/nsources.html.

Table 2.3 Available neutron scattering facilities. Direct links to the corresponding websites are available at http://www.neutron.anl.gov/facilities.html and http://www.ncnr.nist.gov/nsources.html

North and South America	
Centro Atomico Bariloche	Rio Negro, Argentina
Canadian Neutron Beam Centre	Chalk River, Ontario, Canada
High Flux Isotope Reactor (HFIR), Oak Ridge National Laboratory	Oak Ridge, Tennessee, USA
Los Alamos Neutron Science Center (LANSCE)	Los Alamos, New Mexico, USA
Low Energy Neutron Source (LENS), Indiana University	Bloomington, Indiana, USA
McMaster Nuclear Reactor	Hamilton, Ontario, Canada
MITR-II, MIT Nuclear Reactor Laboratory	Cambridge, Massachusetts, USA
NIST Center for Neutron Research	Gaithersburg, Maryland, USA
Peruvian Institute of Nuclear Energy (IPEN)	Lima, Peru
Spallation Neutron Source, Oak Ridge National Laboratory	Oak Ridge, Tennessee, USA
University of Missouri Research Reactor	Columbia, Missouri, USA
Europe	
Budapest Neutron Centre	AEKI, Budapest, Hungary
Berlin Neutron Scattering Center, Helmholtz-Zentrum	Berlin, Germany
Joint Institute of Nuclear Research	Dubna, Russia
FRM-II Research Reactor	Garching, Germany
Institut Laue Langevin	Grenoble, Franc

(continued)

Table 2.3 (continued)

ISIS Pulsed Neutron and Muon Facility	Oxfordshire, UK
JEEP-II Reactor, IFE	Kjeller, Norway
Laboratoire Léon Brillouin	Saclay, France
Ljubljana TRIGA MARK II Research Reactor, J. Stefan Institute	Ljubljana, Slovenia
Nuclear Physics Institute (ASCR)	Rez, Czech Republic
Reactor Institute Delft, Delft University of Technology	Delft, Netherlands
St. Petersburg Nuclear Physics Institute	Gatchina, Russia
Swiss Spallation Neutron Source (SINQ)	Villigen Switzerland
Asia and Australia	
Bragg Institute	Lucas Heights, Australia
High-flux Advanced Neutron Application Reactor (HANARO)	Daecheon, South Korea
Japan Atomic Energy Research Institute (JAERI)	Tokai, Japan
Japan Proton Accelerator Research Complex (J-PARC)	Tokai, Japan
Kyoto University Research Reactor Institute (KURRI)	Kyoto, Japan
Reactor Triga Puspati (RTP), Malaysian Nuclear Agency	Selangor, Malaysia
Planned or under construction	
Austron Spallation Neutron Source	Vienna, Austria
China Advanced Research Reactor (CARR)	Beijing, China
China Spallation Neutron Source (CSNS)	Dongwan, Guangdong, China
European Spallation Source (ESS)	Lund, Sweden

2.3 Photon Sources

Laboratory x-ray generation is based on acceleration of an electron beam towards a metal target by application of positive voltage V. On striking the metal target, the electron's kinetic energy is converted into a photon by two different mechanisms. First, photons are generated when electrons (or any charged particles) decelerate rapidly. This is called Bremsstrahlung ("braking radiation") and yields a continuous spectrum of wavelengths with a lower limit given by $\left(1.24 \times 10^{-6}\right)/V$ [m], where V is the voltage in volts. It follows that generation of x-rays with atomic-size wavelengths is possible with $V \sim 10$ kV [9]. The second process (the fluorescent radiation) is initiated by the incoming particles knocking out electrons from one of the inner shells of an atom in the target and creating a vacancy. The subsequent radiative transition of an electron from an outer shell into the vacancy can produce an x-ray with a characteristic energy equal to the difference in energy between the two electronic shells. For experiments requiring a monochromatic beam, one often uses the K_α line of the generated spectrum that is several orders of magnitude more intense than the bremsstrahlung spectrum [9]. This "vacuum tube" technology has been widely used in laboratories for more than a century. However, the divergence of the x-ray beam and its fixed wavelength (for a given metal target) imposes heat-transfer limitations on the achievable maximum flux of x-rays ($\sim 5 \times 10^{14}$ s^{-1} cm^{-2} for a rotating anode x-ray tube).

Fig. 2.9 Schematic diagram of synchrotron Soleil. The outer circular ring is the synchrotron (i.e., a particle accelerator that brings electrons [light blue beam] to very high speeds). The electrons are accelerated by electric fields in the straight sections between *green squares*. The *red rectangles* are magnets that bend the beam. When the beam is bent, the electrons emit x-rays that are delivered to the various beam lines (the *straight lines* branching out of the synchrotron), which host scientific instruments [10]

The advent of synchrotron x-ray sources offered numerous opportunities for x-ray application in SAS experiments because of enhanced photon flux, low beam divergence, and wavelength tunability. In a synchrotron, the bunches of charged nearly relativistic particles (electrons or positrons) move in a large, closed beam loop containing curved and linear segments, collectively called the storage ring (Fig. 2.9). In curved regions, bending magnets generate a uniform magnetic field perpendicular to the orbit plane. The straight sections contain magnetic devices (magnetic quadrupoles and radio-frequency cavities) that serve to focus and accelerate the electrons to compensate for their gradual slowing down as a result of energy loss in the photon radiation process. The bending magnets are solely responsible for the photon emission in the described setup.

In modern synchrotron facilities, known as third-generation sources, even greater intensity is achieved by use of insertion devices such as wigglers and undulators situated in the straight sections of the storage ring [11]. Undulators provide a way to take maximum advantage of the intrinsic brightness of the synchrotron radiation source. The magnetic structure of today's most common (planar) undulator is an array of closely spaced vertically oriented dipole magnets of alternating polarity. As an electron beam passes longitudinally through the array, its trajectory oscillates in the horizontal plane. Radiation cones emitted at each bend in the trajectory overlap, giving rise to an interference effect. That effect results in a few spectrally narrow peaks in a beam that is highly collimated in both the horizontal and vertical directions, which increases the spectral brightness. Tuning the wavelengths of the harmonics can be accomplished by means of mechanically

Table 2.4 Partial list of high- and intermediate-energy synchrotron sources and their key parameters in decreasing energy order (adopted with modifications from [13])

Name	Location	Energy (GeV)	Perimeter (m)	Current (mA)
SPRING-8	Japan	8	1436	100
APS	US	7	1060	100
ESRF	France	6	844	200
SSRF	China	3.5	396	300
SPEAR3	US	3	240	500
DIAMOND	UK	3	560	300
Australian Synchrotron	Australia	3	216	200
ALBA	Spain	3	269	250
CLS	Canada	2.9	171	500
SOLEIL	France	2.85	354	500
ANKA	Germany	2.5	240	110
PLS	Korea	2.5	281	180
SLS	Switzerland	2.4	240	400
ELECTRA	Italy	2.4	260	320
Nano-Hana	Japan	2	102	300
ALS	US	1.9	197	400
BESSY-II	Germany	1.9	240	270

A full list of light sources and their websites is available at http://www.lightsources.org

adjusting the vertical spacing (gap) between the pole tips. Wigglers are similar to undulators but generally use higher fields and fewer dipoles, which results in the production of a continuous spectrum with a higher flux that extends to shorter wavelengths than those generated by bending magnets. These planar insertion devices produce radiation that is linearly polarized in the horizontal plane, which can be used to examine polarization-dependent and angle-dependent effects. Because of its pulsed nature, synchrotron radiation is especially suitable for examining time-dependent phenomena.

The flux of x-rays generated by third-generation synchrotron sources is more than 9 (wiggler) or 14 (undulator) orders of magnitude higher than that generated by a rotating anode x-ray source [12]. The enormous level of fluxes brings in numerous technical and experimental challenges: the primary optical components, such as monochromator crystals, need to be cooled cryogenically, and organic (especially biological) samples should be exposed to only to a severely attenuated beam to prevent radiation damage. A list of the high- and intermediate-energy photon sources is given in Table 2.4. The number of synchrotron radiation facilities is growing rapidly. As a result, the internet is the most reliable source of up-to-date information on facilities around the world. Useful sites include the following:

- http://www-ssrl.slac.stanford.edu/sr_sources.html
- http://xdb.lbl.gov/Section2/Sec_2-3.html
- http://www.lightsources.org

References

1. Mason, T.E., Arai, M., Clausen, K.N.: Next generation of neutron sources. MRS Bull. **28**, 923 (2003)
2. http://en.wikipedia.org/wiki/High_Flux_Isotope_Reactor
3. Vogel, S.C., Priesmeyer, G.: Neutron production, neutron facilities and neutron instrumentation. Rev. Miner. Geochem. **63**, 27 (2006)
4. Goerigk, G., Varga, Z.: Comprehensive upgrade of the high-resolution small-angle neutron scattering instrument KWS-3 at FRM II. J. Appl. Crystallogr. **44**, 337 (2011)
5. Schober, H.: Neutron applications in earth, energy and environmental sciences. In: Schober, L., Liang, R., Rinaldi, H. (eds.) Neutron Scattering Instrumentation. Springer, New York, NY (2010)
6. Charlton, L.A., Barnes, J.M., Gabriel, T.A., Johnson, J.O.: Spallation neutron source moderator design. Phys. Res. A **411**, 494 (1998)
7. Arai, M., Crawford, K.: Imaging and applications. A reference for the imaging community. In: Anderson, I.S., McGreevy, R., Bilheux, H.Z. (eds.) Neutron Sources and Facilities. Springer, New York, NY (2009)
8. Bauer, G.S.: Physics and technology of spallation neutron sources. Phys. Res. A **463**, 505 (2001)
9. Sivia, D.S.: Elementary Scattering Theory for X-Ray and Neutron Users. Oxford University Press, New York, NY (2011)
10. http://commons.wikimedia.org/wiki/File:Sch%C3%A9ma_de_principe_du_synchrotron.jpg. Copyright © EPSIM 3D/JF Santarelli, Synchrotron Soleil
11. Als-Nielsen, J., McMorrow, D.: Elements of Modern X-Ray Physics. Wiley, New York, NY (2011)
12. Mills, D.M. (ed.): Third-Generation Hard X-Ray Synchrotron Radiation Sources: Source Properties, Optics, and Experimental Techniques. Wiley, New York, NY (2002)
13. Bilderback, D.H., Elleaume, P., Weckert, E.: Review of third and next generation synchrotron light sources. J. Phys. B. Atom. Mol. Opt. Phys. **38**, S773 (2005)

Further Reading

Roe, R.J.: Methods of X-Ray and Neutron Scattering in Polymer Science. Oxford University Press, New York, NY (2000)

Furrer, A., Mesot, J., Strassle, T.: Neutron Scattering in Condensed Matter Physics. World Scientific, London (2009)

Imae, I., Kanaya, T., Furusaka, M., Torikai, N.: Neutrons in Soft Matter. Wiley, Hoboken, NJ (2011)

Some useful web-based resources may be found at:

Carpenter, J.M.: Neutron sources for materials research. http://neutrons.ornl.gov/conf/nxs2010/pdf/lectures/Sources_Carpenter%20ORNL.pdf

Rhyne, J.: Neutron sources. http://neutrons.ornl.gov/conf/nxs2011/pdf/lectures/Scattering%20Facilities_Overview_2011-p.Rhyne.pdf

Mills, D.M., Lang, J.: Synchrotron radiation: properties and production. http://neutrons.ornl.gov/conf/nxs2011/pdf/lectures/D.M.Mills-J.Lang-SourceProperties2011.pdf

Blanchet, C.: SAXS and SANS facilities and experimental practice. http://www.embl-hamburg.de/biosaxs/courses/embo2012ccmb/slides/05-blanchet-facilities-experimental-practice.pdf

Bunker, G.: Synchrotron radiation sources and optics. http://acaschool.iit.edu/lectures07/Howard_SynchrRad.pdf

X-ray data booklet. http://xdb.lbl.gov/

Chapter 3
Constant Flux and Time-of-Flight Instrumentation

Use of small-angle and ultra-small angle scattering instruments in a tandem provides a unique opportunity of exploring sample structure on the length scales between Angstroms and microns in a single experiment. This chapter considers major types and components of small-angle neutron and x-ray scattering instruments. Constant flux instrumentation is discussed based on the General Purpose SANS diffractometer (HFIR, ORNL) and the double-crystal USANS instrument (NCNR, NIST). Operational principles and major components of time-of-flight SAS instrumentation are considered based on the Extended Q SANS and the Time-Of-Flight USANS instruments (SNS, ORNL). X-ray SAS instrumentation is discussed based on SAXS and USAXS diffractometers available at APS, ANL. Advantages and disadvantages of the constant flux and time-of-flight instruments for various types of experiments and different systems are discussed and illustrative examples of the scattering data obtained from various samples using SANS/USANS and SAXS/USAXS instruments are presented.

3.1 General Purpose SANS Instrument at HFIR, ORNL

In the early 2000s, a series of upgrades was undertaken at HFIR, including replacement of the beryllium reflector, installation of a supercritical hydrogen cold source (T ~20°K) in the horizontal beam 4 (HB-4) tube, and construction of a guide hall that hosts a suite of cold source instruments including a GP-SANS instrument on cold guide 2 (CG2) and a biological SANS instrument (Bio-SANS) on cold guide 3 (CG3) that is optimized for the study of biological systems. In addition to the SANS facilities, a suite of other instrumentation has been installed (see the current guide hall layout in Fig. 3.1).

Y.B. Melnichenko, *Small-Angle Scattering from Confined and Interfacial Fluids*, DOI 10.1007/978-3-319-01104-2_3

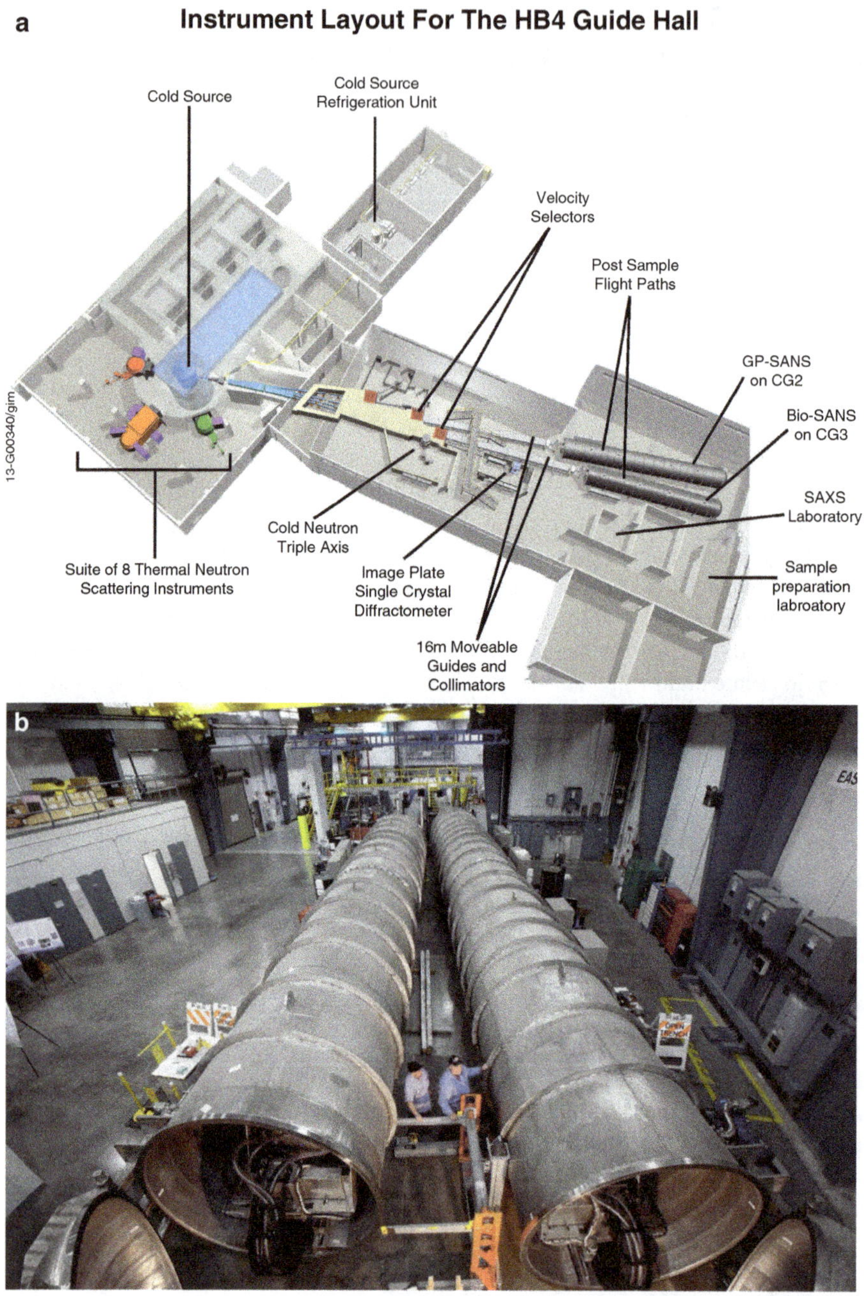

Fig. 3.1 Instrument layout at HFIR (**a**) and two SANS tanks with open lids (**b**)

The GP-SANS instrument [1] can be divided into four major components: the delivery system that brings neutrons to the sample, the sample area, the post sample flight path/detection system, and the instrument control and data handling electronics.

The neutron delivery system consists of the cold source, main guides, velocity selector, and removable guide sections. The energy spectrum of neutrons in thermal equilibrium with the water moderator kept at a constant temperature T is described by the Maxwellian distribution

$$\Phi(\lambda) \sim \frac{1}{\lambda^2} \exp\left(-\frac{h^2}{2k_B T m \lambda^2}\right) \tag{3.1}$$

where λ and m are the wavelength and the mass of the neutron, respectively, and k_B is the Boltzmann constant. The temperature of the water moderator is usually close to room temperature. Hence, the corresponding neutrons, with a maximum peak flux around λ ~ 1 Å, are called thermal neutrons (Fig. 3.2). SANS experiments require neutrons with a considerably longer wavelength, and the Maxwellian energy spectrum can be shifted towards lower energies (longer wavelengths) by passing them through an additional moderator (i.e., cold source) near the reactor core that is maintained at a much lower temperature. The cold source used in HB-4 is situated close (<40 cm) to the center of the HFIR core. It represents an aluminum vessel filled with 505 mL of supercritical hydrogen that is refrigerated to ~20 K at a pressure of 14.5 bar (Fig. 3.3). The cold source produces a Maxwellian wavelength distribution with wavelengths in the range of 4–20 Å.

With an approximately circular area of 78 cm^2, this source feeds neutrons to all four cold guides in the HFIR guide hall.

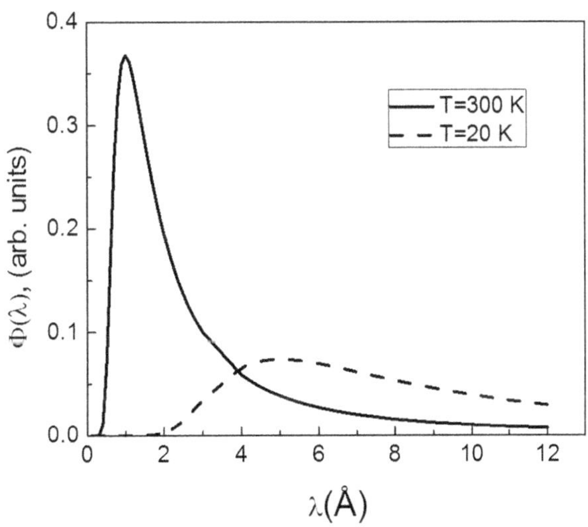

Fig. 3.2 Wavelength distribution of neutrons from cold (20 K) and thermal (300 K) neutrons

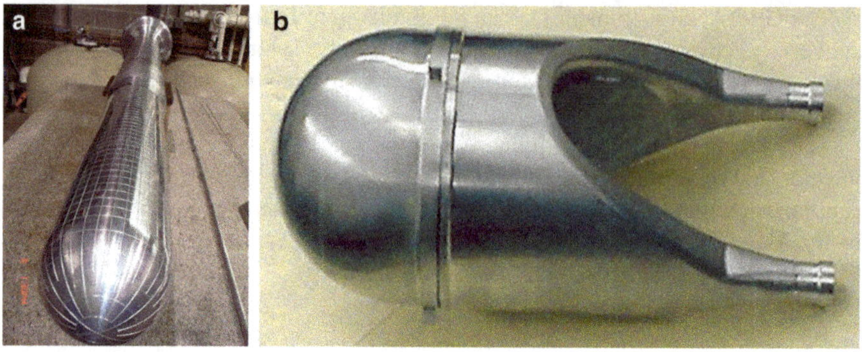

Fig. 3.3 Images of the horizontal beam 4 (HB-4) beam tube (**a**) and a cold source (**b**). The diameter of the cold source is close to 10 cm

The neutron flux at an instrument positioned at a distance (r) from the neutron source will be dramatically reduced according to the r^{-2} law for isotropic radiation. This reduction can be significantly decreased by using neutron guides. A neutron guide works via total reflection of neutrons from the guide's glass walls, which are covered with a smooth single layer or multilayer of a total-reflecting material such as nickel, or the even more efficient ^{58}Ni. Along such total-reflecting surfaces, neutrons can be guided from the source to the instrument with almost no loss. The angular acceptance of a neutron guide can be further increased by reflection from a so-called supermirror, which is composed of a sequence of layers of variable thickness with alternating high positive (e.g., nickel) and high negative (e.g., titanium) SLD. These multilayers form an artificial one-dimensional (1D) lattice that could increase the critical angle of nickel by a factor of five or more. Two straight sections of nickel-coated neutron guides (4 and 16.9 m long) with a beam deviator positioned between them are used to deliver the neutron beam from the cold source to the velocity selector. The supermirror beam deviator, consisting of two flat supermirrors, is used to separate out and shield the fast neutron and γ-ray components of the original beam coming from the reactor, thus minimizing the undesirable radiation background at the SANS instruments.

Monochromatization is essential for any continuous SANS instrument. It can be achieved by using either the Bragg diffraction from crystals or velocity selectors. A neutron velocity selector is a time-gating device that allows neutrons of desired velocity (and corresponding wavelength λ) to pass, while absorbing neutrons that are moving too fast or too slow. As a result, a white incident neutron spectral distribution is filtered into a much more monochromatic distribution with the mean wavelength $\overline{\lambda}$ and a relatively narrow wavelength spread (full width at half maximum [FWHM]) $\Delta\lambda$.

Velocity selectors are of either the solid drum type (with helical slots) or the multidisc type. In drum-based selectors, neutrons are allowed to travel through helical slots covered with neutron-absorbing material (e.g., gadolinium or ^{10}B). For a given rotation speed, only neutrons having the "right" velocity are transmitted through the device. In a multidisc selector, thin neutron-absorbing discs are mounted on an axle at pre-calculated intervals and helical pathways are determined by windows cut into the peripheries of the discs. The main advantage of such selectors is that they are lighter than their drum-based counterparts and are therefore easier and safer to drive to higher speeds, which allows for achieving shorter wavelengths with better wavelength resolution [2]. The GP-SANS is equipped with such multidisc neutron velocity selectors, which were developed and manufactured by Mirrotron [3] (Fig. 3.4). Their performance was determined by directly measuring the TOF distribution of the neutrons transmitted by the selectors as a function of rotation speed and angle of inclination to the beam collimation. Table 3.1 presents some representative parameters for the velocity selector at the GP SANS.

The velocity selector is inserted between the main guides and 16 m of removable guides/collimation. The incident neutron beam monitor, used for data correction, is housed in the velocity selector shielding. Five beam attenuators, mounted after the

Fig. 3.4 Images of the Mirrotron velocity selector: (**a**) selector installed in the radiation shielding box (side view), (**b**) front view of the selector showing the neutron inlet window, and (**c**) a rotor with neutron-absorbing gadolinium disks with perforated slots

Table 3.1 Characteristics of the Mirrotron multidisc neutron velocity selector

Transmission	75 %
Resolution	13 %
Resolution range	8–45 %
Wavelength λ_{min} at $\Delta\lambda/\lambda = 13$ %	3.8 Å
Wavelength range	2.7–20 Å

selector but before the beam monitor, are provided to limit the flux on the detector with an attenuation factor between 3 and 10,000.

A total of 16 m of variable collimation is provided after the velocity selector by eight removable neutron guide sections. By inserting guides into the beam line, the effective source-to-sample distance can be changed within the range of 1.9–17.45 m to vary the beam divergence and flux on the sample. Without evacuation, the beam intensity would be reduced by ~5 % per meter because of absorption and scattering from the dry air. To avoid neutron losses, the guides are placed inside a housing that forms a continuous evacuated volume for the incident beam. Each section has an external motor-driven linear actuator that controls the positioning of a transverse-motion table inside to one of three positions. The tables have neutron-guide elements, 1.9 m long with 4.0 × 4.0 cm cross sections in the first position (Fig. 3.5), and 4.0 cm diameter beam apertures in the second position. The third position is currently unoccupied but could house future optical device upgrades. An exception is the first guide section, which contains 4.0 and 2.0 cm diameter beam apertures at the second and third positions, respectively. The last guide inserted is always immediately followed by a source-defining aperture in the next section. The source apertures are made from sandwitched x-ray and neutron absorbing materials.

The ultimate collimation of the incident beam is defined by the choice of circular beam aperture at the sample position. The apertures represent epoxy castings backed with boron nitride followed by cadmium and having tapered knife-edge openings.

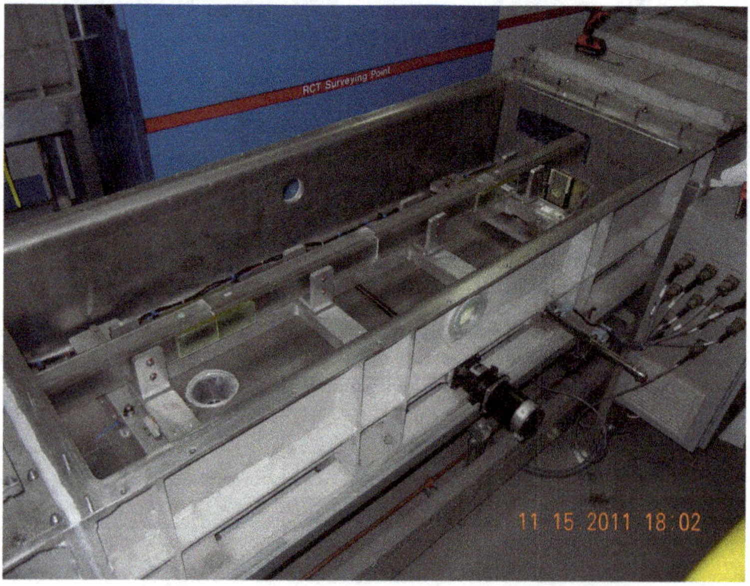

Fig. 3.5 Image of a neutron guide with the top lid removed

The sample area has a 2 m diameter footprint between the movable guide sections and the postsample flight path and detection system. Sample positioning, including horizontal and vertical rotation, is provided through a rail-mounted Huber goniometer suitable for handling heavy loads such as high-field magnets. This arrangement affords maximum versatility for the instrument and provides a robust platform onto which newly developed sample environments can be installed. In a standard configuration, in order to minimize air scattering background, an evacuated beam extender is used to deliver neutrons up close to the sample changer usually located next to the entrance of the SANS tank.

The postsample vacuum vessel is constructed from (nonmagnetic) stainless steel, can be evacuated to a pressure of <0.25 Tor, and has a diameter of 2.5 m and a length of 20.6 m. A carriage mounted on a rail system enables the detector to move along the length of the vacuum vessel, providing sample-to-detector distances (SDDs) in the range of 1.2–20.9 m for the GP-SANS. The detector carriage in the post-sample flight path allows the detector to be moved by up to 45 cm perpendicular to the incident beam. The two-dimensional (2D), position-sensitive detector has an active area of 1 m^2, thereby providing a broad dynamic Q-range to the instrument ($0.001 < Q < 1.0$ Å^{-1}). Most of the existing SANS facilities cover a Q-range that is usually limited by ~Q_{max} ~ 0.6 Å^{-1}. Such a Q-range is more than adequate for studying most soft matter systems with relatively large scattering objects. The extended Q-range, reaching 1 Å^{-1}, available at the GP-SANS provides a unique opportunity to explore the size and shape of sub-nanometer objects, which is especially important for investigation of the structure and adsorption properties of micropores, available, for example, in activated carbons (Sect. 7.3.1).

Scattered neutrons are registered using a 2D, position-sensitive neutron detector that employs 192 ^{3}He tubes, supplied by GE (http://www.gepower.com). Each tube has an active length of 1.04 m and a diameter of 8 mm (Fig. 3.6). The gas pressure in each tube is 22.1 bar, ~20 bar of which is ^{3}He (1 bar = 100,000 Pa). The detector is electronically divided along its length into 256 pixels of ~4.3 mm. To increase the counting efficiency of the whole detector, the tubes are arranged in two planes. Within each plane, the tubes are separated by an 11 mm center-to-center distance. The two planes are 8.2 mm apart, and the tubes between the planes are offset laterally by 5.5 mm to give an effective width of 1.05 m. The detector can safely handle high neutron fluxes. The limits set on global count rate (neutrons per entire detector area), and local count rate (neutrons per pixel) are 2,000,000 and 2000, respectively. Using 8 mm ^{3}He tubes greatly increases maintenance flexibility. The detector can be operated with a few nonfunctioning tubes, which can be replaced between the experiments.

The instrument control and data acquisition systems have borrowed much of the technology already in use at other instruments at HFIR. User interaction with the instrument is through the Spectrometer and Instrument Control Environment (*SpICE*) software, developed at ORNL [4]. *SpICE* is built on *LabVIEW* (National Instruments Corporation, Austin, Texas, USA) and provides a simple

Fig. 3.6 Image of the ^{3}He detector (front view)

command-driven interface for performing all necessary functions of the instrument. *SpICE* can also run user-written macros, and scripting capabilities are provided through the Python programming language (http://www.python.org). The user interacts through a graphical user interface (GUI) developed specifically for SANS instruments at HFIR, which provides an easy-to-use interface with integrated real-time feedback. The raw data are saved in XML files containing pertinent instrument-configuration information, and a template style sheet is available for viewing this information in tabular format using a web browser. Data reduction and initial data analysis are implemented in *Igor Pro* (Wavemetrics Inc., Lake Oswego, Oregon, USA).

Typical neutron flux values at the sample position of the GP-SANS instrument, measured using the source-defining slit 4 cm in diameter, are shown in Table 3.2.

The design of GP-SANS is similar to the design of other constant flux pinhole SANS instruments such as the NG3 and NG7 30 m SANS instruments at the NIST Center for Neutron Research (Gaithersburg, Maryland, USA); D11, D22, and D33 at the ILL (Grenoble, France); KWS-1 and KWS-2 at the FRM-II Research Reactor (Garching, Germany); and the V4 SANS at Helmholtz-Zentrum (Berlin, Germany). More details about the design and operation of these instruments, as well as references to instrumentation-specific articles, can be found at the corresponding websites (Table 2.3).

Table 3.2 Typical values of the neutron flux (neutrons s^{-1} cm^{-2}) at the sample for several representative configurations of the GP-SANS instrument with a wavelength resolution of $\Delta\lambda/\lambda = 13\,\%$

λ (Å)	No. of guides = 0	No. of guides = 4	No. of guides = 8
4.2	3.0×10^6	8.5×10^6	3.3×10^7
4.8	2.7×10^6	7.9×10^6	3.8×10^7
6.0	1.8×10^6	5.3×10^6	3.0×10^7
7.2	1.1×10^6	3.1×10^6	1.9×10^7
8.4	6.0×10^5	1.8×10^6	1.2×10^7
9.6	3.3×10^5	1.0×10^6	7.4×10^6
12.0	1.1×10^5	3.4×10^6	3.0×10^6
14.4	4.3×10^4	1.0×10^5	6.8×10^6
19.0	9.7×10^3	2.9×10^4	3.4×10^5

3.2 Extended Q SANS Instrument at SNS, ORNL

The pulsed nature of the spallation source at SNS makes it possible to compute the wavelength of scattered neutrons. To do this, the time interval between the proton pulse and the moment when neutrons reach the detector is recorded. From the known length of the total flight path, the neutron velocity (and wavelength) can be determined. By the gating the signal at the neutron detector, scattering data are acquired for a number of time intervals (equivalent to neutron wavelengths in increasing order). Therefore, the entire "white" neutron beam can be used in the scattering experiment without monochromatization. The range of scattering vectors for the TOF EQ-SANS is similar to the range of constant flux instruments. The principal benefit of using the TOF EQ-SANS is that the white incident beam allows for covering the full Q-range in a single measurement without time-consuming reconfiguring of the instrument (moving the detector, changing the number of neutron guides, etc.). This is especially useful for studying fast kinetics processes over a large dynamic range of scattering vectors. The disadvantages of using TOF vs. constant flux SANS include the relatively low flux of long-wavelength neutrons, which makes it difficult to reach the low-Q part of the spectrum, as well as the relatively low Q-resolution. Both parameters depend on the duration of the neutron pulse, as well as on the total physical length of the instrument.

The EQ-SANS is located on beam line 6 at SNS (Fig. 2.8). The instrument uses cold neutrons that come from a coupled supercritical hydrogen moderator. Similar to the GP-SANS, the EQ-SANS operates in the pinhole geometry and its main components are shown in Fig. 3.7. The instrument is designed to minimize transport loss for thermal and cold neutrons and to avoid a direct line of sight of the moderator [5]. As discussed earlier, SNS produces ~1 GeV fast neutrons when the proton pulse hits the mercury target. After moderation, thermal and cold neutrons are separated from the remaining fast neutrons by deflecting them using curved beam benders made of glass coated with a supermirror multilayer. The beam

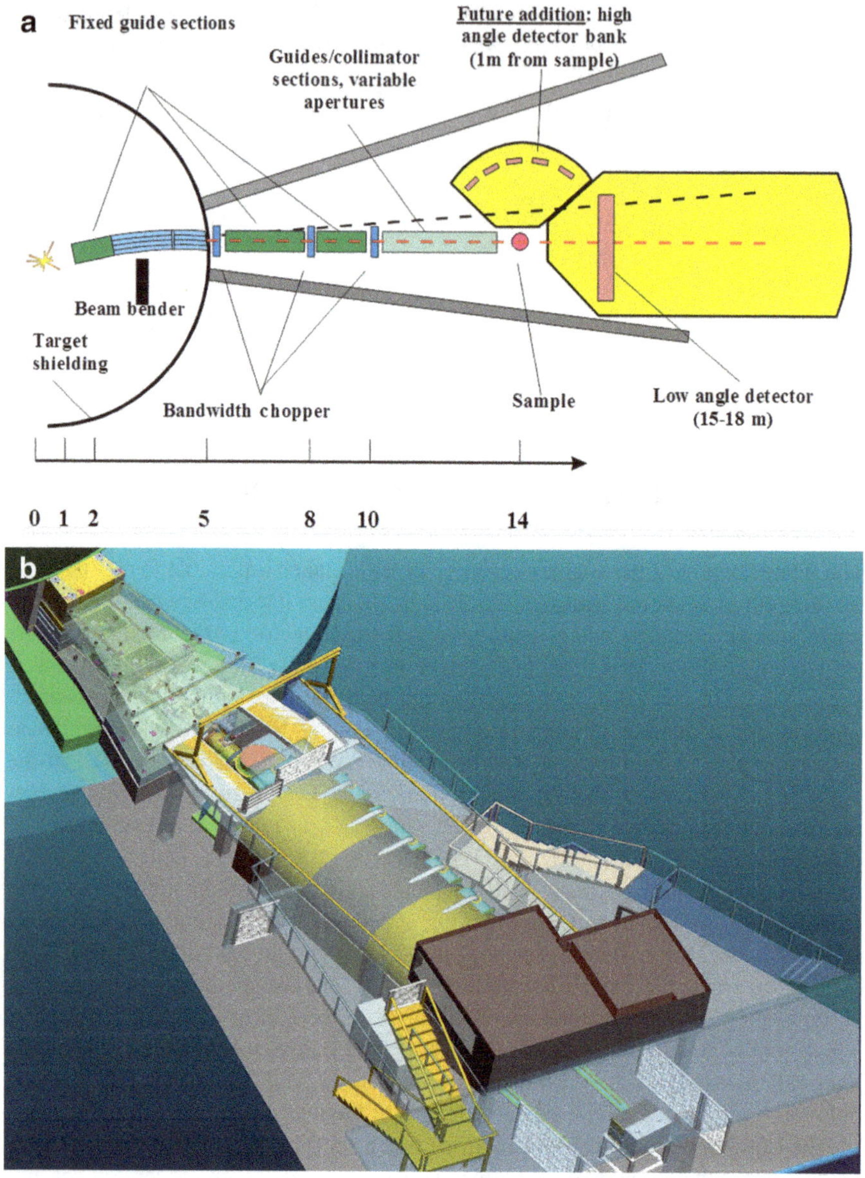

Fig. 3.7 Layout (a) and 3D model (b) of the EQ-SANS instrument

enters a curved multichannel beam bender located in the primary beam shutter and the bulk shield insert of the target monolith. The bender has a bending radius of 95 m and a total length of 3.3 m, which deflect the beam by 2° from the initial direction. After the bender, sections of straight neutron guides follow with gaps between them to accommodate three bandwidth choppers that are used to change the wavelength range in the neutron frames and switch between the "pulse rejection" and the "frame skipping" mode of operation [5].

The secondary shutter used during experiments is a 1 m long rotating drum that has four open and four closed positions. Currently, a neutron guide is installed in one of the open positions, and the other channels are reserved for future upgrades. The collimation system consists of three motorized multiple-slit wheels and a manually controlled sample slit that defines the sample beam size. Six additional downstream guard slits are used to reduce the instrument background by removing residual fast neutrons. Finally, a W-shaped sapphire frame overlap window is installed. Each of the sapphire blades in the window is inclined at a 2° angle with respect to the beam axis. This window is used to reflect neutrons with $\lambda > 33$ Å, which, in combination with the frame definition choppers, ensures a clean beam spectrum with the desirable wavelength range of neutrons.

The neutron detector for the EQ-SANS is similar to that of the GP-SANS (Sect. 3.1). The detector has a total area of 1×1 m^2 and is located in the evacuated scattering tank. The two planes of tubes are arranged on an arc with a radius of 5 m centered towards the sample position. To reach higher Q values in the future, upgrade space has been reserved for installing a high-angle detector bank to cover scattering angles of 30–150°. These high-angle detectors will allow the instrument to simultaneously monitor low-angle scattering and high-angle diffraction, which could be useful for studies of fast kinetic processes such as fluid penetration and adsorption in soil and rocks characterized by extremely wide distributions of pore sizes. The EQ-SANS instrument control and data reduction systems are similar to those used at GP-SANS.

The first scattering data obtained at EQ-SANS, using a dendrimer solution in D_2O, are shown in Fig. 3.8 [6]. The scattering data from the same sample measured using GP-SANS is also shown for comparison. The Q-range covered by both instruments was $0.005 < Q < 0.35$ Å^{-1} (fixed SDD of 4 m; two neutron bands of 2–5.6 Å and 8.9–12.9 Å). For strongly scattering samples, the lowest reliable Q_{min} is ~0.001 Å^{-1} at SDD = 10 m and $\lambda = 20$ Å. Using 2.2 Å and SDD = 1.5 m, the maximum achievable Q_{max} can be extended to ~1 Å^{-1}.

It should be noted that the operational repetition rate of SNS (60 Hz) is not optimal for SANS instruments. At this rate, the frame bandwidth (i.e., the time interval between two consecutive source pulses) is only ~3 Å for a 20 m long instrument, such as EQ-SANS, which does not make full use of the source spectrum offered by the SNS cold moderator. The much lower repetition rate of the SNS second target station will be better suited for support of more optimized and effective SANS instruments.

The design of the EQ SANS is principally similar to the design of other TOF SANS instruments such as the Low-Q diffractometer at the Lujan Neutron Scattering Center (Los Alamos National Laboratory, New Mexico, USA), the LOQ diffractometer at the ISIS Pulsed Neutron and Muon Facility (Rutherford Appleton Laboratory, UK), and the Hi-SANS diffractometer at J-PARC (Tokai, Japan). More details about the design and operation of these instruments, as well as references to instrumentation-specific articles, can be found at the corresponding websites (Table 2.3).

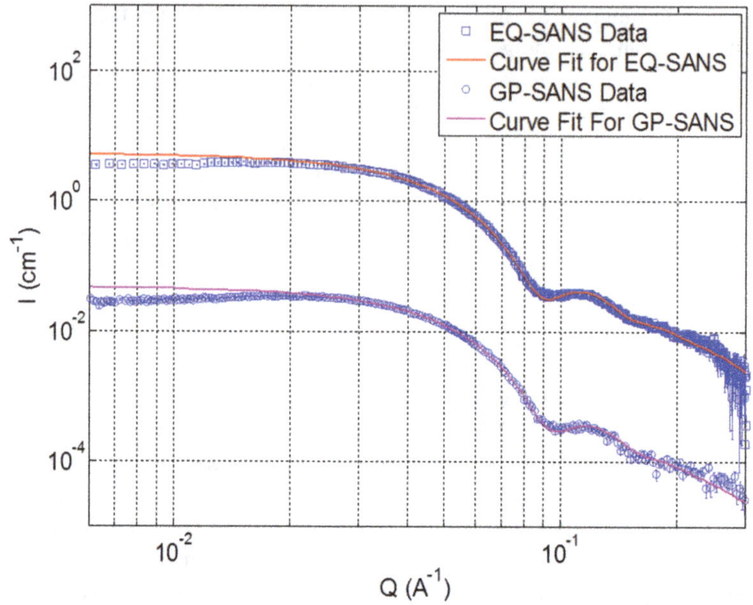

Fig. 3.8 Neutron scattering curves and model fitting of the data collected on EQ-SANS and GP-SANS. The sample is a dendrimers solution with the volume fraction of 0.02 in D_2O. The EQ-SANS data are on an absolute intensity scale, and the GP-SANS data are multiplied by a factor of 1/100. The EQ-SANS data were collected in a single 1 h measurement with an SDD of 4 m, simultaneously using two neutron wavelength bands of 2.0–5.6 Å and 8.9–12.9 Å, respectively. The GP-SANS data are a combination of two sequential measurements of 10 min at SDD = 2 m and 40 min at SDD = 10 m with the neutron wavelength of 4.75 Å [6]. Published with IUCr's copyright permission

3.3 Perfect Crystal USANS Instrument at NIST

Conventional pinhole SANS techniques typically cover a Q-range between $Q_{min} = 10^{-3} < \mathrm{Q} < Q_{max} = 1$ Å^{-1}, which permits the measurement of maximum diffraction distance $D_{max} \sim 2\pi/Q_{min} \sim 6000$ Å. In reality, many systems have structural inhomogeneities with much larger sizes (e.g., aggregates, agglomerates, and large pores), and structural characterization of such objects is out of the resolution window of conventional SANS. Extension of the SAS capability to characterize larger objects with dimensions up to few micrometers became possible due to development of ultra small-angle x-ray scattering (USAXS) and ultra-small-angle neutron scattering (USANS) instruments that are now available at major user facilities around the world. Combining USAXS/USANS and SAXS/SANS data can provide complete structural information on the morphology of strongly polydisperse, multi-hierarchical systems. These techniques have been widely used for structural studies in geology, petrology, archeology, and a variety of soft matter materials.

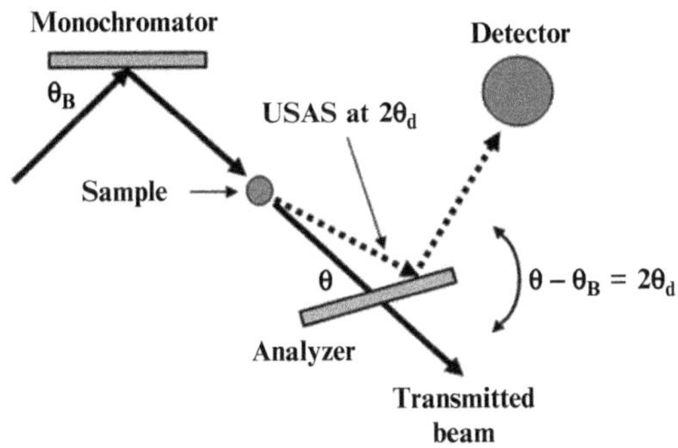

Fig. 3.9 Optical schematic of a double-crystal diffractometer with single-reflection crystals, and the geometry of the USAS experiment

Resolution limits inherent to SANS can be overcome based on the fact that the Bragg reflection of radiation from a single crystal is mostly confined to a very narrow angular range (usually less than several arc seconds). The reflectivity functions cannot be observed in a direct diffraction experiment but can be measured using a parallel double-crystal assembly that is the basis of all double-crystal diffractometers (DCDs). A schematic of the simplest DCD with two identical single-reflection crystals is shown in Fig. 3.9. A sample is placed in between the upstream perfect crystal monochromator (usually silicon or germanium), reflecting radiation at the Bragg angle Θ_B and the identical downstream crystal analyzer. The analyzer crystal is rotated (rocked) with respect to the transmitted beam and provides a reflectivity function or a "rocking curve" (i.e., the transmitted intensity as a function of the angle). The scattering from the sample will represent the difference between the rocking curves measured with and without the sample.

The performance of a single-reflection DCD is strongly limited by the background from the wings of the rocking curve that in most cases obscures the weaker SAS background. Therefore, only ultra small-angle scattering (USAS) from strongly scattering objects can be studied using these instruments. The wing intensity can be decreased and the instrument sensitivity increased using *multiple* reflections from crystal pairs, which allow for increasing the signal-to-noise ratio by more than several orders of magnitude [7]. Multiple reflections before and after the sample suppress the wings of the rocking curve, which enables the measurement of the scattering from smaller structures at angles farther away from the rocking curve center. This advancement stimulated the first USAXS studies of condensed matter [8].

The first attempts to adopt Bonse-Hart methodology to neutron scattering revealed numerous complications related to the much higher penetration power of neutrons compared with x-rays. Rocking curves measured from nearly transparent to neutrons silicon channel-cut crystals showed strong contamination from back-face reflections. In addition, some other sources of parasitic scattering such as edge-face diffraction and surface-induced scattering had to be eliminated. Major progress was achieved when the classical Bonse-Hart scheme was modified by cutting the crystal and inserting a neutron-absorbing material (cadmium) between adjacent reflectors in the middle of the long wall of a triple-reflection, channel-cut silicon

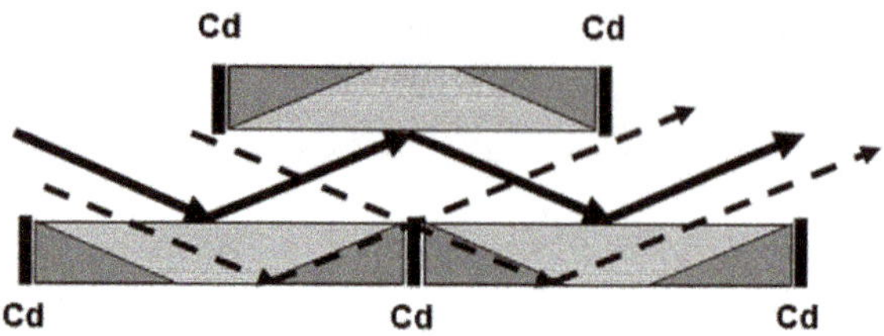

Fig. 3.10 The triple-bounce channel-cut crystal modified for neutron radiation. The cadmium absorber in the middle of the long wall filters out parasitic scattering, whereas the absorbers on the crystal edges help to eliminate edge-face diffraction

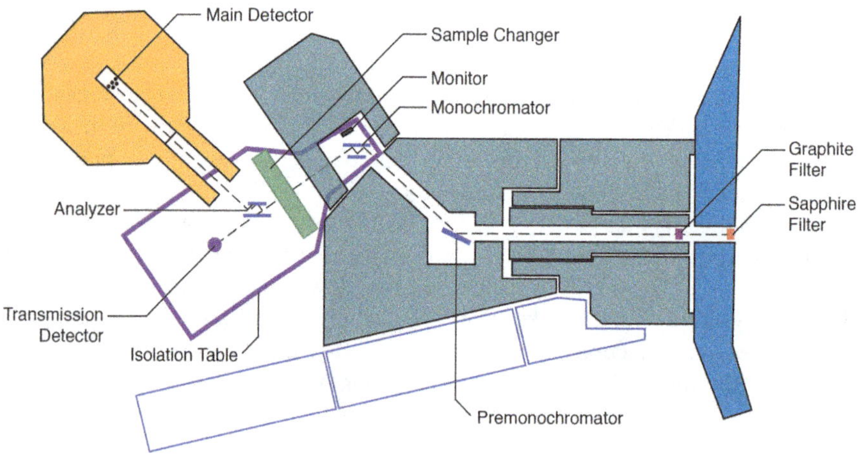

Fig. 3.11 Schematic layout of the perfect-crystal diffractometer (BT-5 USANS) showing the main instrument components, as described in the text. Adopted with modifications from http://www.nist.gov/ncnr/bt5-usans-ultra-small-angle-neutron-scattering.cfm

crystal [9] (Fig. 3.10). This neutron absorber serves to filter out the parasitic scattering originated from the first single reflection and allows only the useful neutrons to pass through the trapezoidal zones reflecting from the front and back diffractive surfaces. This modification, along with other improvements and modifications, increased signal-to-noise ratio by about four orders of magnitude, and made the Bonse-Hart technique fully adoptable for neutrons [10].

The first fully optimized USANS instrument became operational at NIST in 2000 [11]. The BT-5 USANS (Perfect Crystal Diffractometer for USANS) is a thermal neutron instrument located on thermal beam tube BT-5 of the NIST reactor. Figure 3.11 shows the layout of the instrument. The monochromator and the analyzer are channel-cut, triple-reflection silicon (220) crystals set up at the Bragg angle $\Theta_B = 38.4°$. Resolution for this type of instrument is independent of sample size; hence, large crystals are used to provide a beam cross section of up to 5×5 cm, which defines maximum sample sizes that can be used for experiments.

Both crystals are mounted on an anti-vibration table. Mounted directly behind the monochromator is a beam monitor of the fission chamber type for monitoring possible fluctuations of the beam intensity. There are two sample positions supported independently from the table. For routine use, a translation slide is used to move multiposition sample holders. To accommodate large auxiliary equipment, such as furnaces, cryostats, or magnets, the sample area can be enlarged by moving both the analyzer and the detector out on precision slides, which opens up a 58 cm diameter secondary sample table. The secondary position has the beam intensity reduced by a factor of 0.6 due to the beam defocusing.

The instrument operates at wavelength $\lambda = 2.38$ Å ($\Delta\lambda/\lambda = 6$ %), close to the maximum of the Maxwellian distribution of the thermal neutron spectrum. The typical Q-range of this instrument is $3 \times 10^{-5} < Q < 5 \times 10^{-3}$ Å^{-1}, and its sensitivity, or signal-to-noise ratio, is ~2.5×10^{6} for $Q \geq 5 \times 10^{-4}$ Å^{-1}. The increased total neutron flux at the sample (~1.7×10^{4} Å^{-1} n/cm^{2} s) was achieved by incorporating a 2D focusing pyrolytic graphite (PG002) premonochromator, which provided a gain factor of ~6 without narrowing the available Q-range, losing resolution or sensitivity.

In a typical experiment, a series of data sets is collected over merging angular intervals. The lowest-angle set is rocked through the main beam to define the zero scattering angle, and than the data for larger scattering angles are recorded. The sets of files corresponding to scattering from the sample or empty container are combined, sorted, and shifted to zero angle. Sample transmissions are automatically calculated, and the multiple scattering probability can be estimated from the ratio of the wide-angle and zero-angle sample transmissions [12].

In most cases, it is desirable to combine USANS and SANS data taken from the same sample to be able to analyze the entire scattering curve simultaneously. Scattering data in a pinhole SANS and Bonse-Hart USANS experiment are taken in different geometries that require different instrumental smearing methods for slit-smeared USANS and pinhole-smeared SANS data. Desmearing of the USANS data can be performed using Lake's iterative method [13]. The slit-smeared data can be interactively masked and smoothed before desmearing, if desired. Often the best results are obtained without smoothing, which can introduce additional artifacts. When combining of SANS and USANS data is not required, desmearing is typically not necessary since the slit-smeared USANS data can be used "as is" by fitting to appropriately smeared structural models. Typical combined USANS/SANS data sets measured from a shale sample are shown in Fig. 3.12.

The design of BT-5 USANS is principally similar to the design of other USANS instruments such as the Kookaburra USANS at the Australian Nuclear Science and Technology Organization (ANSTO), Australia; the KIST-USANS at the High-Flux Advanced Neutron Application Reactor (HANARO), Korea; and the V12 bent-crystal diffractometer at the Berlin Neutron Scattering Center (BENSC), Germany. The CRG-C thermal neutron interferometer S18 at the ILL, as well as the

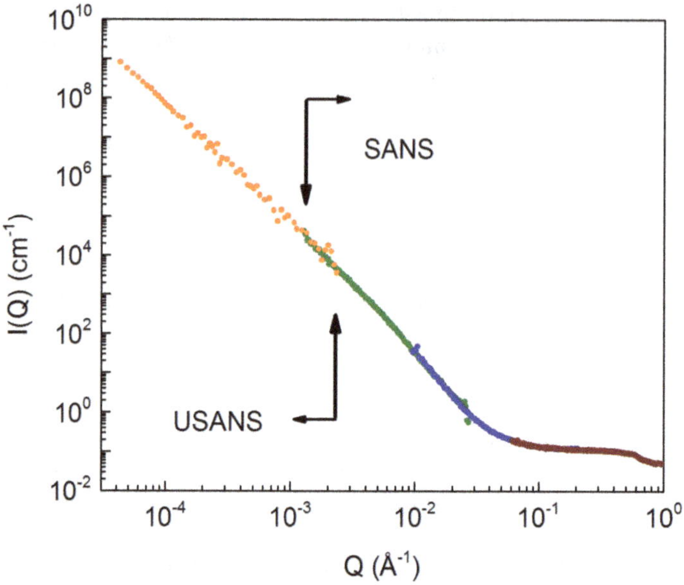

Fig. 3.12 Combined USANS and SANS data from Barnett shale. SANS data taken at three different SDDs. The corresponding Q-ranges are marked in different *colors*

MORPHEUS diffractometer at PSI in Switzerland, can be reconfigured and used in the USANS mode. More details about the design and operation of these instruments, as well as references to the instrumentation-specific articles, can be found at the corresponding websites shown in Tables 2.1 and 2.2.

3.4 Time-of-Flight USANS at SNS

The performance of existing reactor-based Bonse-Hart double-crystal diffractometers is nearly optimal. It has, however, an inherent limitation related to the fixed wavelength originating from the Bragg reflection from a crystal. As a result, the Q-resolution is also fixed and optimized for the highest Q-resolution ($Q_{min} \sim 2 \times 10^{-5}$ Å^{-1}). The USANS data overlap with the data taken using conventional SANS only for strongly scattering samples (e.g., rocks; Fig. 3.12), leaving a measurable gap in the range $\sim 8 \times 10^{-4}$ Å$^{-1} < \sim 4 \times 10^{-3}$ Å^{-1} for weakly scattering objects. In some cases this mismatch can seriously compromise the data analysis, especially if the sample has diffraction maxima in this Q-range.

The presence of high-order Bragg reflections in the spectrum of a reactor-based DCD can be ignored because of their negligibly small contribution in the intensity. By contrast, the discrete nature of the multiwavelength pulses generated by spallation sources offers an opportunity to optimize the flux-on-sample spectrum from

Table 3.3 Bragg diffraction parameters for the Si(220) family of crystallographic planes in the back scattering geometry at Bragg angle $\Theta_B = 70^\circ$ and the calculated TOF-USANS instrument resolution as a function of wavelength λ [14]

n	Si(h,k,l)	d (Å)	λ (Å)	Q_{min} (10^{-6} Å^{-1})
1	220	1.92	3.609	45.1
2	440	0.96	1.804	20.6
3	660	0.64	1.203	11.8
4	880	0.48	0.902	7.14

the first Bragg reflection and simultaneously boost the instrument resolution at high Q-values, making use of the normally extremely low-intensity high-order reflections [14]. Calculations show that the total flux gain factor for the SNS liquid hydrogen moderator and the source power of 2 MW could be as high as ~30 [10]. This significant gain factor could help to eliminate the gap between measurable USANS and SANS data for weakly scattering samples.

Table 3.3 summarizes the diffraction parameters for the first four Bragg reflections from the Si(220) family of crystallographic planes at $\Theta_B = 70^0$ chosen to be used in the SNS TOF-USANS instrument. The last column also gives the Q-resolution calculated for each wavelength. As can be seen, the discrete multiwavelength spectrum allows for achieving the Q-resolution of the conventional (reactor-based) DCD, such as the BT-5 USANS at NIST, for the second-order reflection. The availability of the discrete multiwavelength spectrum of Bragg diffraction increases the flux on sample in the range $Q_{min} > 4.5 \times 10^{-5}$ Å^{-1} and at the same time extends the USANS Q-range to the smaller Q-values. The TOF approach makes the USANS instrument resolution adjustable and allows to cover the entire Q-range in one run by performing an angular scan of the analyzer crystal.

3.5 SAXS and USAXS Instruments at APS, ANL

Similar to SANS, SAXS data contain information about size, shape, volume, and surface area of scattering objects, and a pinhole SAXS camera delivers structural information on length scales between 10 and 1000 Å. Structures with larger dimensions can be resolved using USAXS techniques that are based on the original Bonse-Hart double-crystal configuration, which uses multiple reflections from perfect crystals to reach a region of scattering angles Q that is generally hidden behind the beam stop in the pinhole SAXS instrument. The APS USAXS/SAXS instrument extends the capabilities of SAXS by covering more than 10 decades in scattering intensity and spanning more than 4 decades of scattering vector, which increases the size resolution range from few Angstroms up to few micrometers. The unique capability that distinguishes this instrument from other USAXS instruments available at major x-ray synchrotron facilities is that it allows for collecting

SAS data over the entire Q-range in a single scan [15, 16]. Additional important features include 2D-collimated USAXS, USAXS imaging, and anomalous USAXS measurements near the absorption edge. With the use of high-resolution, high-precision slits, the APS USAXS/SAXS can utilize beam with much smaller cross sections which allows to study microstructural variations in different sample areas [17].

A schematic of the APS USAXS/SAXS instrument is shown in Fig. 3.13. This instrument receives photons from an APS undulator, which provides continuous access to x-rays in the energy range from 3.2 to 80 keV, corresponding to the range of wavelengths from 4.1 to 0.155 Å. The photons pass through either a Si(111) or Si (220) monochromator that is capable of tuning the photon energy between 8 and 35 keV. The instrument can be operated in either 1D (Fig. 3.13a) or 2D (Fig. 3.13b) collimation geometry. For isotropically scattering samples, 1D-collimated USAXS mode is preferable. Its advantage is high x-ray throughput, a large sampling area (1 mm high × 2 mm wide), large Q-range ($10^{-4} \leq Q \leq 1$ Å^{-1}), and large scattering intensity range (up to 9 decades). As the analyzer crystal only collimates scattering in the vertical direction, this geometry introduces smearing along the horizontal axis, which requires a numerical desmearing of raw data [13]. The 2D-collimated USAXS mode makes desmearing unnecessary; however, it reduces the dynamic range of x-ray intensity by 2 decades, narrows the available Q-range to $10^{-4} \leq Q \leq 0.1$ Å^{-1}, and requires use of a smaller beam (0.4 × 0.4 mm). With an azimuthal rotational sample stage, the anisotropic scattering intensity patterns can be recorded by either scanning Q at a fixed azimuthal angle or by scanning the azimuthal angle at a fixed Q.

To meet the need for high-Q scattering measurements from weakly scattering materials, the instrument is equipped with a pinhole camera (PILATUS 100 K) that covers the Q-range between 0.05 and 1.2 Å^{-1}. In conjunction with the USAXS/SAXS range, this allows for measurements of extended Q-range scattering profiles, which is especially useful for characterization of materials with hierarchical structures, such as coal or shale. The schematic of the pinhole SAXS setup is shown in Fig. 3.13c. An example of the combined USAXS/SAXS data taken from metal organic framework (MOF) and zeolite-based CO_2 absorbent is shown in Fig. 3.14.

An *Igor Pro*–based data reduction package (Indra) has been developed for reduction of USAXS/SAXS data [18]. The data reduction consists of several standard steps such as calculation of the sample transmission, absolute calibration, and subtraction of the backgrounds. Correction for slit smearing of 1D-collimated USAXS data, as well as correction for multiple scattering, is included in the package. A tool suite, Irena [18], was developed for the analysis of x-ray and neutron SAS data, within the commercially available *Igor Pro* software. Irena brings together a comprehensive set of tools for materials science, materials physics, chemistry, polymer science, and other fields. In addition to the standard Guinier and Porod fits, the suite includes size distribution determination in the dilute limit, modeling of dilute-limit SAS from multiple noninteracting populations of scattering objects, unified fits, pair distance distribution functions, etc. [18].

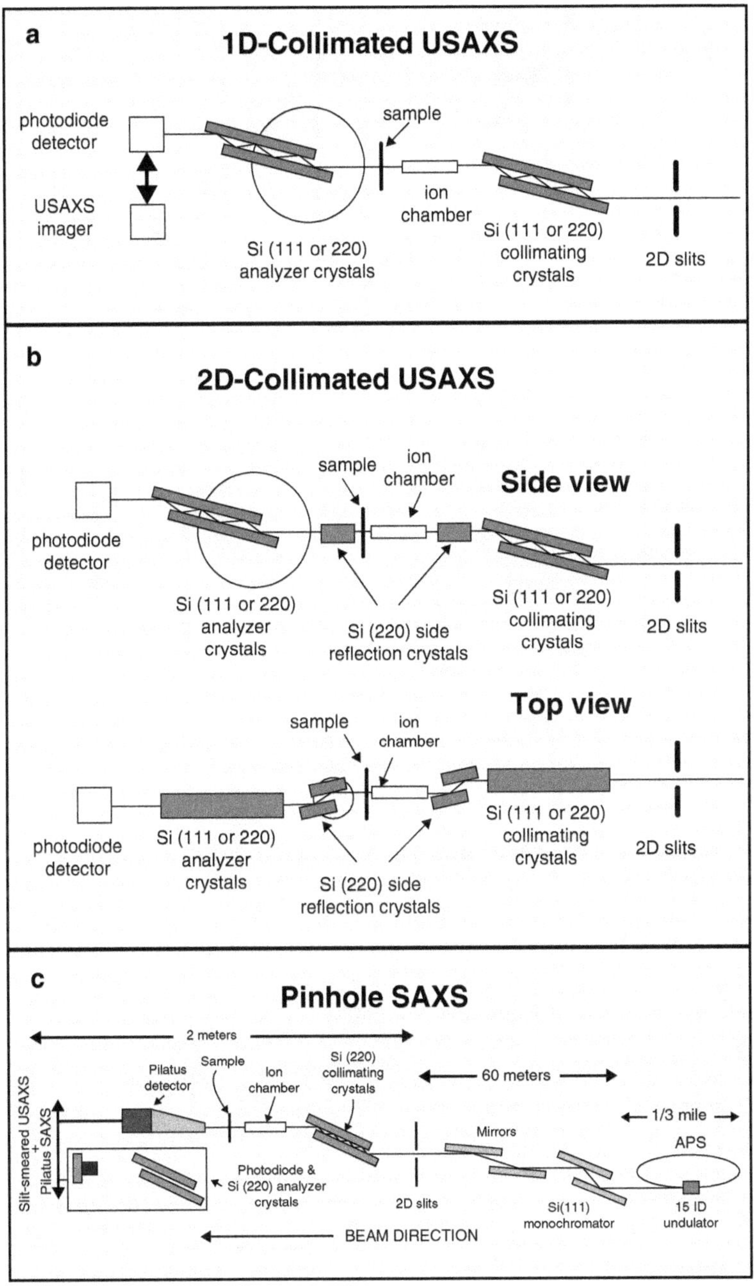

Fig. 3.13 Schematic of the APS USAXS/SAXS instrument in (**a**) 1D-collimated or USAXS-imaging configuration, (**b**) 2D-collimated configuration, and (**c**) pinhole SAXS configuration. Reprinted from [15] with kind permission from Springer Science and Business Media

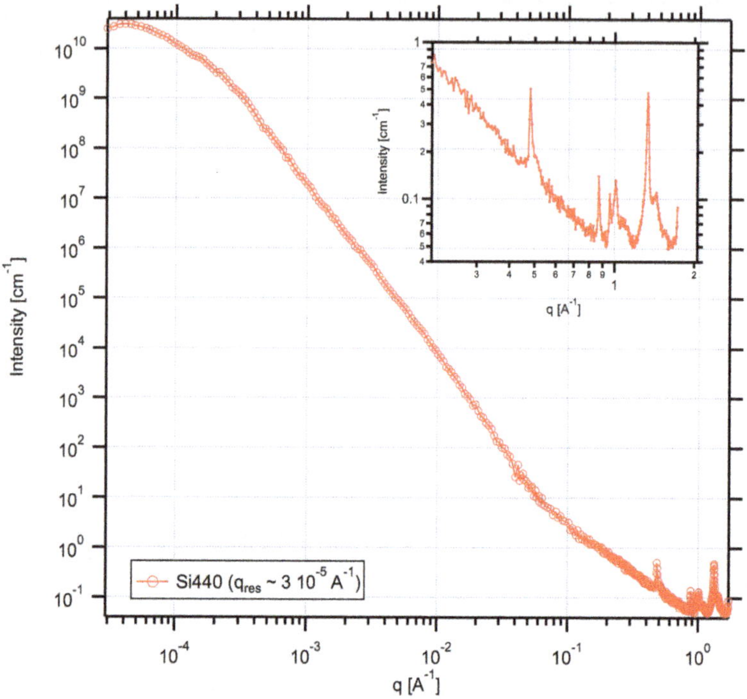

Fig. 3.14 Combined USAXS [Si(440), 21.0 keV, desmeared] and SAXS data from a new MOF- and zeolite-based CO_2-absorbent material (courtesy of J. Ilavsky)

Irena is available for both Windows and Macintosh operating systems and can be downloaded for free from the APS website [19].

The instrument continues to be improved and upgraded. One of the forthcoming upgrades will make use of the high-energy x-rays available in the undulator spectrum delivered to the instrument. Such x-rays have the advantage of higher penetration power, which would allow for studying the structure of thicker samples in complex environments, such as pressure cells. In addition, use of the high-energy x-rays would allow for intensity measurements much closer to the forward direction, decreasing the minimum Q from 10^{-4} to $\sim 2 \times 10^{-5}$ Å^{-1}, and thus making it possible to probe structural objects with sizes up to ~30 μm. More details about instrument performance, as well as examples of recent research using the APS USAXS/SAXS instrument, can be found in recent reviews [15, 16]. Pinhole SAXS and Bonse-Hart USAXS instruments are currently available at major synchrotron sources all around the world. Because of their high flux, resolution, and versatility, these instruments are in high demand for researchers working in all areas of materials science, physics, chemistry, and biology.

References

1. Wignall, G.D., Littrell, K.C., Heller, W.T., Melnichenko, Y.B., Bailey, K.M., Lynn, G.W., Myles, D.A., Urban, V.S., Buchanan, M.V., Selby, D.L., Butler, P.D.: The 40 m general purpose small-angle neutron scattering instrument at Oak Ridge National Laboratory. J. Appl. Crystallogr. **45**, 990 (2012)
2. Hammouda, B.: Multidisc neutron velocity selectors. Nucl. Instr. Meth. Phys. Res. **A321**, 275 (1992)
3. Rosta, L.: Neutron physical properties of a multidisc velocity selector. Phys. B **156–157**, 615 (1989)
4. Lumsden, M.D., Robertson, J.L., Yethiraj, M.: SPICE—spectrometer and instrument control environment. Phys. B **385–386**, 1336 (2006)
5. Zhao, J.K., Gao, C.Y., Liu, D.: The extended Q-range small-angle neutron scattering diffractometer at the SNS. J. Appl. Crystallogr. **43**, 1068 (2010)
6. Liu, D., Hong, K., Gao, C.Y., Melnichenko, Y., Littrell, K., Smith, G., Zhao, J.: First data acquired on the extended Q-range small-angle neutron scattering (EQ-SANS) diffractometer at the spallation neutron source. J. Appl. Crystallogr. **44**, 1120 (2011)
7. Bonse, U., Hart, M.: Tailless X-ray single-crystal reflection curves obtained by multiple reflections. Phys. Lett. **7**, 238 (1965)
8. Bonse, U., Hart, M.: Small-angle X-ray scattering by spherical particles of polystyrene and polyvinyltoluene. Z. Phys. **189**, 151 (1966)
9. Agamalian, M., Wignall, G.D., Triolo, R.: Optimization of a Bonse-Hart ultra-small-angle neutron scattering facility by elimination of the rocking-curve wings. J. Appl. Crystallogr. **30**, 345 (1997)
10. Agamalian, M.: Bonse-Hart USANS instrument. In: Imae, T., Kanaya, T., Furusaka, M., Torikai, N. (eds.) Neutrons in Soft Matter. John Wiley & Sons, New York (2011)
11. Barker, J.G., Glinka, C.J., Moyer, J.J., Kim, M.H., Drews, A.R., Agamalian, M.M.: Design and performance of a thermal-neutron double-crystal diffractometer for USANS at NIST. J. Appl. Crystallogr. **38**, 1004 (2005)
12. Kline, S.R.: Reduction and analysis of SANS and USANS data using Igor Pro. J. Appl. Crystallogr. **39**, 895 (2006)
13. Lake, J.A.: An iterative method of slit-correcting small angle X-ray data. Acta. Crystallogr. **23**, 191 (1967)
14. Carpenter, J.M., Agamalian, M., Littrell, K.C., Thyagarajan, P., Rehm, C.: Time-of-flight implementation of an ultra-small-angle neutron scattering instrument. J. Appl. Crystallogr. **36**, 763 (2003)
15. Ilavsky, J., Zhang, F., Allen, A.J., Levine, L.E., Jemian, P.R., Long, G.G.: Ultra-small-angle X-ray scattering instrument at the advanced photon source: history, recent development, and current status. Metall. Mater. Trans. A **44A**, 68 (2013)
16. Ilavsky, J., Jemian, P.R., Allen, A.J., Zhang, F., Levine, L.E., Long, G.G.: Ultra-small-angle X-ray scattering at the advanced photon source. J. Appl. Crystallogr. **42**, 469 (2009)
17. Radlinski, A.P., Busbridge, T.L., MacA Gray, E., Blach, T., Cheng, G., Melnichenko, Y.B., Cookson, D.J., Mastalerz, M., Esterle, J.: Dynamic micromapping of CO_2 sorption in coal. Langmuir **25**(2385) (2009)
18. Ilavsky, J., Jemian, P.R.: Irena: tool suite for modeling and analysis of small-angle scattering. J. Appl. Crystallogr. **42**, 347 (2009)
19. http://usaxs.xray.aps.anl.gov/staff/ilavsky/irena.html

Further Reading

Higgins, J.S., Benoit, H.C.: Polymers and Neutron Scattering. Clarendon, Oxford (1994)

Pethrick, R.A., Dawkins, J.V. (eds.): Modern Techniques for Polymer Characterization. Wiley, Chichester, New York (1999)

Roe, R.J.: Methods of X-Ray and Neutron Scattering in Polymer Science. Oxford University Press, New York, Oxford (2000)

Lindner, P., Zemb, T. (eds.): Neutron, X-Ray and Light Scattering: Introduction to an Investigative Tool for Colloidal and Polymeric Systems. North-Holland, Amsterdam, Oxford, New York, Tokyo (1991)

Furrer, A., Mesot, J., Strassle, T.: Neutron Scattering in Condensed Matter Physics. World Scientific, New Jersey, London, Singapore (2009)

Imae, T., Kanaya, T., Furusaka, M., Torikai, N. (eds.): Neutrons in Soft Matter. Wiley, Hoboken, New Jersey (2011)

Borsali, R., Pecora, R. (eds.): Soft-Matter Characterization. Springer, New York (2008)

Chapter 4
Sample Environment

Sample environment is the collective name given to the pieces of equipment that hold or contain the sample and allow the variation of external parameters, such as pressure, temperature, magnetic field, shear rate, etc. This chapter considers different types of sample cells and containers used for the structural studies of porous materials at ambient conditions as well as for investigation of the fluid adsorption in pores at ambient and elevated pressure and temperature. Particular attention is given to description of the design and operation principles of high-pressure cells used for the SAS studies of the confined supercritical fluids and gases under pressure using different types of radiation (neutrons and x-rays).

4.1 Sample Cells for Ambient Conditions

The workhorses of SANS and USANS instruments are computer-controlled motorized multiple-position sample changer racks, which also provide a means of regulating and maintaining sample temperature. Such racks are mounted on horizontal translation stages, which allow for precise positioning of each sample in the beam. The sample changer available at the GP-SANS instrument at HFIR, ORNL (Fig. 4.1a), is made of ultrathermal conductive copper, the temperature of which is controlled by Peltier elements. The sample changer provides both cooling (up to ~ −20 °C) and heating (up to 150 °C) capability, with temperature stability and gradients better than ±0.01 °C. The changer has 12 sample positions, split into 4 sets of 3 samples. The temperature of each set can be programmed and controlled independently by the instrument computer. The sample changer can be used with standard quartz Hellma cells (flight path lengths of 1, 2, 5, and 10 mm), as well as with cells fitted with demountable quartz windows (Fig. 4.1b). To provide optimal transmission, the flight path length of demountable cells can be varied continuously between 0.1 and 12 mm using Teflon or stainless steel spacers. The EQ-SANS

Y.B. Melnichenko, *Small-Angle Scattering from Confined and Interfacial Fluids*, DOI 10.1007/978-3-319-01104-2_4

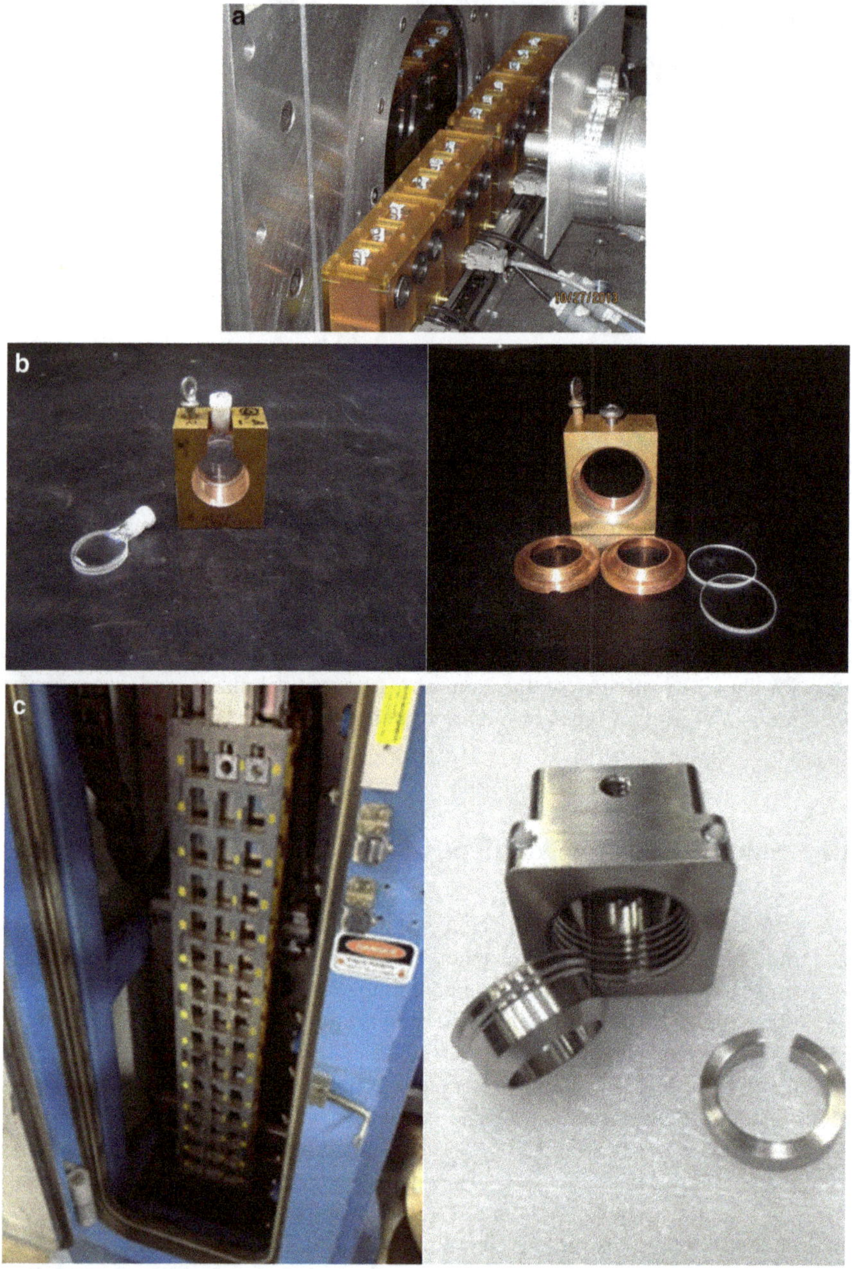

Fig. 4.1 (**a**) Peltier element-controlled 12-position sample changer on a translation stage of the GP-SANS, HFIR. (**b**) Sample holder for Hellma cells (*left*) and demountable cells (*right*). (**c**) Water-bath-controlled 45-position changer and titanium sample holder used at EQ-SANS, SNS

instrument at SNS is equipped with a water-bath-controlled sample changer that can accommodate 45 samples in titanium cell holders (Fig. 4.1c).

Because of the limited transparency to x-rays, there is a limited choice of materials that can be used for fabricating x-ray cells. A multiposition sample changer usually consists of a large plate with a number of holes. Samples in the form of foils or powders are placed between two layers of Scotch or Kapton polyimide adhesive tape and mounted on top of the plate holes (up to 30 samples can be used at the USAXS/SAXS instrument at APS). Computer-controlled stepping motors can position samples in the beam with a precision of about 10 μm, which allows 2D sample position scanning. Liquid sample holders can be made of very thin (~125 μm) glass microscopy cover slips with spacers that provide sample confinement between the slips. Use of glass as a window material requires usage of higher x-ray energies (above 10–11 keV). Windows for x-ray cells for chemically benign liquids can be made of Kapton film. Minimum sample volume of the smallest container is ~50 μL (8 mm diameter opening, 1 mm thick).

The solid and liquid sample holders described herein have been widely used to study the structure of pores and pore interconnectivity in rocks using contrast-matching liquid solvents [1, 2]. The first SANS vapor adsorption experiments were conducted by preparing several samples of a porous solid, bringing them in equilibrium with a vapor at different fixed relative pressures P/P_0 (P_0 saturation vapor pressure), sealing the cells and measuring SANS patterns as a function of P/P_0 [3, 4]. The first experimental setup (SANS adsorption or SANSADSO) allowing *in situ* SANS measurements of the equilibrium as well as kinetic aspects of vapor adsorption and condensation in pores was built in the end of the 1990s at BENSC, Germany [5]. The sample holder for this setup was designed as a metal tube fitted with Viton-sealed sapphire windows (1 mm thick, 15 mm in diameter) that could accommodate quartz cells for powder samples with the path length ranging from 1 to 3 mm. The tube temperature was controlled using a water bath thermostat. It was connected to a vapor generator, which allowed *in situ* SANS measurements from porous media as a function of the relative pressure P/P_0 later dubbed as "neutron porosimetry".

4.2 SANS High-Pressure Cells

In situ SAS studies of adsorption of pressurized fluids require using temperature-controlled high-pressure cells that provide low-level parasitic scattering from the vessel windows. Because of the high transparency to neutrons, windows for SANS high-pressure cells can be made from metal or crystalline materials (aluminum, titanium, or sapphire). The first SANS high-pressure high-temperature gas-liquid cells with the pressure exceeding 1 Kbar were built for studies of soft matter and phase behavior of complex fluids [6–11]. Fig. 4.2 shows the ORNL-1 cell, which is 133.4 mm long with an outside diameter (OD) of 76.1 mm. The cell body was made of stainless steel alloy 304 L and the retaining end caps—of a high-strength steel

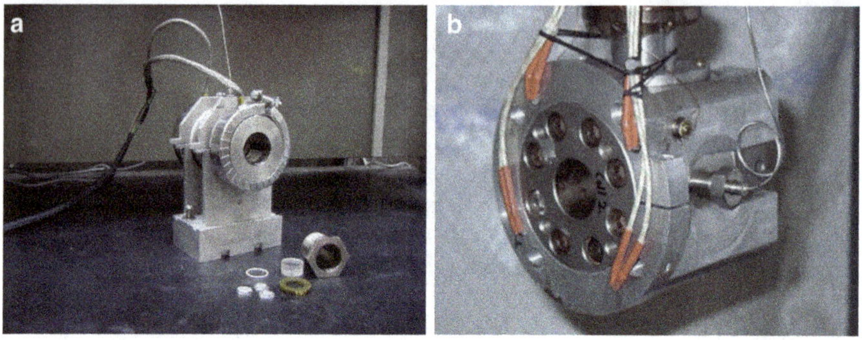

Fig. 4.2 (**a**) ORNL-1 high-pressure cell [6] with the removed outlet sapphire window, endcap, brass washer, and C-seal. The image also shows demountable aluminum powder sample holders with caps having 0.3 mm diameter holes to allow fluid access to a sample. (**b**) Assembled McHugh high-pressure cell. Both cells have a high Q-limit of ~0.2 Å^{-1} and are available at NIST and HFIR (ORNL) for high-pressure studies of scattering objects with sizes ≥30 Å

alloy (Nitronic 60), to minimize the probability of galling. The cell is fitted with sapphire windows (25.4 diameter and 12.7 mm thickness), which are sealed using silver-plated metal C-type seals. The internal diameter of the cell is 17 mm, and the maximum flight path between the internal window surfaces (20 mm) can be reduced using sapphire spacers. The temperature is measured using a resistance temperature detector (RTD) and is controlled to within ±0.1 °C. The cell is rated for 1 Kbar at 250 °C.

The McHugh high-pressure cell is 94 mm long and 76 mm in diameter; the cell body, constructed of a high-strength steel alloy (Nitronic 50), is 94 mm long and has a 76 mm OD (Fig. 4.2b). The cell is fitted with sapphire windows (19.1 OD and 19.1 mm thickness), which are sealed with elastomeric O-rings. The path length between the inner surfaces of the windows is adjustable between 1 and 5 mm using spacers. The sample temperature is measured with a calibrated thermocouple. An aluminum band heater with four resistance heating cartridges and four cooling lines is used to maintain the cell temperature with an accuracy of ±0.5 °C. The cell is rated for 2 Kbar at 200 °C.

The retaining endcaps of both ORNL-1 and McHugh cells machined with a 30° included-angle cone allow a neutron beam to exit the scattering cell at a maximum angle of 15° relative to the outer window surface. The theoretical achievable value of the maximum scattering vector calculated using (1.4) for $\lambda = 4.7$ Å ($Q_{max} \sim 0.69$ Å^{-1}) is in practice much smaller for both cells ($Q_{max} \sim 0.2$ Å^{-1}) because samples are mounted behind a thick sapphire window, deep in the cell body. This value of Q_{max} corresponds to the minimum detectable size of the scattering objects of $2\pi/Q_{max} \sim R_{min} \sim 30$ Å, which is sufficient for studies of most polymer-supercritical fluid mixtures. However, it is not satisfactory for investigation of nanometer and subnanometer objects (e.g., interlamellar spacing in anthracite coal or pores in activated carbons). Both types of high-pressure cells are available to users of the SANS and USANS facilities at ORNL and NIST.

The conceptual design of the second generation of high-pressure cells (ORNL-2) is similar to the ORNL-1 cell. These cells are shorter (76.2 mm long, 76.2 mm OD) because of a shorter maximum flight path (10 mm), which reduces thermal mass and accelerates the heating and cooling process. They are designed to use the same metal seals and entry sapphire window as those used with the ORNL-1 cell. The exit sapphire window is larger than for ORNL-1 cells (25.4 diameter and 12.7 mm thickness), which allows for increasing the maximum scattering angle and thus the value of Q_{max} to ~ 0.6 Å^{-1} (R_{min} ~ 15 Å). A set of four ORNL-2 cells allows for measuring SANS from four different samples at the same pressure and four different temperatures independently controlled by the instrument computer (Fig. 4.3). An option for independent pressure control for each cell is under development.

Further increase of the maximum scattering angle and corresponding Q_{max} is not possible within the conceptual design of the ORNL-1 and ORNL-2 cells. It was therefore decided to design the cell body out of titanium and eliminate the restriction on the detector-side acceptance angle by using a metal "dome" exit path, as shown in Fig. 4.4. The body of the third-generation high-pressure cell ("Dome cell") is made of titanium and has a length of 53.3 mm and an OD of 58.4 mm. The cell is sealed with a single sealing nut and the same metal C-type seals as those used for the ORNL-1 and ORNL-2 cells. Solid porous samples can be placed directly in the cell. To avoid moving powder particles during fluid injection or discharge, powder samples are placed in thin-wall aluminum sample holders. These holders and/or solid samples are supported by a sapphire hemispherical insert that minimizes the parasitic volume, which otherwise would be filled with the injected fluid. The temperature of the cell is controlled by four 25 W heater cartridges.

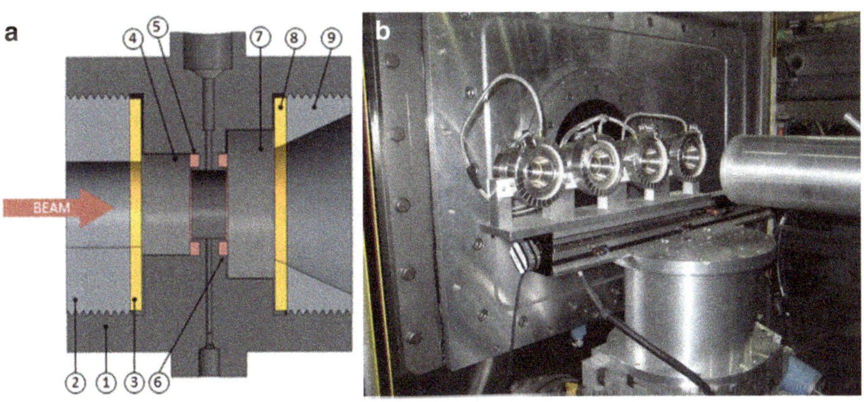

Fig. 4.3 (**a**) Cross-sectional view of the ORNL-2 cell. Labels: (*1*) cell body, (*2*) entry sealing nut, (*3*) and (*8*) brass washer, (*4*) entry sapphire window, (*5*) and (*6*) metal C-seals, (*7*) exit sapphire window, (*9*) exit sealing nut. Image courtesy of J. Carmichael, ORNL. (**b**) A set of four ORNL-2 high-pressure cells (P_{lim} = 1 kbar, T_{lim} = 200 °C) on a translation stage in the beam of GP-SANS, HFIR. The cells were rotated by 180° to show the wider exit windows that allow for increasing the maximum scattering angle and the value of Q_{max} to ~ 0.6 Å^{-1}

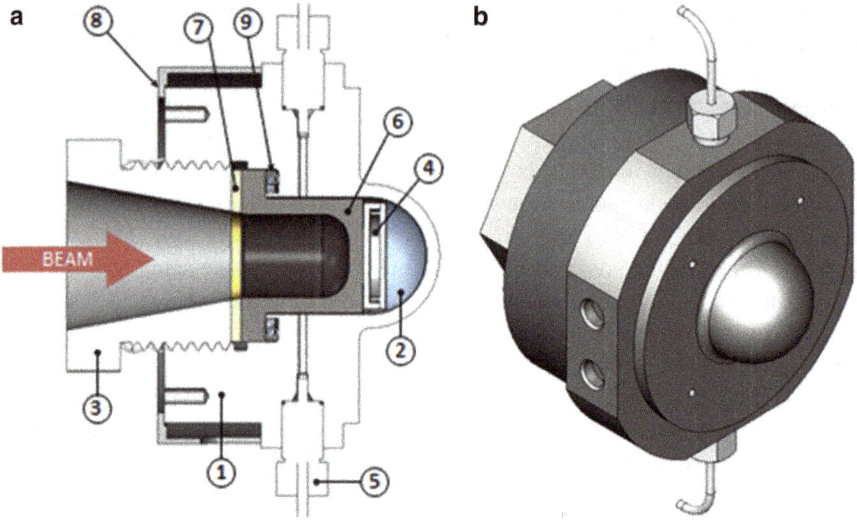

Fig. 4.4 (**a**) Cross-sectional view of the ORNL Dome cell: (*1*) cell body, (*2*) sapphire hemisphere spacer, (*3*) sealing nut, (*4*) aluminum sample holder, (*5*) pressure adapters, (*6*) titanium sealing insert, (*7*) brass plate, (*8*) wire retainer, and (*9*) metallized c-ring seal. (**b**) Isometric view of the complete cell. Images courtesy of J. Carmichael (ORNL, USA)

The thickness of the titanium dome is 1.9 mm; such design increases the scattering angle up to ~180°. The cell is rated for a maximum pressure and temperature of 1 Kbar and 300°C, respectively. A set of four Dome cells was built and mounted on a single translation stage, which allows for measuring SANS from four different samples at the same pressure and four different temperatures independently controlled by the instrument computer. An option of independent pressure control of each cell is under development. Tests have shown that scattering from the empty Dome cell in the high Q-range ($Q > 0.05$ Å^{-1}) is flat, which is ideal for studying small scattering objects with sizes of <125 Å. In the low Q-limit, scattering from an empty titanium cell is comparable to that of aluminum and higher than that of sapphire only by a factor of ~2.

The described SANS high-pressure are usually used for studies of the fluid adsorption and condensation at temperatures above the room temperature. However, in many cases it is desirable to expand the temperature range down to cryogenic conditions (e.g., for studies of hydrogen or helium adsorption near their critical gas-liquid temperatures) [12, 13]. A cross section and image of a cryogenic SANS high-pressure cell that meets these requirements is shown in Fig. 4.5. The cell body is made of 7075-T6 aluminum and is designed to fit into a 50 mm bore of a cryostat equipped with sapphire windows to minimize parasitic scattering. On top of the cell there is a pressure feed and a connector used to mount the cell onto the cryogenic stick. The inner diameter (ID) of the sample area is 20 mm, and the maximum internal flight path of the cell (5 mm) can be adjusted using sapphire spacers. The cell has a round insert made of neutron absorbing

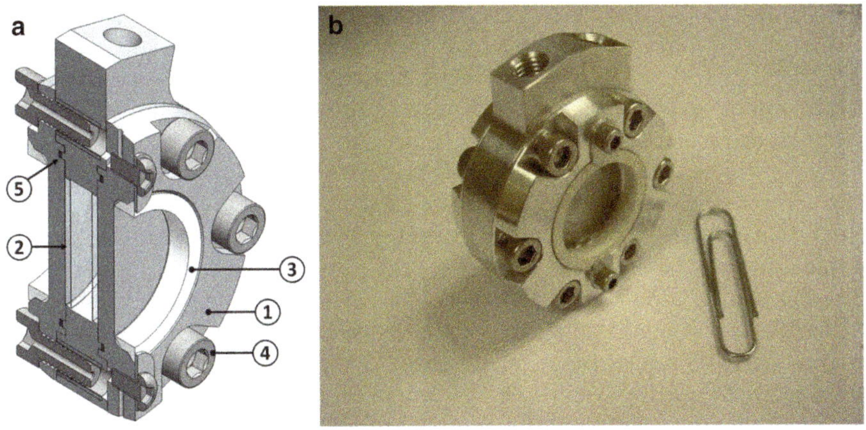

Fig. 4.5 (**a**) Cross-sectional view of the ORNL cryogenic high-pressure cell: (*1*) cell body, (*2*) aluminum window, (*3*) boron nitride insert/shield, (*4*) retaining bolts, (*5*) indium seal. (**b**) Cryogenic high-pressure cell. Images courtesy of J. Carmichael (ORNL, USA)

material boron nitride, which helps to align the cell in the cryostat against neutron beam. The cell has three different sets of aluminum windows, with the thicknesses T_{Al} optimized for three pressure ranges: $T_{Al} = 3.05$ mm for P_{max} 200 bar, $T_{Al} = 1.91$ mm for P_{max} 100 bar, and $T_{Al} = 1.22$ mm for P_{max} 50 bar. The cell mounted in the cryostat cannot be placed close to the entry silicon window of the SANS tank; therefore, the value of Q_{max} is limited to ~0.35 Å^{-1} ($R_{min} \sim 20$ Å).

A number of other SANS high-pressure cells (see, for example, [14, 15] and references therein) are available to users at various facilities such as NIST, USA; ILL, France; PSI, Switzerland; Laboratoire Leon Brillouin, France; and Helmholtz Zentrum, Germany. Each high-pressure cell is unique and has its advantages and disadvantages for a particular type of experiment. The suite of available high-pressure cells and their capabilities is continuously expanding, and potential users are advised to consult local instrument scientists and facility sample environment websites for updated information.

4.3 SAXS High-Pressure Cells

A key difference between SANS and SAXS high-pressure cells is the design and material used for fabrication of the cell windows. A good SAXS window must satisfy several, often contradictory, requirements. The window material should be relatively transparent to x-rays and provide low-level background SAS compared with the sample under investigation. At the same time, it must be capable of withstanding high pressures without significant deformation. The window scattering should be symmetrical about the beam center, especially in experiments where azimuthal averaging of the 2D scattering pattern is performed. Ideally, the SAXS window should also provide uniform scattering that is insensitive to slight changes in the cell position if the cell is removed and put back in the beam after loading a sample. Finally, the window should be chemically inert to the loaded samples. The strong absorption of x-rays by matter imposes severe limitations on a number of materials that can be used for fabrication of windows for SAXS high-pressure cells. The only materials that satisfy to a reasonable extent at least some of the aforementioned requirements are low-Z diamond and beryllium.

Despite their much higher cost, diamonds are often considered to be the preferred material for SAXS windows because of their high hardness, very low adhesion, unmatched thermal conductivity, high melting temperature, high dielectric breakdown strength, radiation hardness, and unique chemical inertness [16]. A number of SAXS studies have shown that the least parasitic scattering is observed from nitrogen-free diamonds, classified as type IIa. However, such diamonds are very rare in nature and make up only 1–2 % of all natural diamonds. Recent developments in microwave plasma-assisted chemical vapor deposition (MPCVD) synthesis have enabled high growth rates of single-crystal diamonds, reducing their cost and changing the prospects for their application as window material for SAXS. The type IIa diamonds can easy sustain multi-megabar pressures. Recent synchrotron SAXS studies have confirmed their extremely low level of parasitic scattering, which was found to be comparable with the scattering measured from a 7.5 μm thick Kapton film [17].

Figure 4.6a shows the schematic of a beryllium window cell designed for high-energy x-ray diffraction studies of hydrogen-metal systems in the transmission mode [18]. The cell body consists of two stainless-steel flanges, sealed by an O-ring. The thickness of the sample is defined by the separation of the beryllium windows and is thus fixed for a particular set of flanges. In the two available versions, the sample thicknesses are 0.5 and 1.0 mm, aimed at transition metals and light elements, respectively. The cell flanges were designed for a working

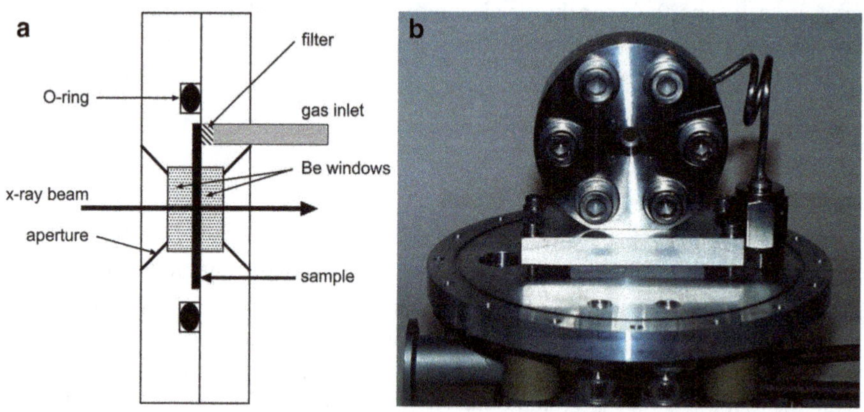

Fig. 4.6 (**a**) Design concept of the beryllium window x-ray cell for studies under gas pressure in transmission geometry. (**b**) X-ray pressure cell, without windows, mounted on a height-adjustable stage. Images courtesy of Evan MacA. Gray (Griffith University, Australia)

pressure rating of 100 bar at 300 °C. Figure 4.6b shows the cell (without its vacuum shroud, beryllium windows, and lead flange masks) mounted on a stage that bolts directly to a Huber 410 goniometer head. With a close-coupled gas-handling system attached, the whole cell assembly can be rotated to any angle within the acceptance angle designed into the flanges. Rotation about the beam axis is also possible, contributing to good powder averaging inside the cell. The cell is mounted on a plate beneath which a Peltier-effect heat pump can be fitted for temperature control near room temperature. Alternatively, the cell can be held by a cradle, which permits its thermal isolation from the base plate by ceramic insulating posts. The cell is then heated with small cartridge heaters set into its cradle.

The cross section of one of the most advanced SAXS high-pressure cells built so far is shown in Fig. 4.7a. This high-pressure cell was designed to accommodate a wide variety of samples for a broad range of experiments. Key design parameters include the following [19]:

- The cell is suitable for static and pressure jump experiments up to 5 Kbar at temperatures between −20 and 120 °C.
- The x-ray windows remain constantly in place during sample changes to allow for accurate subtraction of background scattering from the high-pressure windows, an essential data processing step when investigating samples that show diffuse scattering.
- The cell requires minimal intervention during a typical synchrotron experimental run.

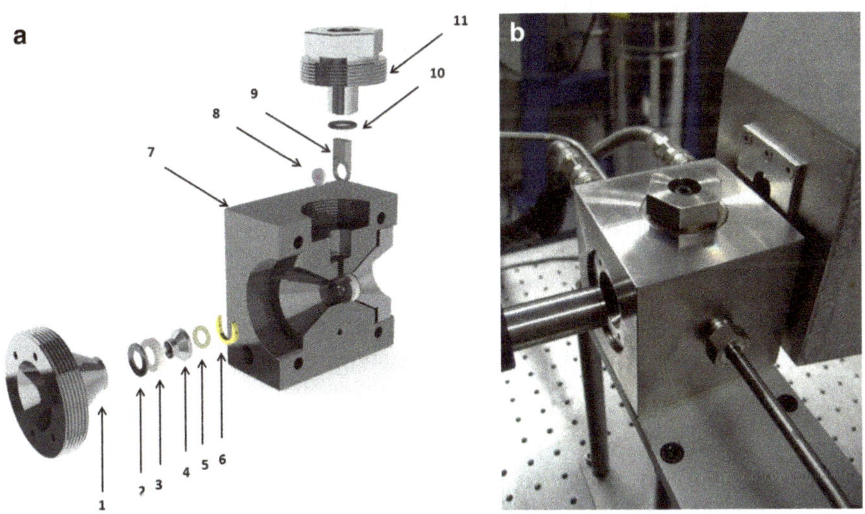

Fig. 4.7 (**a**) Cross section of the diamond window x-ray high-pressure cell: (*1*) window plug, (*2*) anti-extrusion washer, (*3*) window seal, (*4*) window support, (*5*) diamond window, (*6*) window spacer and sample carrier guide, (*7*) cell body, (*8*) sample Teflon carrier, (*9*) sample loader, (*10*) sealing O-ring, and (*11*) sample plug. (**b**) Cell positioned on beam line I22 at the Diamond Light Source. Both images courtesy of N. J. Brooks (Imperial College, UK)

- The sample is self-contained to avoid cell contamination and preclude the need for frequent cleaning.
- The sample is easy to load and unload to minimize downtime.
- The cell allows access to both small-angle and limited wide-angle x-ray scattering.

The basic geometry of the cell consists of two perpendicular bores: one along the x-ray path running through the length of the cell and sealed at both ends by x-ray windows and the other for loading the sample (Fig. 4.7). This loading port is a major advantage over previous pressure cells, where the windows have to be removed to load samples. Removing the windows is a major problem as even minor rotation of the cell body around the x-ray beam can significantly affect background scattering, making it difficult to accurately subtract it from experimental data. The two bores intersect via a rectangular slot to minimize dead volumes, reduce surface areas, and hence minimize stresses. A smaller additional intersecting bore, running perpendicular to both the x-ray path and sample port, is used to connect the cell to the high-pressure network on one side and to a bleed valve on the other. The bleed valve allows any trapped air in the cell to be purged after changing the sample. An ID of 14 mm was chosen for both bores, and the rectangular sample slot was machined with a 10 × 4 mm cross section.

The x-ray windows for the high-pressure cell are 1 mm thick, 5 mm diameter, flat single-crystal chemical-vapor deposition type IIa diamonds (EasyLab Technologies, Ltd., UK). Although diamonds have a relatively low transmission for CuK_{α} x-rays (1.54 Å at 8 keV), this pressure cell was developed specifically for use at a synchrotron beam line where a continuously variable x-ray energy is available in the range of 3.7–20 keV (3.4–1.6 Å). Operation with 17 keV (0.73 Å) x-rays allows for ~62 % transmission through 2 × 1 mm of diamond. Additionally, the use of 17 keV x-rays could help reduce radiation damage to biological samples by decreasing x-ray absorption relative to 8 keV. The windows consist of circular plates mounted on supports with a circular aperture of 2.5 mm, allowing a scattering angle of up to 21°. The calculated window failure pressure is 1235 Mpa (12.35 Kbar), significantly higher than the designed cell operation pressure of 500 MPa (5 Kbar). The maximum deflection at the window center at this pressure is 0.859 μm.

There are four different types of high-pressure seals used in the pressure cell. A Poulter-type seal is used to seal each window against its window support. Bridgman-type seals are used between the window supports and the cell body. A simple O-ring seal can be used as an alternative sample closure for pressures below 4 Kbar. Cone-to-cone couplings are employed in commercial high-pressure fittings and have been used to connect the cell to the high-pressure network equipped with a remotely controlled pressure generation system. A custom-designed high-pressure reservoir with a volume of 50 cm^3 adds significant volume to the network and allows for performing pressure jumps.

The described setup was designed specifically for use at beam line I22 at the Diamond Light Source, UK. It has been fully integrated with the I22 beam line control systems and is available to users for high-pressure experiments. Diamond window SAXS high-pressure cells are also available at the Cornell High Energy Synchrotron Source (CHESS) G1 station, USA [20]; the BL-15A station at the High Energy Accelerator Research Organization (KEK), Japan [21], and some other facilities.

References

1. Gethner, J.S.: The determination of the void structure of macroporous coals by small-angle neutron scattering: void geometry and structure in Illinois No-6 bituminous coal. J. Appl. Phys. **59**, 1068 (1986)
2. Hall, P.J., Antxustegi, M., Ruiz, W.: Contrast-matching small-angle neutron scattering evidence for the absence of a connected pore system in Pittsburgh No. 8 coal. Fuel **77**, 1663 (1998)
3. Lin, M.Y., Sinha, S.K., Huang, J.S., Abeles, B., Johnson, J.W., Drake, J.M., Glinka, C.: A SANS study of capillary condensation in porous media. J. Mater. Res. Soc. Proc. **166**, 449 (1990)
4. Ramsay, J.D.F., Wing, G.: Small angle neutron scattering investigations of water sorption in porous silica and ceria gels. J. Colloid Interface Sci. **141**, 475 (1991)
5. Hoinkis, E.: Small angle neutron scattering study of C_6D_6 condensation in a mesoporous glass. Langmuir **12**, 4299 (1996)
6. Melnichenko, Y.B., Kiran, E., Wignall, G.D., Heath, K.D., Salanival, S., Cochran, H.D., Stamm, M.: Pressure and temperature-induced transitions in solutions of poly(dimethyl siloxane) in supercritical carbon dioxide. Macromolecules **32**, 5344 (1999)
7. DiNoia, T.P., Kirby, C.F., van Zanten, J.H., McHugh, M.A.: SANS study of polymer-supercritical fluid solutions: transitions from liquid to supercritical fluid solvent quality. Macromolecules **33**, 6321 (2000)
8. Gabke, A., Kraineva, J., Kohling, R., Winter, R.: Using pressure in combination with X-ray and neutron scattering techniques for studying the structure, stability, and phase behavior of soft condensed matter and biomolecular systems. J. Phys. Condens. Matter **17**, S3077 (2005)
9. Takeno, H., Nagao, M., Nakayama, Y., Hasegawa, H., Hashimoto, T., Seto, H., Imai, M.: High pressure cell for small-angle neutron and light scattering studies of phase transitions in complex liquids. Polymer J. **29**, 931 (1997)
10. Pepi, G., Baroni, P.: A new high pressure cell with sapphire windows for small-angle scattering. J. Appl. Crystallogr. **36**, 814 (2003)
11. Bonetti, M., Calmettes, P.: Sapphire-anvil cell for small-angle neutron scattering measurements in large-volume liquid samples up to 530 MPa. Rev. Sci. Instrum. **76**, 043903 (2005)
12. Morris, J.R., Contescu, C.I., Chisholm, M.F., Cooper, V.R., Guo, J., He, L., Ihm, Y., Mamontov, E., Melnichenko, Y.B., Olsen, R., Pennycook, S.J., Stone, M., Zhang, H., Gallego, N.C.: Modern approaches to studying gas adsorption in nanoporous carbons. J. Mater. Chem. A **1**, 9341 (2013)
13. Prisk, T.R., Pantalei, C., Kaiser, H., Sokol, P.E.: Confinement-driven phase separation of quantum liquid mixtures. Phys. Rev. Lett. **109**, 075301 (2012)
14. Klosterman, M., Foster, T., Schweins, R., Lindner, P., Glatter, O., Strey, R., Sottmann, T.: Microstructure of supercritical CO_2-in-water microemulsions: a systematic contrast variation study. Phys. Chem. Chem. Phys. **13**, 20289 (2011)
15. Kohlbrecher, J., Bollhadler, A., Vavrin, R.: A high pressure cell for small angle neutron scattering up to 500 MPa in combination with light scattering to investigate liquid samples. Rev. Sci. Instrum. **78**, 125101 (2007)

16. Asmussen, J., Reinhard, D. (eds.): Diamond Films Handbook. Marcel Dekker, New York (2002)
17. Wang, S., Meng, Y., Ando, N., Tate, M., Krasnicki, S., Yan, C., Liang, Q., Lai, J., Mao, H., Gruner, S.M., Hempley, R.J.: Single-crystal CVD diamonds as small-angle X-ray scattering windows for high-pressure research. J. Appl. Crystallogr. **45**, 453 (2012)
18. Mac, E., Gray, A., Cookson, D.J., Blach, T.P.: X-ray diffraction cell for studying solid-gas reactions under gas pressures to 100 bar. J. Appl. Crystallogr. **39**, 850 (2006)
19. Brooks, N.J., Gauthe, B.L.L.E., Terill, N.J., Rogers, S.E., Templer, R.H., Ces, O., Seddon, J. M.: Automated pressure cell for pressure jump x-ray diffraction. Rev. Sci. Instrum. **81**, 064103 (2010)
20. Ando, N., Chenevier, P., Novak, M., Tate, M.W., Gruner, S.M.: High hydrostatic pressure small-angle X-ray scattering cell for protein solution studies featuring diamond windows and disposable sample cells. J. Appl. Crystallogr. **41**, 167 (2007)
21. Tanaka, Y., Takahashi, Y., Morita, T., Nishikawa, K.: Sample holder for small-angle X-ray scattering measurements and density fluctuation of supercritical Xenon. Jap. J. Appl. Phys. **47**, 334 (2008)

Further Reading

Praise, J.: High pressure studies. Rev. Min. Geochem. **63**, 205 (2006)
Klotz, S.: Techniques in High Pressure Neutron Scattering. CRC Press, Boca Raton, London, New York (2012)

Chapter 5
Practical Aspects of Planning and Conducting SAS Experiments

This chapter begins with tips on writing a successful beam time proposal followed by the guidelines on the optimization of the instrument configuration that allow maximizing output of the high quality experimental data. Methods of the instrument and sample background subtraction are described along with the choice of optimal sample thickness and other parameters that permit minimizing contribution of multiple scattering. Placing SAS data on absolute scale is important as a sensitive indicator of the applicability of the structural model chosen for the SAS data analysis. This can be achieved by using absolute standards or direct beam measurements. Accurate estimation of the effective thickness of powder samples is required for placing SAS data on absolute scale. Several methods of such estimation using different physical parameters are described along with the contrast variation methods using liquid solutions of protonated and deuterium substituted solvents and pressurized gases as contrast matching media.

5.1 Applying for Beam Time

The vast majority of neutron and photon sources operate as "user facilities" that provide access to neutron research instruments for researchers from industry, universities, and government agencies. These facilities are operated with different modes of access. At SNS and HFIR, access is available via a general user program. Calls for proposals are usually issued twice a year, and beam time is allocated on a competitive basis after a peer-review process in which proposals are reviewed for scientific merit. A fraction of total available beam time (~25 % at SNS and HFIR) is allocated for scientific research conducted by instrument personnel and for instrument and sample environment upgrades, etc. A limited amount of beam time might be provided via a "rapid access" route for conducting urgent experiments or feasibility measurements.

Y.B. Melnichenko, *Small-Angle Scattering from Confined and Interfacial Fluids*, DOI 10.1007/978-3-319-01104-2_5

There is no access charge for researchers whose results are made available to the scientific community by being published in the open scientific literature. A fraction of beam time may be used for confidential, proprietary research, which can be performed only under full cost recovery. Most user facilities have financial assistance programs covering travel, room and board, etc. to encourage new users to acquire their first experience in neutron and x-ray scattering.

SAS instruments are in great demand and are usually oversubscribed by a factor of 2–3. Because of the strong competition for beam time, developing good proposal writing skills is essential. Some tips that might help first-time users obtain beam time follow.

One of the most important factors is choosing the "right" facility for a particular SAS experiment, including the type of radiation (neutrons vs. x-rays), power (brilliance), and the availability of the relevant sample environment. The higher radiation flux provided by synchrotron sources permits better statistics to be obtained in a shorter time, which could be critical for studying weakly scattering samples or for experiments where it is necessary to study, for instance, a large number of samples at numerous temperatures or pressure increments. Synchrotron or TOF instruments are usually preferable for kinetic or time-resolved experiments. Both TOF and constant flux SANS are methods of choice if contrast variation or zero average contrast conditions should be used during experiments. Other important factors, such as experience, established collaboration with a local scientist, or the availability of funds for travel, should also be considered.

After choosing a facility and a particular SAS instrument, potential users should contact instrument staff, who are available to answer questions and provide details about recent instrument and sample environment upgrades that might not yet be listed on instrument fact sheets or elsewhere online. Instrument scientists (aka local contacts) can advise on the feasibility of proposed experiments, help to estimate and justify the amount of requested beam time, and help to address why their specific facility is the best choice to meet specific experimental requirements, as well as comment on a statement of research. Such specific information is important because it could help reviewers fully understand and better evaluate the proposal and hence facilitate allocating beam time for the proposed experiment.

Most user facilities offer downloadable proposal templates and advice on what specific information should be included. An important component of any proposal is a clear articulation of scientific goal. Science at user facilities is diverse, and reviewers cover broad areas. Therefore, the scientific context of the proposal should be understandable to nonexperts in your field. Other key elements of a successful proposal include the following:

- Providing background information explaining why the proposed experiment is important.
- Placing your research plan in the context of what others are doing and including references to the literature.
- Supplying sufficient details about your samples and their availability.

- Explaining planned data analysis methods, including a brief description of the structural models (form and structure factors), as appropriate.
- Explaining why you need this particular user facility and instrument and why this experiment cannot be conducted elsewhere.
- Justifying the amount of beam time requested by estimating the sample run time for each instrument configuration, number of samples, temperatures, pressures, etc.

More tips for writing a successful beam time proposal can be found in [1].

5.2 Choice of the Instrument Configuration

The basic geometry of an SAS experiment is similar for x-rays and neutrons, and its optimization depends crucially on a number of parameters such as type and available amount of material studied, number of samples and measurements (e.g., at different temperatures, pressures, and composition), required sample environment, and Q resolution. Therefore, the instrument configuration should be carefully evaluated and optimized in each experiment to obtain the best quality data. A particular instrument configuration is defined by parameters such as the wavelength, SDD, the sample-source distance (SSD), and the detector offset perpendicular to the incident beam. Also important is the right choice of the source aperture (i.e., aperture that defines the size of the incident beam after the last neutron guide), the sample aperture (i.e., the aperture that defines the size of the beam actually illuminating the sample), and the size of the beam trap. Characteristics of major instrument components currently available at the ORNL GP-SANS are considered below:

Source aperture. Such apertures are located at the end of each guide; they represent a movable neutron "source" that is positioned at the last guide section inserted into the beam. The most-used source aperture is 4.0 cm in diameter, with the exception that in the highest resolution configuration, when all the guide sections have been removed from the beam, a 2.0 cm aperture is available to further reduce the minimum Q-value to $Q < 0.001$ Å^{-1}.

Sample aperture. In SANS it is common, but not mandatory, to use an incident beam with a circular cross section with a diameter between 2 and 20 mm. The choice of beam size is always a compromise since use of larger apertures increases flux proportional to the area of the beam but decreases Q-resolution. A standard set of round and rectangular sample apertures made of highly efficient neutron and γ-ray absorbing materials is available with sizes between 2 and 20 mm, which defines the corresponding minimum sample width of between 4 and 22 mm.

Sample-source distance. By inserting guide sections into the beam line, the effective SSD may be changed between 2 and 18 m to vary the resolution and the flux on sample.

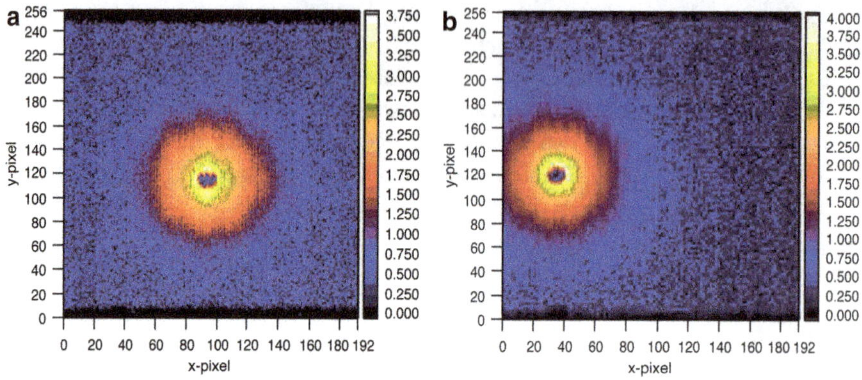

Fig. 5.1 Two-dimensional image of the GP-SANS detector with (**a**) zero and (**b**) 40 cm offset against the beam

Sample-detector distance. A carriage mounted on a rail system inside the evacuated detector tank enables the detector to translate along the length of the vessel, providing SDDs of ≈1.2–20.9 m.

Dynamic range and detector offset. Offsetting the detector relatively to the beam on the fixed-wavelength instruments provides the advantage of increasing the dynamic Q range (i.e., achieving a larger range of scattering vectors that can be measured by one configuration of the instrument; see Fig. 5.1 for an illustration). A maximum detector offset of 450 mm is available at the GP-SANS instrument. TOF SANS instruments have a much greater dynamic range as they use a polychromatic beam of neutrons. The dynamic range actually accessible by both types of instruments is generally quite similar (see Chap. 3).

Beam trap. The beam trap (also known as the beam stop) is usually a round piece of neutron-absorbing material that is used to protect the detector from direct unattenuated beam. Its size should be sufficient to cover the beam image on the detector. The beam-trap sizes available at GP-SANS are 10.1, 7.6, and 5.0 cm in diameter.

Wavelength (λ) and its resolution (Δλ/λ). Average wavelengths of $4.2 < \lambda < 20$ Å are available for GP-SANS. The velocity selector can provide a broad range of wavelength resolutions ($8 < \Delta\lambda/\lambda < 45$ %) depending on the tilt angle of the selector against the beam. The resolution of the selectors used at GP-SANS at zero tilt angle, corresponding to the maximum transmission coefficient, is $\Delta\lambda/\lambda$ ~ 13 %. Higher Q- and λ-resolution might be achieved at longer wavelengths and finite tilt angles, respectively, at the cost of lower flux.

A frequently chosen set of configurations on the GP-SANS includes two SDD settings, one of 20.9 and one of 4.8 m. With an offset of the detector of 400 mm and $\lambda = 6$ Å the covered Q-range is from ~2.9×10^{-3} to 0.62 Å^{-1}. For studies of small, subnanometer objects/pores, one more SDD of 1.2 m is usually added, which increases a value of Q_{max} to ~1 Å^{-1}. A lower Q limit can be reduced down to

Q ~0.0015 Å^{-1} by choosing longer wavelength (e.g., 12 Å) neutrons. Use of the focusing lenses and longer wavelength $\lambda \approx 20$ Å helps to further reduce Q_{min} down to ~4 × 10^{-4} Å^{-1} at the cost of about 20-fold reduction of the flux on the sample and a significant increase of the contribution of multiple scattering. Although focusing lenses help to increase the Q resolution by a factor of ~3.6, the resulting Q_{min} still remains an order of magnitude lower than that achievable by USANS. Therefore, USANS represents a preferable complimentary to SANS technique for studying the structure of multi hierarchical systems.

Unless it is absolutely necessary for a specific sample, it is recommended to keep the wavelength constant and vary the SDD to cover the needed Q-range. This helps avoid repeating the transmission, calibration, and detector sensitivity measurements at each additional wavelength, as these parameters are λ dependent [2]. The choice of the SSD is always a compromise between the size of the direct beam (defined by the sample aperture) and the flux. Usually, an SSD matching the SDD is used ("symmetric instrument configuration"). However, for strongly scattering samples—like hydrogenated solutions or nanoporous carbons—measured at very short SDDs, use of SSD much longer than SDD may be advantageous as such asymmetric configuration reduces the flux on the sample and hence eliminates the necessity of inserting a beam attenuator to avoid detector saturation and damage. In addition, use of the longer SSD increases resolution in the high Q-limit and therefore improves the data quality.

Figure 5.2 illustrates the geometry of a typical SAS experiment. A round source aperture defines the size of the effective neutron source with diameter D_1. A sample

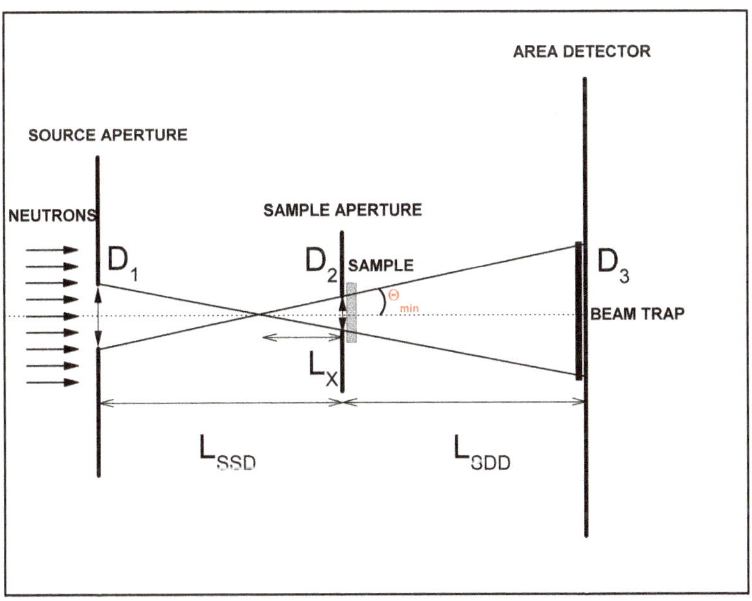

Fig. 5.2 Schematic of the SAS setup

aperture with diameter D_2 defines the size of the beam actually irradiating the sample. The image of the beam in the plane of the area detector is a circle with diameter D_3. The distance between the source aperture and the sample is L_{SSD}, and the distance between the sample and the detector is L_{SDD}. Beams touching the edges of the source and sample apertures intersect at some distance from the sample L_X. Simple geometrical considerations show that the value of L_X can be calculated as

$$L_X = L_{SSD}\frac{D_2}{D_2 + D_1}. \tag{5.1}$$

The diameter of the beam image on the detector is

$$D_3 = D_1\frac{L_{SDD} + L_X}{L_{SSD} - L_X}. \tag{5.2}$$

Equations (5.1) and (5.2) can be used to calculate the diameter of the beam on the detector and thus the minimum size of the beam trap required to block the direct beam, if the sizes of all apertures, as well as L_{SSD} and L_{SDD}, are known.

In most instances the sizes of the beam trap and source aperture are known and one needs to calculate the maximum size of the sample aperture D_2 that will produce a beam image completely covered by the beam trap for a certain combination of L_{SSD} and L_{SDD}. In this case, it is convenient to rewrite Eqs. (5.1) and (5.2) as follows:

$$L_X = \frac{D_3 L_{SSD} - D_1 L_{SDD}}{D_1 + D_3}. \tag{5.3}$$

$$D_2 = \frac{L_X D_1}{L_{SSD} - L_X}. \tag{5.4}$$

For the longest L_{SDD} of 20.9 m, symmetric instrument configuration ($L_{SSD} = L_{SDD}$), source aperture size $D_1 = 4$ cm, and beam trap size $D_3 = 7.6$ cm one obtains $L_X \cong 649$ cm and $D_2 \cong 0.18$ cm. Hence, at these instrument parameters the standard Hellma quartz cells with an ID of 2 cm can be used as sample holders. The minimum scattering angle for this configuration is $arcsin[(D_3/2)/(L_X + L_{SDD})] \cong 0.08°$, and the corresponding Q_{min} available are ~1.47×10^{-3} and ~2.9×10^{-3} Å^{-1} for $\lambda = 12$ Å and 6 Å, respectively. A few other examples of configuration calculations are given subsequently.

In many experiments it is necessary to measure accurately the size of the scattering objects with the radius of gyration R_g, and the order of magnitude of R_g is known apriori from independent measurements. To a first approximation, the L_{SDD} is related to R_g as [3]

$$L_{SDD}(\text{in cm}) \approx 50\,(R_g/\lambda), \tag{5.5}$$

where both R_g and λ are in Angstroms. If the radius of gyration of the scattering objects in the sample is expected to be close to $R_g = 100$ Å, the adequate L_{SDD} should be of the order of ~833 cm. For the symmetric configuration ($L_{SSD} = L_{SDD}$), $D_1 = 4$ cm, $D_2 = 0.18$ cm (close to the maximum sample size available), we calculate from Eq. (5.3) $L_X = 833\ (0.18/(0.18+4)) \cong 36$ cm. From Eq. (5.2), the diameter of the beam image on the detector is $D_3 \cong 4.4$ cm; therefore, a beam trap with a diameter of 5 cm can be used for this configuration with the achievable Q_{min} for $\lambda = 6$ Å $(4\pi/6) \times 2.5/(833+36) = 0.006$ Å^{-1}. In this geometry, the maximum Q-value is $Q_{max} = 0.125$ Å^{-1}, with the beam centered on the detector. If a wider dynamic Q-range is desired, the detector can be offset by up to 450 mm to increase Q_{max} up to ~0.24 Å^{-1}.

In some experiments it is necessary to resolve Bragg peaks emerging from the scattering from ordered structures in a sample. For a sample representing hexagonally packed cylinders with diameter and length equal to 1000 Å and a center-to-center distance between the cylinders of $D = 2000$ Å, the expected position of the peaks can be calculated as $Q_{\text{peak}} = 4\pi/\left(d\sqrt{3}\right) = 0.0036$ Å^{-1}. The expected FWHM of these peaks is ~0.002 Å^{-1}, and thus the value of Q_{min} required to measure the shape of the peaks should be on the order of 0.001 Å^{-1}. Scattering from such samples is usually weak, which prohibits using the smallest source aperture ($D_1 = 2$ cm). The better strategy of reaching the required Q-resolution is to optimize the instrument configuration around the smallest beam stop available at GP-SANS ($D_3 = 5$ cm). Using Eqs. (5.1)–(5.4), calculations can be made for the longest available $L_{SDD} = 20.9$ m and symmetric configuration: $L_X = 232$ cm. Thus, the required size of the sample aperture is $D_2 = 0.5$ cm. As shown in Fig. 5.3, the chosen configuration is adequate for the sample in question. Positioning the detector with the beam in the center (zero offset) is beneficial in this case as it facilitates resolving weak secondary peaks arising from scattering from the hexagonal lattice.

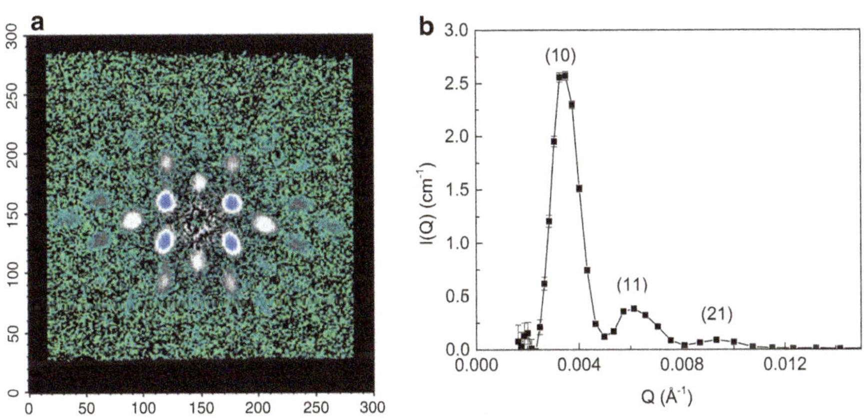

Fig. 5.3 Two-dimensional SANS pattern from hexagonally packed cylindrical pores in silica wafers with axes parallel to the beam (**a**) and corresponding radially averaged cross section showing three resolved peaks (**b**)

5.3 Detector Sensitivity and Instrument Backgrounds

Before beginning an SAS experiment, the local contact will help align the sample changer and beam trap against the beam. An indication of correct sample environment alignment is the absence of "parasitic" scattering (flares, hot spots, etc.) that can be checked by monitoring the 2D scattering pattern on the instrument computer. The position of the beam trap on the detector will vary with the SDD, collimation, and wavelength since neutron beam tends to fall under gravity. Therefore, the alignment has to be checked for each configuration, especially for long wavelengths and large SDDs. The alignment is usually performed using strongly forward-scattering material (graphite, porous aluminum, etc.) and by monitoring the shadow of the beam trap on the detector. Special visualization software is usually available to monitor vertical and horizontal slices of the 2D scattering pattern, and the beam trap is considered to be properly aligned when the same number of neutrons is observed on the beam trap edges (symmetric scattering is observed). After alignment, the beam trap is removed from the beam, and the intensity of the properly attenuated direct beam is measured through an empty position. The beam center is then calculated with standard routines and is used afterwards for radial averaging of the scattering patterns. The integrated number of neutrons in the direct beam and the attenuation factor can be used to calculate the flux for a given instrument configuration and to place data on absolute scale (as described subsequently). This measurement will also be used for transmission calculations, as discussed in Sect. 5.4.

In addition to scattering from the sample, the detector records data from various backgrounds, which must be subtracted to give the scattering cross section of the material under study. Even with the beam closed, the detector will record a signal arising from the general radiation background in the guide hall, cosmic rays, and the electronic noise of the counting system. This is called the "dark current" (aka "blocked beam") background that is normally measured by blocking the beam with the strongest attenuator available. Typical dark current backgrounds are ~5–10 Hz, and most of the signal comes from electronic noise.

The next background to measure is the "empty beam" or "empty cell" background, which records the extra scattering signal measured when the beam is allowed to impinge on the beam stop, with no sample mounted. For more complex experiments involving pressure cells, cryostats, furnaces, etc., this signal is enhanced by the extra scattering from ancillary equipment (e.g., windows of a pressure cell). It should be carefully measured before the actual material under study is loaded.

Additionally, the variation in the sensitivity of the various pixels of the detector must be accounted for, along with a geometrical correction to ensure that each pixel is given a proper normalization [4]. The spatial variation of the detector efficiency (ε) is measured using a predominantly incoherent scatterer. At ORNL, this "sensitivity correction" function is provided by the local contact and is measured from a poly(methyl methacrylate) (PMMA) sheet ~3.5 cm thick. It is sometimes referred to

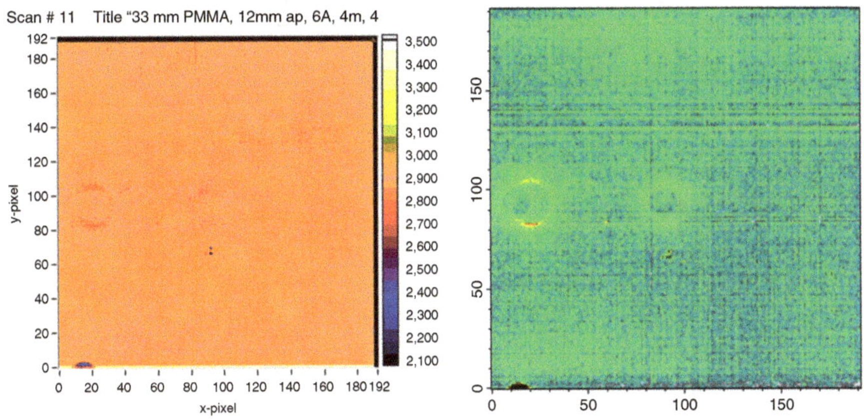

Fig. 5.4 (**a**) Two-dimensional image of the Ordella detector after measuring scattering from a 33 mm thick brick of PMMA for 6 h. (**b**) Pixel sensitivity calculated based on the measured flood pattern and used for actual data reduction

as the "flood pattern" and has an angle-independent intensity in the Q-range measured. Thus, to a good approximation, any variation in the measured signal may be attributed to the detector efficiency and used in the data analysis software to correct for this effect (see Fig. 5.4). This efficiency correction averages unity when summed over the whole detector, and any deviations from this value reflect local variation of the detector efficiency. Pixels with less than 50 % of the average sensitivity correction (e.g., partially shaded pixels near the beam trap) are excluded from data analysis. Finally, the signal from the sample is measured, and the net scattering is corrected for the detector efficiency on a pixel-by-pixel basis.

5.4 Optimal Sample Thickness, Transmission, and Multiple Scattering

Optimal sample thickness depends on the total cross section (i.e., the sum of the absorption and coherent and incoherent cross sections); see Eq. (1.19). The interaction between neutrons and materials is generally much weaker compared with that between x-rays and materials. Because of the relatively low fluxes available for SANS instruments compared with SAXS instruments, thicker samples are required for neutron experiments. For x-rays, typical samples range in thickness from 0.01 to 1 mm, whereas neutron samples range from 0.5 to 5 mm. The scattering intensity $I(\Theta)$ is proportional to $L\,e^{-\mu L} = L\,e^{-L/\Lambda}$, where L is sample thickness, μ is the linear attenuation coefficient, and $\Lambda = \mu^{-1}$ is the mean free path of radiation in the sample. The maximum of function $I(\Theta)$ is reached at ~$1/e$, corresponding to $T \sim 0.37$ %, which defines the theoretical optimum for the sample thickness. In practice, however, higher transmissions (typically 50 ~ 90 %) are preferable as they help to

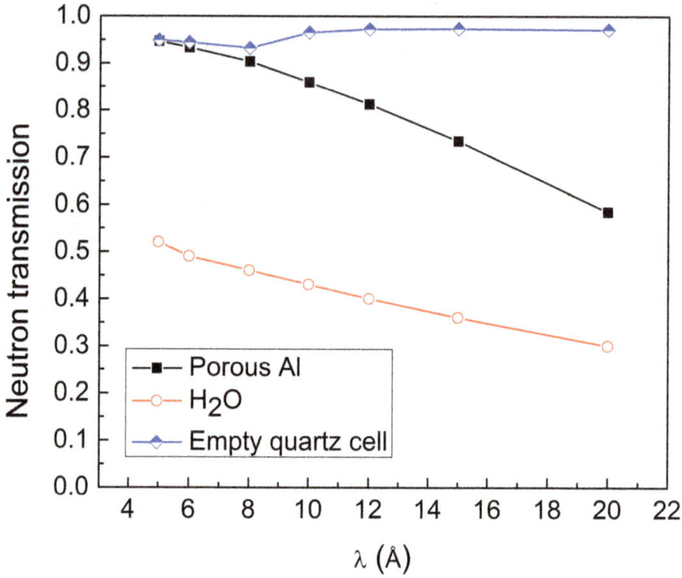

Fig. 5.5 Variation of the transmission of two of the SANS standard samples as a function of λ: porous aluminum (Al-4, thickness 11.14 mm), and 1 mm water contained in a quartz cell. Also shown is the transmission of an empty quartz cell that appears to be virtually independent of the neutron wavelength

minimize multiple scattering or absorption ($L << \Lambda$). Another important parameter that influences the transmission value is the radiation wavelength, as illustrated by Fig. 5.5. The variation of transmission with λ should be taken into account when conducting an experiment.

The transmission is measured by comparing the number of neutrons counted with and without a sample in the direct beam. The main detector can be used for this purpose with the beam trap moved out of the beam and an attenuator inserted in the beam upstream of the sample to reduce the overall intensity to a safe level. Alternatively, an additional small detector can be placed in the center of the beam stop, which allows for continuous monitoring of the transmission. The latter method is preferred for kinetics types of experiments.

If a sample is placed in a container (e.g., a liquid polymer solution in a Hellma cell or a sample of coal in a high-pressure cell), the scattering from the container should be subtracted. In such a case it is convenient to use the run with the empty container as an "empty beam" measurement, and sample transmissions should be measured against the number of counts with the container in the beam. It follows from Table 1.1 that the transmission of most hydrogen-containing materials is mainly defined by the amount of the incoherently scattering hydrogen atoms. Transmission of such samples and thus coherent signal-to-noise ratio is usually reduced by using deuterium-substituted components (e.g., D_2O instead of H_2O or deuterated methane, CD_4, instead of "normal" methane, CH_4). It should be also emphasized that the transmission value is a wavelength-dependent parameter. If multiple wavelengths are used, measurements should be made at each λ to properly normalize scattering data and place them on absolute scale.

Multiple scattering occurs when radiation gets scattered more than once in the sample; therefore, its probability increases with the sample thickness. Multiple scattering effects lead to data smearing and should be minimized to facilitate data interpretation. The probability of multiple scattering increases for strongly scattering $I(Q \to 0) \geq 10^4\ \mathrm{cm}^{-1}$ and attenuating samples ($T < 40\ \%$). According to [5], the contribution of *coherent* multiple scattering reaches ~2 % for the transmission values of ~0.8 and will further increase for strongly absorbing and incoherently scattering samples. Until now there has been no general method of correcting data for multiple scattering effects, although several empirical approaches have been proposed [6–10]. The probability of multiple scattering varies with sample thickness and wavelength as $L\lambda^2$. Possible ways of decreasing its contribution include use of thinner samples and shorter wavelengths, as well as reduction of scattering signal by decreasing the neutron contrast in the sample (e.g., by using fully or partially deuterated compounds in a deuterated matrix).

5.5 Subtraction of the Sample Background

No general rule can be given as to what is a sample background, as its definition can vary in different experiments. In experiments with extremely dilute solutions or dispersions it is usual to consider scattering from the solvent or dispersion medium as background. If the concentration ϕ of the dispersed phase increases, it begins to occupy a significant volume of the sample and a reduced amount of the scattering from solvent, ~(1-ϕ), should be subtracted. In more concentrated hydrogen-containing samples, the background is mostly generated by incoherent scattering from hydrogen, which gives rise to the flat Q-independent scattering (see Fig. 5.6a, [11]). In solid porous samples, such as coal or shale, background might be a sum of incoherent scattering and coherent background that arises from scattering from very small pores. Some researchers subtract both backgrounds before analyzing the residual scattering; while in many instances, structural characterization of subnanometer porosity itself is the aim of the experiment and therefore incoherent scattering only should be determined and subtracted before structural analysis in the high Q-limit. Figure 5.6b shows SANS data obtained from a shale sample IL-1 [12]. The incoherent background can be estimated using the tabulated (bound) incoherent scattering length of hydrogen $b_{inc} = 2.52 \times 10^{-12}$ cm and the number density N_H of hydrogen atoms in the sample:

$$\left[\frac{d\Sigma}{d\Omega}(Q)\right]_{inc} = N_H (b_{inc})^2 = N_H \frac{\sigma_{inc}}{4\pi} = N_H \cdot \frac{79.7 \cdot 10^{-24} cm^2}{4\pi}. \tag{5.6}$$

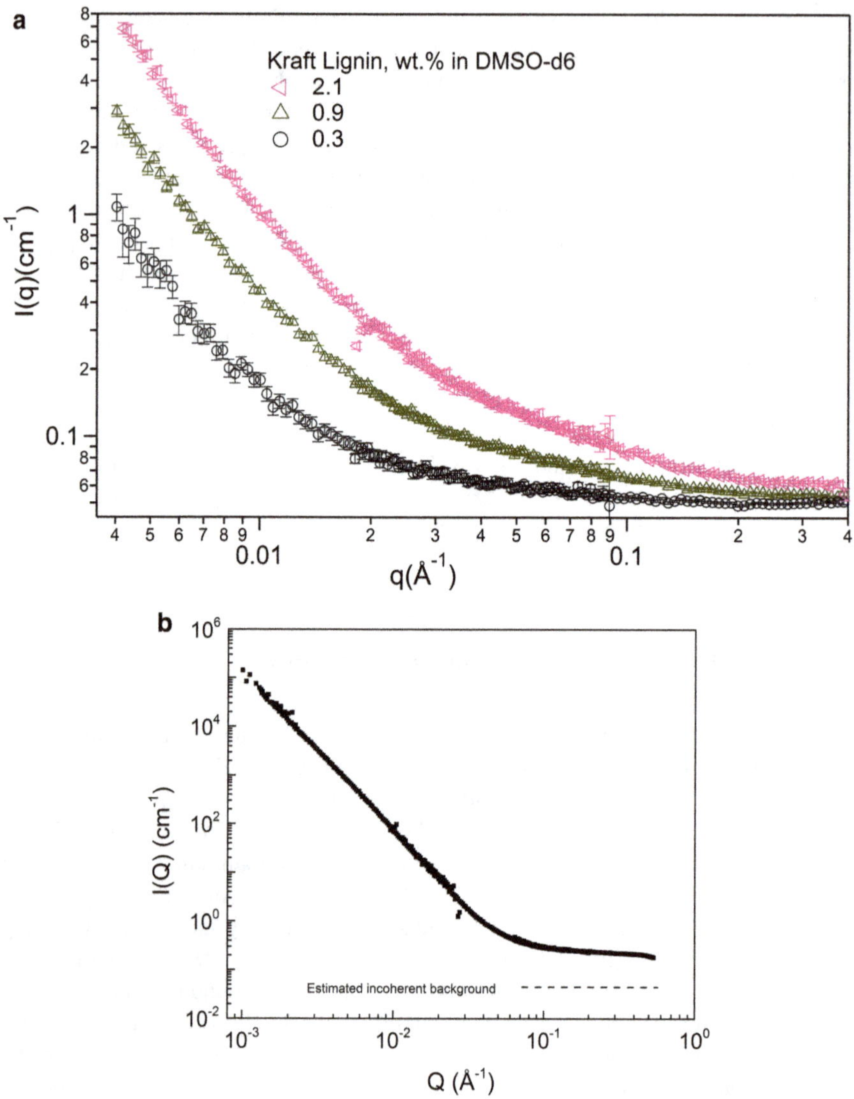

Fig. 5.6 (**a**) SANS data from Kraft lignin solutions in DMSO-d$_6$ at different lignin concentrations shown in the inset. Flat scattering at $Q > 0.2$ Å^{-1} is the result of incoherent scattering from hydrogen in the lignin. Its amplitude ($I_{inc} \sim 0.5$–0.7 cm^{-1}) correlares with the hydrogen/polymer content in solution. Reprinted with permission from [11]. Copyright 2012 American Chemical Society. (**b**) SANS from New Albany shale sample IL-1. The dashed line shows an estimated incoherent background of $I_{inc} \sim 0.047$ cm^{-1} based on the hydrogen content in the sample using Eq. (5.6). Reprinted with permission from [12]. Copyright 2015 American Chemical Society

The scattering length for bound atoms gives a value of incoherent scattering that is smaller than real b_{inc}; however, it still might provide a reasonable estimate of the incoherent background if the hydrogen content is known. Using values of the shale density $\rho_{SHALE} = 2.55$ g/cm^3 and weight concentration of hydrogen $\rho_H = 0.005$ g/cm^3 [12], we calculate $N_H = (\rho_{SHALE} \cdot \rho_H)\ N_A/1 \approx 7.7 \times 10^{21}$ and thus $(d\Sigma/d\Omega)_{inc} \sim 0.049$ cm^{-1}. As shown in Fig. 5.6b, the estimated flat background is by a factor of ~5 lower than the measured scattering signal in the limit of high Q, which leads to the conclusion that in this Q-range SANS is generated by coherent scattering from small pores rather than by incoherent background from hydrogen. Another way of estimating incoherent background in case the scattering varies as $Q^{-\alpha}$ and does not reach plateau in the high Q-limit is to fit the data in the high Q-limit to the equation:

$$\frac{d\Sigma}{d\Omega}(Q) = A \cdot Q^{-\alpha} + \left[\frac{d\Sigma}{d\Omega}(Q)\right]_{inc}. \tag{5.7}$$

The value of the background $\left[\frac{d\Sigma}{d\Omega}(Q)\right]_{inc}$ is accurately determined from a linear plot of $Q^{\alpha} + \frac{d\Sigma}{d\Omega}(Q)$ as a function of Q^{α}:

$$Q^{\alpha}\frac{d\Sigma}{d\Omega}(Q) = A + Q^{\alpha}\left[\frac{d\Sigma}{d\Omega}(Q)\right]_{inc}, \tag{5.8}$$

where $\left[\frac{d\Sigma}{d\Omega}(Q)\right]_{inc}$ is the slope and A is the intercept. In many cases this simple empirical method gives reasonable results.

5.6 Data Acquisition Time, Masking, and Radial Averaging

The right choice of acquisition time per sample at each instrument configuration is an important part of developing a data collection strategy and for optimum use of allocated beam time. In most cases longer acquisition time is required at longer SDDs. Count rates at short SDDs are usually much higher, and the measurements can be completed within few minutes. However, longer counting times might be required if the sample contains significant amount of hydrogen. The parameter that is used for estimating acquisition time is the so-called "bare" count rate (CR), which represents the difference between the total number of counts registered by the detector per second with beam passing through the sample and through the empty sample container (e.g., the Hellma cell in case of a liquid sample or high-pressure cell). Short measurements of ~10 s give a good estimate of count rate for each sample and container, and the time required to accumulate the desired total number of counts (N_{tot}) is simply N_{tot}/CR. Acceptable statistics are usually achieved on the GP-SANS if $N_{tot} \sim 10^6$. The error bars in the measurements are given by Poisson statistics $\delta(N) = \sqrt{N}$, and therefore the fractional error is

$\delta(N)/N = 1/\sqrt{N}$. It follows from this equation that to improve the fractional error by a factor of X, one needs to count a factor of X^2 more neutrons. For instance, improving error bars from ~4 % to ~2 % requires a four times longer counting time. The counting time required for measuring the backgrounds (dark current and empty cell runs) should provide the same quality statistics as those from a sample. There is no point in collecting exceptionally good statistics on a sample if background is measured with poor statistics since resulting errors are added in quadrature.

Currently, SANS and SAXS data are collected on position-sensitive detectors that generate 2D data sets for each sample. The detector efficiency measurement will remove most artifacts caused by the detector. In high-pressure experiments, scattering data at the shortest wavelengths could show parasitic scattering in the form of spots coming from inhomogeneities in the cell window and the position of such spots could be pressure and temperature dependent. Data from such pixels and spots can be masked out using available software, which excludes systematic errors from data analysis.

In many instances scattering from samples is isotropic, that is, circularly symmetric around the beam center. Such 2D data sets can be radially averaged and reduced to 1D data sets depending only on a scalar Q (see Fig. 5.7). Such averaging is performed after the reduction and masking of "bad" pixels. This averaging is

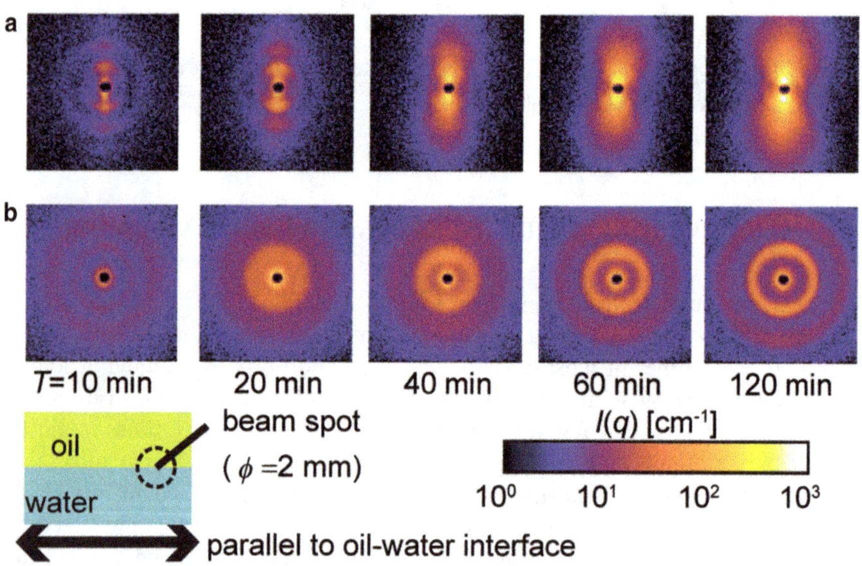

Fig. 5.7 Time evolution of 2D SANS patterns from a droplet of oil floating on the oil–water interface. (**a**) Moving interface at non-equilibrium conditions. (**b**) Static interface. Scattering from the droplet in static interface is isotropic, and 2D patterns can be radially averaged against the beam center to produce $I(Q)$ patterns. In case of the moving interface, scattering is highly anisotropic and can be analyzed in different Q-directions (wedges) showed as straight lines. Images courtesy of Yutaka Sumino (Tokyo University of Science, Japan)

conducted against the center of the incident beam, the coordinates of which are determined during a "direct beam" run, as explained in Sect. 5.4. For anisotropically scattering samples, 2D patterns cannot be circularly averaged and the variation of the scattering as a function of scattering vector can be analyzed only along specific directions of $\boldsymbol{Q}$ (see Fig. 5.7).

5.7 Absolute Calibration

Nowadays it is a common practice to place all intensity data on absolute scale in the form of a differential scattering cross section per unit volume $d\Sigma/d\Omega(Q)$. Although the use of absolute units is not required for measurement of spatial dimensions (e.g., determining the radius of gyration R_g of a particle or a polymer coil), it is essential, for example, for calculating pore volume and the internal surface area of porous systems. In addition, it forms a valuable diagnostic tool for the detection of artifacts. Because the cross section varies as the sixth power of the scatter linear dimension, it represents a sensitive indicator of whether an appropriate structural model has been chosen. For example, scattering studies of colloidal micellar solutions can be modeled by core-shell spherical micelles as a function of a set of parameters describing the particle structure and interactions. On an arbitrary intensity scale, it is possible to produce excellent fits of the particle shape, which could be *in error by as much as 3 to 4 orders of magnitude in intensity* [13]. Thus, absolute calibration allows such artifacts to be recognized, and the model parameters might be restricted to those that reproduce the observed cross section.

This behavior illustrates the point referred to earlier that the intensity is extremely sensitive to the particle or molecular dimensions and even an approximate (~10 %) absolute calibration is sufficient to reveal the presence of such artifacts.

When a coherent beam with the flux Φ_0 (s^{-1} cm^{-2}) illuminates a sample with transmission T, the measured intensity $I(Q)$ (s^{-1}) scattered in the direction $\boldsymbol{Q}$ within a solid angle $\Delta\Omega$ is related to the differential cross section per unit volume $d\Sigma/d\Omega(Q)$ as [14, 15]:

$$\frac{d\Sigma}{d\Omega}(Q) = \frac{1}{V}\frac{d\sigma}{d\Omega}(Q) = \frac{I(Q)}{\Phi_0 \Delta\Omega(Q)\varepsilon TSL} = \frac{I(Q)}{I_0 \Delta\Omega(Q)\varepsilon TL}, \tag{5.9}$$

where $V = SL$, the illuminated volume of the sample defined by the sample aperture with the area S and sample thickness L; ε is the detector efficiency; and $I_0 = \Phi_0 S$ is the incident neutron intensity (number of neutrons per second at the sample position for a given area S). In high-pressure, as well as soft matter, experiments, the samples are contained in a high-pressure vessel or a quartz cell that can contribute significantly to the measured scattering signal. After correcting for the dark current background, the scattering from the empty cell (*EC*) should be subtracted from the measured intensity:

$$\frac{d\Sigma}{d\Omega}(Q) = \frac{1}{LI_0}\left[\frac{[I(Q)]_{MEASURED}}{I_0\Delta\Omega(Q)\varepsilon T_{SAMPLE}t_{SAMPLE}} - \frac{I(Q)_{EC}}{I_0\Delta\Omega(Q)\varepsilon T_{EC}t_{EC}}\right]. \tag{5.10}$$

If measurements are conducted with "free standing" samples (no sample container), the measured scattering should be corrected to the scattering from air and the transmissions of samples in a cell should be calculated with respect to the empty beam. If the scattering from an *EC* is chosen as "empty beam," the transmissions are measured with respect to the *EC* and T_{EC} is set to unity.

With the known sample aperture size and sample thickness (which define illuminated sample volume), time of measurement, and sample transmission, after correcting for background the data can be normalized using Eq. (5.10) to give a coherent cross section, which is an intensive property independent of the sample dimensions. A solid angle in Eq. (5.10) is calculated as $\Delta\Omega = \Delta A/(L_{SDD})$, where ΔA is a cross section of the detector area. The only remaining unknown parameter in Eq. (5.10) is the incident intensity I_0, which can be measured using the "direct beam" method. This approach is based on determining the intensity of the direct beam I_{att} (counts per second) defined by a sample aperture with the area *S* and attenuated by a calibrated attenuator with the attenuation factor *F*, in which case $I_0\ (n/s) = F \cdot I_{att}$.

Another method used to normalize the data is to use calibrated samples or "absolute standards," which are samples with a well-known differential scattering cross section $(d\Sigma/d\Omega)_{standard}$. Absolute scaling is performed by measuring $I_{standard}$ at an identical instrument configuration and the same sample aperture, after which the absolute cross section of the sample is calculated as

$$\left(\frac{\mathrm{d}\Sigma}{\mathrm{d}\Omega}\right)_{\text{sample}} = I_{sample}\frac{\left(\frac{\mathrm{d}\Sigma}{\mathrm{d}\Omega}\right)_{standard}}{I_{standard}} \equiv I_{sample}K_{ABS}, \tag{5.11}$$

where K_{ABS} is the absolute scaling factor.

An absolute standard should scatter with a reproducible cross section and be stable with time. A primary standard should ideally possess a calculable cross section, whereas secondary standards are obtained by cross calibration. The scattering signal should be strong to facilitate accurate measurements in a short time. In many cases it is desirable to have standards that can be used over a wide range of scattering vectors to allow for calibration at different instrument configurations. One of the accepted primary standards is a single crystal of mechanically unstressed, hydrogen-free vanadium metal between 1 and 3 mm thick. Its scattering is entirely incoherent; however, the disadvantage of vanadium is that its cross section is very small, which requires a long acquisition time to accumulate reasonable statistics. In addition, if left in long contact with air, vanadium can absorb a significant amount of hydrogen, which will significantly alter the scattering cross section. Incoherent scattering from water is much higher than from vanadium, and for this reason it is adopted as a secondary standard at many SANS facilities, among other incoherent scatterers such as polymers (e.g., PMMA). However, it should be remembered that use of water is complicated by the fact that its scattering is both temperature and wavelength dependent. In addition, neutron scattering from water is not entirely angular independent and tends to peak in the forward direction because of inelastic scattering effects. Despite some complications, use of incoherent standards is pretty common for short SDDs (relatively high values of the scattering vector), where scattering is flat within the experimental error (see Fig. 5.8a).

At long SDDs (low scattering angles), the flat incoherent scattering signal is relatively low, and this requires development of a different set of coherently scattering standards. Various materials can be used for this purpose, and several absolutely calibrated samples are available to help users perform normalization to absolute scale, as listed in Table 5.1, along with the suggested Q-ranges and SDDs. Figure 5.8b gives examples of how standards can be used to calculate the absolute scaling factor K_{ABS} and how to use Eq. (5.11) for placing experimentally measured SAS intensities on absolute scale. Assume that the intensity from the 1 mm H_2O standard was measured in the sample position. The average measured intensity $\bar{I}_{standard} = 0.033\ s^{-1}$. For the 1 mm H_2O standard, $d\Sigma/d\Omega_{standard} = 0.94\ cm^{-1}$ (see Table 5.1) and thus $K_{ABS} = 0.94\ cm^{-1}/0.033\ s^{-1} \approx 28.48\ cm^{-1}\ s^{-1}$. For the Al-4 standard, $d\Sigma/d\Omega_{standard}(0) = 130\ cm^{-1}$ (see Table 5.1). The measured intensity $I(0)_{standard} = 122.16\ s^{-1}$, and therefore $K_{ABS} = 1.064\ cm^{-1}\ s^{-1}$.

5.8 Instrument Resolution

On most SAS instruments, the major contributors to the resolution of the scattering vector ($\Delta Q/Q$) are the wavelength spread $\Delta\lambda/\lambda$, the incident beam divergence, and the spatial resolution of the detector. Each SAS instrument has limited Q-resolution that could lead to the smoothing of the form factor oscillations, broadening of the observed maxima, and decreasing the slope in log$I$–log$Q$ representation. The wavelength distribution is determined by the monochromator or the velocity selector. The beam divergence can be estimated from the instrument configuration and size of the source and sample apertures.

The Q-resolution at a given Q can be written according to a Taylor expansion as [2]:

$$\Delta Q = -Q\left(\frac{\Delta\lambda}{\lambda}\right) + \left(\frac{4\pi}{\lambda}\right)\cos\Theta\Delta\Theta. \tag{5.12}$$

Therefore

$$\Delta Q^2 = Q^2\left[\left(\frac{1}{2\sqrt{2ln2}}\frac{\Delta\lambda}{\lambda}\right)^2\right] + \left[\left(\frac{4\pi}{\lambda}\right)^2 - Q^2\right]\Delta\Theta^2. \tag{5.13}$$

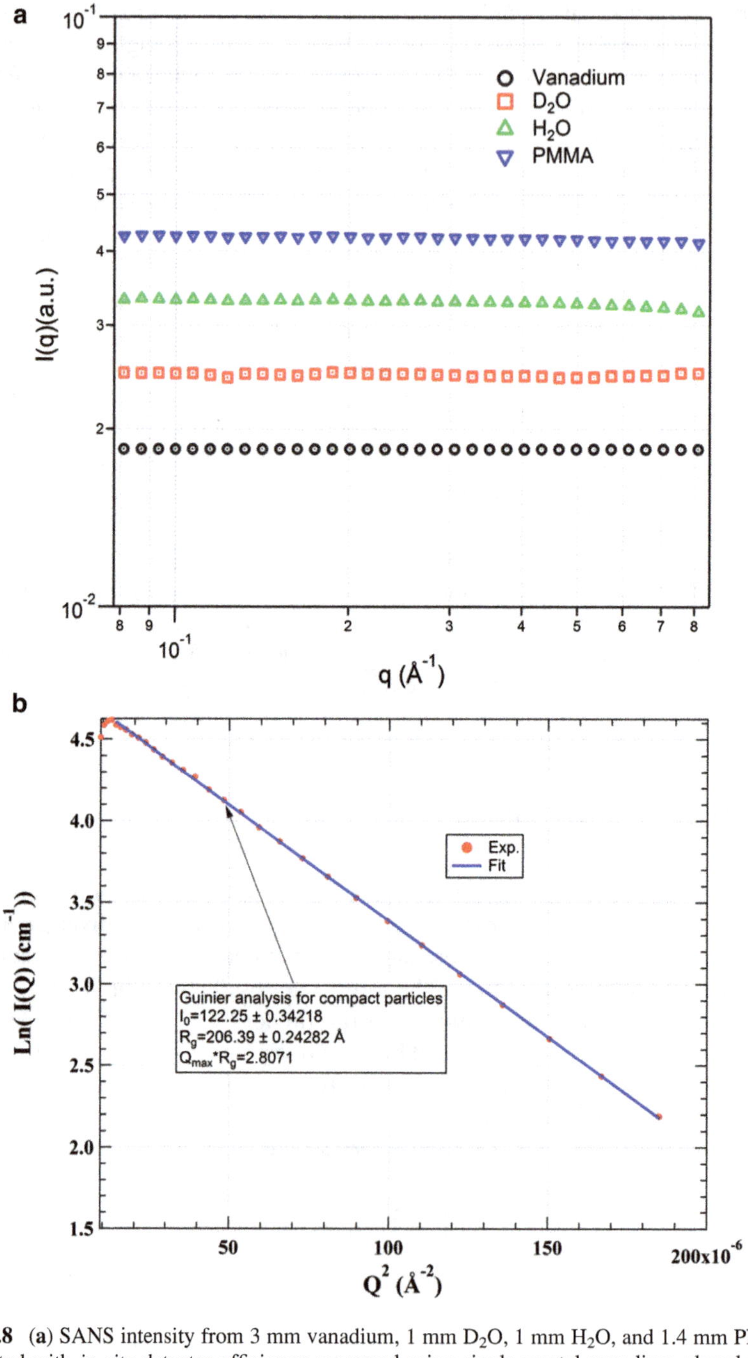

Fig. 5.8 (**a**) SANS intensity from 3 mm vanadium, 1 mm D_2O, 1 mm H_2O, and 1.4 mm PMMA corrected with *in situ* detector efficiency measured using single-crystal vanadium placed in the sample position. Data shifted for the sake of clarity. (**b**) Guinier plot for the Al-4 standard

Table 5.1 Examples of precalibrated scattering standards available at ORNL SANS facilities

Standard	T (λ)	L, mm	$d\Sigma/d\Omega_{standard}$, cm^{-1}	Parameter, Å	Range of SDD, m	Equation
Porasil A	0.915 (4.75)	1.0	47.5 ± 2	$\xi = 20.9 \pm 1$	1–4	Debye-Bueche
Porasil B	0.871 (4.75)	1.81	450 ± 30	$\xi = 43.2 \pm 2$	4–8	Debye-Bueche
Urupunga 4	0.612 (7)	1.1	–	Slope	4–20	Slope
Al-4 (1)	0.954 (4.75)	11.14	130 ± 7	$R_g = 203 \pm 3$	10–20	Guinier
Al-4 (2)	0.96 (4.75) 0.935 (8) 0.917 (6)	9.6	122 ± 5	$R_g = 201 \pm 3$	10–20	Guinier
D30	0.66	1.25	63.7 ± 2	R_g 73.2 ± 2	4–8	Debye
1 mm H_2O	0.53 (4.75)	1	0.94 ± 0.02	Isotropic	1–8	Average
1 mm D_2O	0.88 (4.75)	1	0.058 ± 0.002	Isotropic	1–3	Average
PMMA	0.467 (5)	1.4	0.895 ± 0.005	Isotropic	1–8	Average

T is the (wavelength-dependent) sample transmission, *L* is the sample thickness, and $d\Sigma/d\Omega(0)$ is the intercept in the $Q \rightarrow 0$ limit (coherently scattering samples) or an average intensity (incoherently scattering samples). Equations used for the data fitting are discussed in Chap. 6

$\Delta\lambda/\lambda$ is related to the FWHM value of the triangular function describing the wavelength distribution, and $\Delta\Theta$ is related to the width of the direct beam. From Eq. (5.13) it follows that ΔQ^2 depends on the wavelength spread and also on the Q-value. The contribution of $\Delta\Theta$ is approximately constant with Q. At low $Q < 10^{-2}$ Å^{-1}, the uncertainty is dominated by the distribution of the angle. At large Q-values, ΔQ is mostly defined by the wavelength spread.

The resolution function $R(Q',Q,\Delta Q)$ describes the distribution of the Q vectors at a given instrument configuration. The experimentally measured cross section $\frac{d\Sigma}{d\Omega}(Q)_{exp}$ is the real neutron cross section $\frac{d\Sigma}{d\Omega}(Q)$ Eq. (1.23) smeared by the resolution function $R(Q',Q,\Delta Q)$. In many instances it is adequate to approximate the resolution function of an instrument by a Gaussian with the argument of ΔQ^2. In this case

$$\frac{d\Sigma}{d\Omega}(Q)_{exp} = \int_{Q'-\infty}^{Q'+\infty} \frac{1}{\Delta Q\sqrt{2}} exp\left[-\frac{(Q'-Q)^2}{2(\Delta Q)^2}\right]\frac{d\Sigma}{d\Omega}(Q)dQ'. \tag{5.14}$$

Various procedures for incorporating resolution in the data analysis have been suggested based on the convolution of structural models with the instrument resolution function of constant flux [2, 16–19] as well as TOF SAS instruments [20].

Thus far it was assumed that all scattering objects that are present in a sample are identical in size and shape, which is known as monodispersity. If there is variation, or polydispersity, but no correlation in the locations and orientations of the objects, the resultant differential cross section is a weighted average of the scattering from individual constituents. Thus, the form factor is smeared with the size distribution function $G(\mu,r,\sigma)$

$$\frac{d\Sigma}{d\Omega}(Q)_{poly} = \int_0^\infty G(\mu, r, \sigma)\frac{d\Sigma}{d\Omega}(Q)dr, \tag{5.15}$$

where σ is the parameter characterizing the width of the distribution around a maximum value μ. $G(\mu, r, \sigma)$ is usually approximated by the following distribution functions:

$$G_G(\mu, r, \sigma) = \frac{1}{\sigma\sqrt{2\pi}} exp\left[-\frac{(r-\mu)^2}{2\sigma^2}\right] \quad \text{(Gaussian)}, \tag{5.16}$$

$$G_{LN}(\mu, r, \sigma) = \frac{1}{r\sigma\sqrt{2\pi}} exp\left[-\frac{(lnr-\mu)^2}{2\sigma^2}\right] \quad \text{(log-normal)}, \tag{5.17}$$

$$G_S(\mu, r, \sigma) = \left[\frac{Z+1}{\mu}\right]^{Z+1} \frac{r^Z}{\Gamma(Z+1)} \exp\left(-\frac{(Z+1)r}{\mu}\right) \text{(Schultz)}, \tag{5.18}$$

where $Z = \frac{\mu^2}{\sigma^2} - 1$ and $\Gamma(Z) = (Z-1)!$ is the gamma function.

The variance of these distribution functions is σ^2 (Gaussian), $[exp(\sigma^2 - 1)]\cdot exp(2\mu + \sigma^2)$ (log-normal), and $\mu^2/(Z+1)$ (Schultz). These equations give similar numerical values (Fig. 5.9), and the difference between them usually cannot be resolved in an SAS experiment.

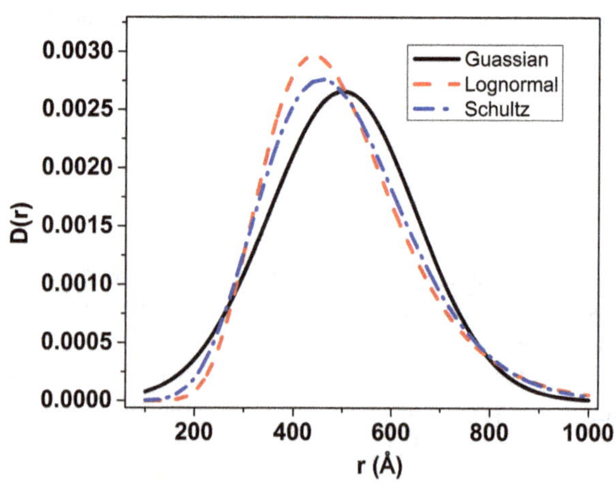

Fig. 5.9 Size distribution functions (Eqs. (5.16)–(5.18)) for the average size μ = 500 Å and a standard deviation of 150 Å

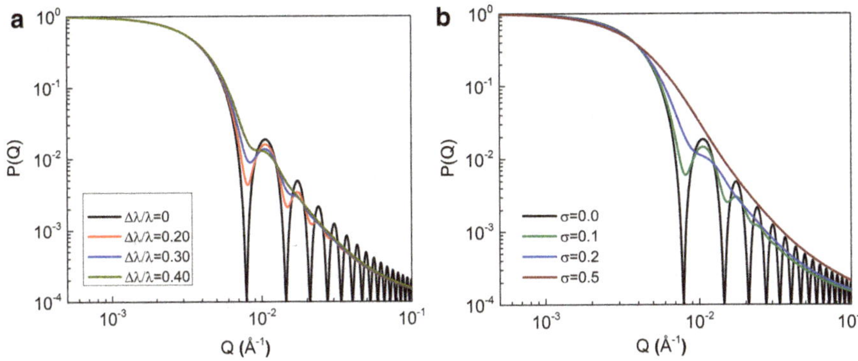

Fig. 5.10 Effect of the wavelength resolution $\Delta\lambda/\lambda$ (**a**) and the polydispersity index σ (**b**) on the ideal scattering from a monodisperse core-shell cylinder

Mathematically, the smearing functions (Eqs. (5.14) and (5.15)) are similar, and both the polydispersity and instrument resolution contribute to the smearing of the true cross section, as illustrated in Fig. 5.10.

5.9 Effective Thickness of Powder Samples

Powders formed of porous pellets/particulates of micrometer size are often used in materials science, and SAS has been used for a long time to explore the structure and physical properties (fractal dimension, total porosity, and surface area) of such systems. As mentioned earlier, the optimum sample thickness to minimize the probability of multiple scattering is in the range of 0.01–1 mm for x-rays. However, in many cases when the absorption coefficient is high, to maintain a reasonable x-ray transmission the sample is limited to a thin layer of particulates of unknown thickness [21]. For neutrons, the typical range of sample thickness is between 0.5 and 5 mm, although in some SANS experiments, to increase signal-to-noise ratio, it can be increased up to 20 mm if the sample components are predominantly coherent scatterers, such as dilute solutions of fullerenes in CS_2 [22] and mixtures of fluorosurfactants with supercritical CO_2 [23]. The sample thickness is the parameter that is required for placing SAS data on absolute scale as it enters the calibration correction factor; see Eqs. (5.9) and (5.10). For liquid or gaseous samples, the sample thickness is identical to the internal spacing between the windows of the sample holder (e.g., a quartz cell or high-pressure vessel). The thickness of bulk solid samples can be measured easily. However, the situation is more complicated when the sample is in the form of particulates (powders, granules, fibers, etc.) that do not completely fill the sample holder volume.

Geological samples are seldom available as an oriented core. The most common type of rock sample available from exploration drilling is a "drill cutting," which must be crushed to a coarse powder for SAS experiments. Preparing geological samples for SAS in the form of particulates is actually advantageous because it allows the SAS measurement to be taken from a representative sample of rock (coal or shale) in which components are positioned in various orientations, rather than being limited to one orientation. Typically, geological samples are sieved to obtain reasonably monodispersed particles of 0.5 mm diameter. Such particles are large enough to avoid scattering from interparticle voids in both SAS and USAS domains, which could contribute to scattering from pores and bias data analysis. In addition, the use of sieved quasi-monodispersed spherical particulates reduces the effective thickness of samples by a factor of ~2, which helps to minimize the contribution of multiple scattering [24].

Samples in the form of particulates are usually placed in quartz sample holders (for experiments at ambient conditions) or in aluminum sample holders for high-pressure cells with internal thickness *L*. Apparently, in the case of particulates, the effective sample thickness L_{eff} will be less than the nominal thickness of the sample holder because of the availability of interparticle voids. The effective thickness of the equivalent porous platelets can be calculated using the skeletal, apparent, and bulk density of the sample particulates.

The *true, or skeleton, density* (ρ_{skel}) of a porous material is defined as the ratio of the mass to the volume occupied by the skeleton. In other words, the contribution of the volume of pores (for solid materials) and both the volume of pores and interparticle voids (for particulate materials) should be subtracted when measuring the skeletal density. By definition:

$$\rho_{skel} = \frac{m_{skel}}{V_{skel}}. \tag{5.19}$$

V_{skel} is usually measured by the displacement of air in pores and interparticle voids with a gas or a liquid that does not chemically react with the skeleton. Most commercially available pycnometers use helium as the displacement gas. That is why the true or skeletal density is often called the "helium" density. The use of helium is preferential because it is inert and has a small molecular size, which facilitates penetration in the smallest pores [25].

Another relevant density, *apparent* density (ρ_{app}), is defined as the mass of a skeleton normalized to the skeleton volume. The difference between ρ_{skel} and apparent density is that the mass of skeleton is now normalized by the volume occupied by the skeleton plus the volume occupied by the pores:

$$\rho_{app} = \frac{m_{skel}}{V_{skel} + V_{pore}}. \tag{5.20}$$

The apparent density takes into account the volume occupied by the matrix pores and ignores the volume occupied by interparticle voids present in a particulate material. It can be measured by displacement of a nonwetting liquid (e.g., mercury) that fills interparticle voids but does not penetrate the pores.

The bulk density ρ_{bulk} is the skeleton mass divided by the total sample volume V_{tot} that includes the volume of interparticle voids:

$$\rho_{bulk} = \frac{m_{skel}}{V_{tot}} = \frac{m_{skel}}{V_{skel} + V_{pore} + V_{void}}. \tag{5.21}$$

This density can be measured, for example, by pouring powder into a graduated container (or sample holder) and dividing the powder mass by the container volume V_{tot}. For samples with a regular shape (e.g., disk, sphere, or cube) this density can be estimated by dividing sample mass by the calculated sample volume. If the chemical composition and densities of all skeleton components are exactly known, this parameter can also be evaluated using transmission (x-ray or neutrons) measurements. For this purpose, one first calculates transmission of the sample completely filled with skeleton material (no pores or voids); see Eqs. (1.20–1.22):

$$T_{skel} = \exp\left[-\sum_i N_i \sigma_{tot,i} L\right], \tag{5.22}$$

where N_i is the number density of skeleton component "i" with a total cross section $\sigma_{tot,i}$. The transmission of porous particulate sample T_{sam} will be higher because of the presence of pores and interparticle voids. This reduces the bulk sample density, which can be calculated from

$$\rho_{bulk} = \rho_{skel} \frac{T_{sam}}{T_{skel}}. \tag{5.23}$$

The requirement of knowing the chemical composition might appear difficult to satisfy apriori for complex systems such as geological samples; however, it is perfectly applicable to chemically simpler materials such as porous glasses and ceramics [26].

Finally, *envelope (or geometric) density* calculations can be used to estimate the volume of interparticle voids and thus the effective thickness of a particulate sample. If a sample holder with volume V_{tot} is filled with closely packed spherical particulates with diameter D, the ratio of the total volume of spheres ($N_{sp} \cdot \pi D^3/6$) and V_{tot}, and thus the ratio of the effective thickness of the particulates and the total internal thickness of the sample holder, is ~0.52. This means, for example, that for quasi-spherical monodisperse particulates closely packed in a sample holder with internal thickness L, the effective sample thickness will be ~$L/2$ (i.e., about a factor of two lower that the nominal thickness of the sample holder [27]).

A more accurate estimation of the effective thickness of particulate samples can be reached if the density of the skeleton and sample porosity (V'_{pore} [cm^3/g]) is known from independent measurements. In such a case, the pore volume in a sample with weight m_{skel} is $V_{pore} = V'_{pore} \cdot m_{skel}$, and the apparent density can be calculated using Eq. (5.20):

$$\rho_{app} = \left(\rho_{skel}^{-1} + V'_{pore}\right)^{-1}. \tag{5.24}$$

In its turn, ρ_{bulk} can be determined by weighing the sample loaded in the sample holder with the volume V_{tot} (Eq. (5.21)). The effective sample thickness is then calculated via

$$L_{eff} = L\frac{\rho_{bulk}}{\rho_{app}}, \tag{5.25}$$

where L is the total internal thickness of the sample holder. If the sample porosity is not known, the effective sample thickness can be estimated for samples with relatively low porosity $V_{skel} >> V_{pore}$, in which case $\rho_{app} \to \rho_{skel}$ and

$$L_{eff} \approx L\frac{\rho_{bulk}}{\rho_{skel}}. \tag{5.26}$$

Many geological samples, such as coal and tight gas shales, are characterized by relatively low porosity (generally less than 10 %) [28], and Eq. (5.26) can be used for such samples, if the error caused by neglecting sample porosity can be tolerated. In the opposite limit $V_{skel} << V_{pore}$, characteristic, for example, of silica aerogel with a porosity of 96 % [29], $\rho_{app} = \left(V'_{pore}\right)^{-1}$ and

$$L_{eff} \approx L\rho_{bulk}V'_{pore}. \tag{5.27}$$

To summarize, calculation of the effective sample thickness of particulate materials requires knowing the skeleton/helium density, bulk density, internal thickness, and total volume of the sample holder, as well as mass of the loaded sample. The effective sample thickness, rather than the nominal internal thickness of the sample holder, should be used for calculation of the differential neutron cross section in Eqs. (5.9) and (5.10).

5.10 Contrast Variation with Liquids and Gases

The scattering power of a binary system (e.g., polymers or colloids dispersed in a solvent, pores in activated carbon or silica, etc.) is proportional to the contrast factor $(\Delta\rho^*)^2$ (neutron or x-ray) that is defined as the square of the difference in the SLD of a scattering object and the surrounding medium; see Eq. (1.18). When the scattering objects are dispersed in a solvent, the contrast of the object can be varied by changing the average scattering density $\overline{\rho_2^*}$ of the solvent, which is called the "solvent contrast variation method" [30, 31]. For SANS studies of materials, various types of contrast variation methods have been developed, including label triangulation, inverse contrast variation, triple isotopic substitution, and spin contrast variation methods [32]. In the case of SAXS, the $\overline{\rho_2^*}$ can be varied by adding small molecules (salts, sucrose, glycerol, etc.) to the solvent. However, except for the use of high concentrations of salt (such as NaBr), in many cases it is hard to approach the electron density of an object. In addition, adding molecules to the solvent could introduce significant perturbation in the structure and interactions between the objects.

Contrast variation can be also conducted using anomalous SAXS (ASAXS), which does not require adding additional components to the system. Development of this technique was triggered by the availability of synchrotron radiation sources, which generate photons over a wide energy range. The contrast variation is performed by varying the energy of the incident x-rays. ASAXS experiments can differentiate between various scattering components if the x-ray energy is varied close to the absorption edge of one of the system constituents [33]. As was explained in Sect. 1.4, the absorption edge is characterized by a sharp resonant increase in x-ray absorption with increasing x-ray energy as *K*- or *L*-shell atomic electrons are excited. The increase in absorption is reflected in anomalous variation of both the real and imaginary part of the x-ray scattering form factor. Studying this variation as a function of x-ray energy just below the absorption edge allows for delineating the scattering contribution from various system components according to their atomic composition [34].

In the case of SANS, contrast variation can be conducted based on the isotope dependence of the scattering amplitude of elements (see Table 1.1), for example, by changing the ratio between the nondeuterated (normal) and deuterated solvents. Contrast matching of porous materials using SANS can be used for determination of the SLD of the porous matrix by varying the SLD of the liquid introduced into the pores and finding the value of the average SLD $\left(\overline{\rho_2^*}\right)$ at which the contrast goes to zero. The application of such a contrast variation method is based on the assumption that mixing nondeuterated and deuterated solvents does not affect the solute structure and that the basic scattering functions are invariant upon isotopic substitution.

Suppose that neutron cross section $d\Sigma/d\Omega(Q)$ is measured from a porous SiO_2 with an SLD of the solid skeleton of 3.47×10^{10} cm^{-2} (Table 1.2). Different samples of porous SiO_2 are filled with isotopic mixtures of H_2O (SLD $= -0.56 \times 10^{10}$ cm^{-2}) and D_2O (SLD $= 6.39 \times 10^{10}$ cm^{-2}) of different composition and, therefore, different average SLDs of the mixed solvent $\left(\overline{\rho_2^*}\right)$. Since $d\Sigma/d\Omega(Q) \sim (\Delta\rho^*)^2$ (see Eq. (1.18)), plotting the square root of $d\Sigma/d\Omega(Q)$ (at fixed Q) as a function of $\overline{\rho_2^*}$ should give a linear plot, and plotting $d\Sigma/d\Omega(Q)$ *vs.* $\overline{\rho_2^*}$ should give a parabola, as shown in Fig. 5.11. Under the condition that the SLD of the solid matrix does not fluctuate in space and the incoherent background is properly subtracted, the scattering signal should go to zero at the contrast matching point when the SLD of the isotopic mixture becomes equal to that of SiO_2 $\left(\rho_1^* = \overline{\rho_2^*}\right)$. If the SLD of the skeleton fluctuates from point to point, the scattering signal never

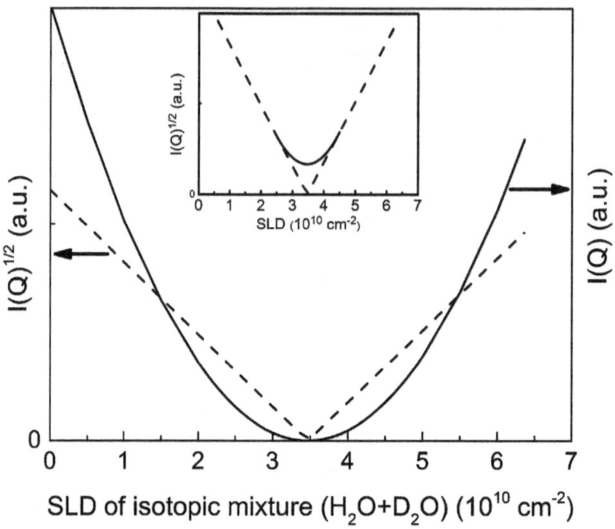

Fig. 5.11 Simulated variation of the scattering intensity at a fixed Q as a function of the SLD of isotopic mixture ($H_2O + D_2O$) with a different concentration of the components filling pores in porous SiO_2 with the SLD $= 3.47 \times 10^{10}$ cm^{-2}. *I(Q)* and $I(Q)^{1/2}$ are parabolic and linear functions of the average SLD of isotopic mixture, respectively. *I(Q)* goes to zero at the contrast matching point, provided the SLD of SiO_2 is constant and all pores are accessible to the mixture. Otherwise, *I(Q)* is not zero in the contrast matching point, as discussed in the inset

reaches zero, even when the average SLD of the skeleton is matched. Even in the absence of fluctuations of the skeletal SLD, a nonzero scattering intensity at the contrast matching point of binary porous materials can be caused by the presence of closed pores. These closed pores are inaccessible to the contrast matching fluid, which results in a residual scattering because of the contrast between the SLD of the matrix and the SLD of the unfilled pores.

The true/helium skeleton density of fine powders is often different from the density of bulk pieces of the same material because in the process of preparing the powders, atoms or molecules located at the surface are often forced out of their equilibrium positions within the solid structure. The percentage of such atoms in bulk materials is negligibly small; however, it increases significantly for small particles. Contrast matching SANS was employed in [35] to measure the skeleton density of activated carbon powder produced from raw coal by steam activation. The experiments were performed using mixtures of normal and deuterated toluene of known composition. The variation of measured SANS intensity *I(Q)* with the average SLD of (h + d) toluene is shown in Fig. 5.12 for several values of *Q*. As expected, the data can be fit to a parabola with the minimum of the function $I\left(\overline{\rho_2^*}\right)$ close to zero. The contrast match point reveals some variation with *Q*: it has shifted to higher values of $\overline{\rho_2^*}$ at high *Q*, which is equivalent to higher values of the apparent carbon density. The corresponding skeletal density is approaching the maximum

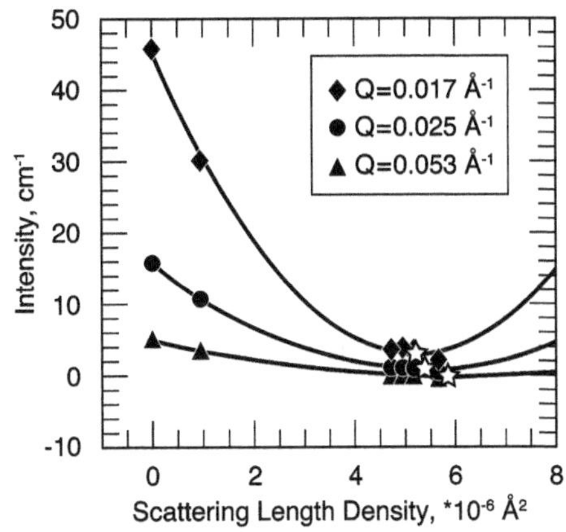

Fig. 5.12 Variation of the SANS signal with the SLD of H/D toluene for a series of Q-values. The star symbols show the fitted minimum scattering position. Reprinted from [35] with permission from Elsevier

value of 1.85 g/cm^3, which is lower than the density of graphite (2.15 g/cm^3) because of the availability of closed pores.

In addition to isotopic substitution, contrast variation can also be performed by filling the pores with nonadsorbing or weakly adsorbing supercritical fluids or pressurized gases, such as CO_2 or deuterated methane (CD_4), and measuring the scattering patterns as a function of pressure (or fluid density). Using D-methane rather than "normal" methane (CH_4) helps to minimize the contribution of incoherent scattering from hydrogen nuclei to the SANS data. An important advantage of using gases or supercritical fluids is their excellent penetrability into porous structure because of their orders-of-magnitude lower viscosity compared with the corresponding liquids. Using Eqs. (1.15) and (1.17), it is easy to show that the SLDs of CO_2 (ρ^*_{CO2}) and CD_4 (ρ^*_{CD4}) vary with their physical densities (ρ_{CO2}), and (ρ_{CD4}) as:

$$\rho^*_{CO2} = (2.49 \cdot \rho_{CO2})10^{10} \text{ cm}^{-2}, \quad \text{and} \tag{5.28}$$

$$\rho^*_{CD4} = (10 \cdot \rho_{CD4})10^{10} \text{cm}^{-2}, \tag{5.29}$$

where ρ_{CO2}, ρ_{CD4} can be calculated using appropriate equations of state of the bulk fluids at arbitrary temperature and pressure. At room temperature and pressure of 1 Kbar, the maximum density of CO_2 is $\rho_{CO2} = 1.14$ g/cm^3 and the maximum density of CD_4 is $\rho_{CD4} = 0.34$ g/cm^3, which means that the SLD can be varied between 0 and ~2.84×10^{10} cm^{-2} (CO_2) and ~3.4×10^{10} cm^{-2} (CD_4). Using the SLD value of silica calculated in Sect. 1.3 $\left(\rho^*_1 = 3.47 \times 10^{10} \text{cm}^{-2}\right)$ and Eq. (1.18) we obtain

$$\frac{d\Sigma}{d\Omega}(Q) \sim [3.47 - 2.49\rho_{CO2}]^2 \cdot 10^{20} cm^{-4}. \tag{5.30}$$

And thus the contrast matching point for this system should be reached at a CO_2 density of ~1.39 g/cm^3.

Contrast variation for a binary system (e.g., Vycor porous glass saturated with supercritical CO_2) was performed at T = 35 °C and T = 80 °C as a function of fluid density [36]. The SANS data reveal a characteristic maximum ("correlation" or "Vycor" peak) at Q_{MAX} ~ 0.02 Å^{-1} [37] (Fig. 5.13a). The amplitude of the maximum $I(Q_{MAX})$ was determined and plotted in $I(Q_{MAX})^{1/2}$ vs. ρ_{CO2} coordinates in Fig. 5.13b, where it can be seen that $I(Q_{MAX})^{1/2}$ is a linear function of ρ_{CO2}. A fit to Eq. (5.30) shows that the contrast matching point is reached at $\rho_{CO2} = 1.36 \pm 0.06$ g/cm^3. This value corresponds to an SLD of 3.39×10^{10} cm^{-2}, which is in good agreement with $\rho^*_{CO2} = 3.35 \times 10^{10}$ cm^{-2}, reported previously in [37] at the Vycor match point. This density also coincides (within experimental error) with $\rho_{CO2} = 1.39$ g/cm^3, which was calculated assuming the density of the solid phase of Vycor glass to be equal to the density of amorphous silica $\rho_{SiO2} = 2.2$ g/cm^3 (see Sect. 1.3). Thus, the data demonstrate that in the range of temperature and fluid density studied, porous Vycor glass saturated with CO_2 represents a truly binary system, with no adsorbed fluid present in the sample.

Contrast variation of a binary system (porous fractal silica with deuterated methane CD_4) was applied in [38] to monitor structural homogeneity and pore accessibility of the silica matrix to D-methane at room temperature T = 23 °C. The skeletal SLD of the studied porous silica (3.47×10^{10} cm^{-2}) was calculated using the density of amorphous silica $\rho_{SiO2} = 1.8$ g/cm^3, and the pressure at which the contrast matching point is reached (P_{CM}) was estimated to be P_{CM} ~ 514 bar. In this experiment both USANS and SANS were measured as a function of pressure and were normalized to the scattering intensity measured in vacuum $I(Q,VAC)$. The resulting ratios of $I(Q,P)/I(Q,VAC)$ are shown in Fig. 5.14 for different values of Q in both USANS and SANS domains. The scattered intensity shows a monotonic decrease as the pressure approaches the calculated P_{ZAC} value followed by a monotonic increase. The nearly zero value of the intensity around P_{CM} for all values of Q indicates the absence of measurable fluctuations of the SLD as well as insignificant amount of closed pores in the silica matrix.

Generally, analysis of the scattering data becomes much more complicated if the sample represents a ternary system. Examples of such systems are a two-component polymer blend with suspended nanoparticles, or a porous matrix with a layer of densified adsorbed fluid on the pore walls with the pore core filled with unadsorbed (bulk-like) fluid. The analysis may be simplified significantly by contrast matching of two out of three components. This leaves only the remaining component "visible" to the radiation, which reduces the problem to the binary system case.

Contrast matching SANS was applied to study the structure of a polymer-templated porous silica SBA-15, the structure of which consisted of a 2D hexagonal array of cylindrical mesopores "decorated" with micropores with interconnectivity [39]. The accessibility of micropores was studied using probe molecules of different size and shape: isotopic mixtures of (1) water, H_2O/D_2O; (2) decane, $C_{10}H_{22}/C_{10}D_{22}$; (3) isooctane, C_8H_{18}/C_8D_{18}; (4) cyclohexane, C_6H_{12}/C_6D_{12}; and (5) tributylamine, $N(C_4H_9)_3/N(C_4D_9)_3$. The average SLD of these isotopic mixtures

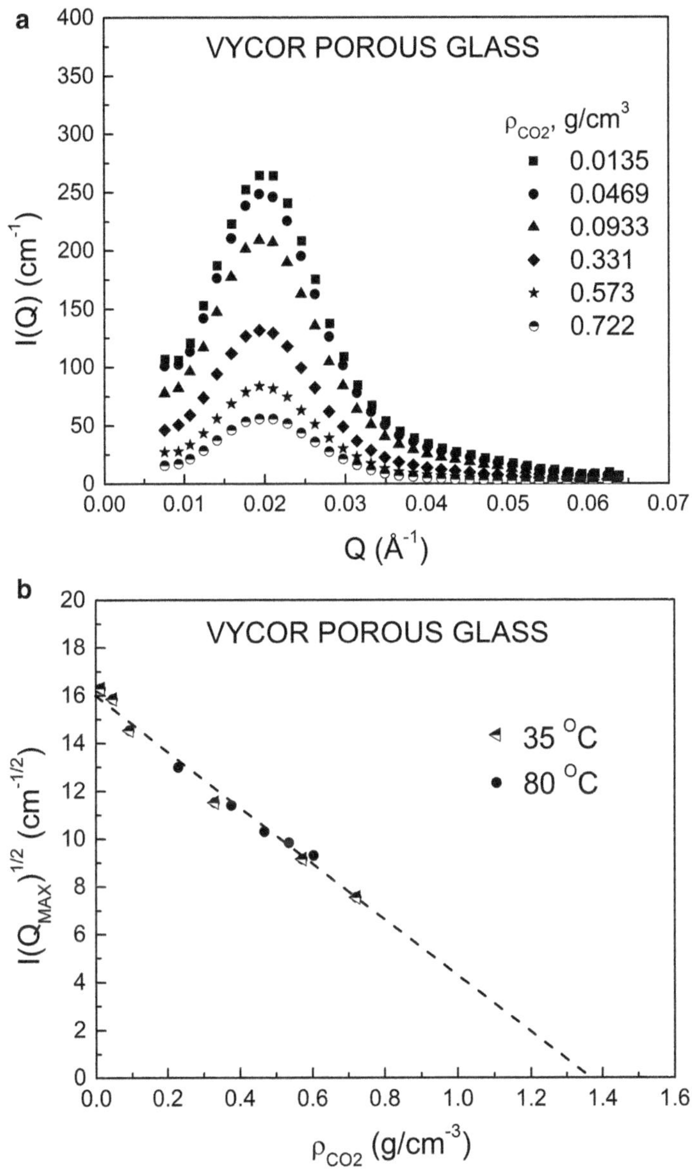

Fig. 5.13 (**a**) Neutron cross section of Vycor porous glass saturated with supercritical CO_2 at T = 35 °C as a function of the fluid density shown in the inset. (**b**) $I(Q_{MAX})^{1/2}$ *vs*. the density of CO_2 at two different temperatures specified in the inset. Reprinted with permission from [36]. Copyright [2006], AIP Publishing LLC

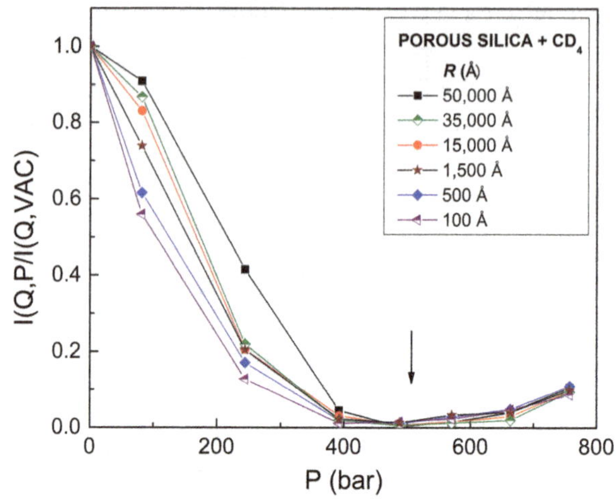

Fig. 5.14 Variation of the normalized cross section as a function of pressure of D-methane for porous fractal silica in pores of different sizes R (see *insets*). The arrow shows calculated pressure $P_{CM} = 514$ bar at which contrast matching point should be reached. Pore sizes estimated using relation $\overline{R} \approx 2.5/Q$ (Sect. 6.10). Reprinted from [38]. Copyright (2012), with permission from Elsevier

was varied to match the SLD of the silica matrix. If all pores are completely accessible to the probe molecule, then coherent scattering should be eliminated. However, if a molecule is too large to fit into micropores, residual scattering should remain in the contrast matching point. As shown in Fig. 5.15, the variation of the amplitude of the main (10) diffraction peak vs. the hydrogen isotopologue mole fraction agrees well with the expected parabolic dependence. For water, decane, and tributylamine, the diffraction intensity from mesopores decreases below detection limits, whereas for cyclohexane and isooctane there is significant intensity in the (10) peak remaining at the contrast match point. This indicates that these molecules are too big to fill the micropores, which gives an estimate for the micropore diameter of between 5.7 and 6.2 Å.

Contrast variation using pressurized fluids can be used to evaluate the physical density of an adsorbed layer in porous matrices of a different nature, which is virtually inaccessible by any other technique. As mentioned previously, the square root of intensity at each Q-value for a binary system should follow a linear function connecting points of the full contrast (pores in vacuum) and contrast-matched pores at a pressure P_{CM} that can be calculated based on the known chemical composition and the skeleton density. Deviations from such linear variation at low pressures are usually interpreted to be the result of a formation of the third component or "phase," which represents a layer of adsorbed fluid with the density exceeding the density of unadsorbed fluid in the pore core. The physical density of this layer can be evaluated by comparing the experimentally measured $I(Q)^{1/2}$ with the calculated linear variation based on the two-component (or two-phase) approximation. Figure 5.16 shows the variation of the cross section from the porous fractal silica as a function of CO_2 density at the value of $Q = 8.75 \times 10^{-3}$ Å^{-1} corresponding to the pore size $R \sim 286$ Å [40]. Measured scattering is equal to the calculated scattering when the densities of the adsorbed and unadsorbed phases are identical, which occurs at $\rho_{CO2} \approx 0.82$ g/cm^3. At higher pressures the measured $I(Q)^{1/2}$ falls

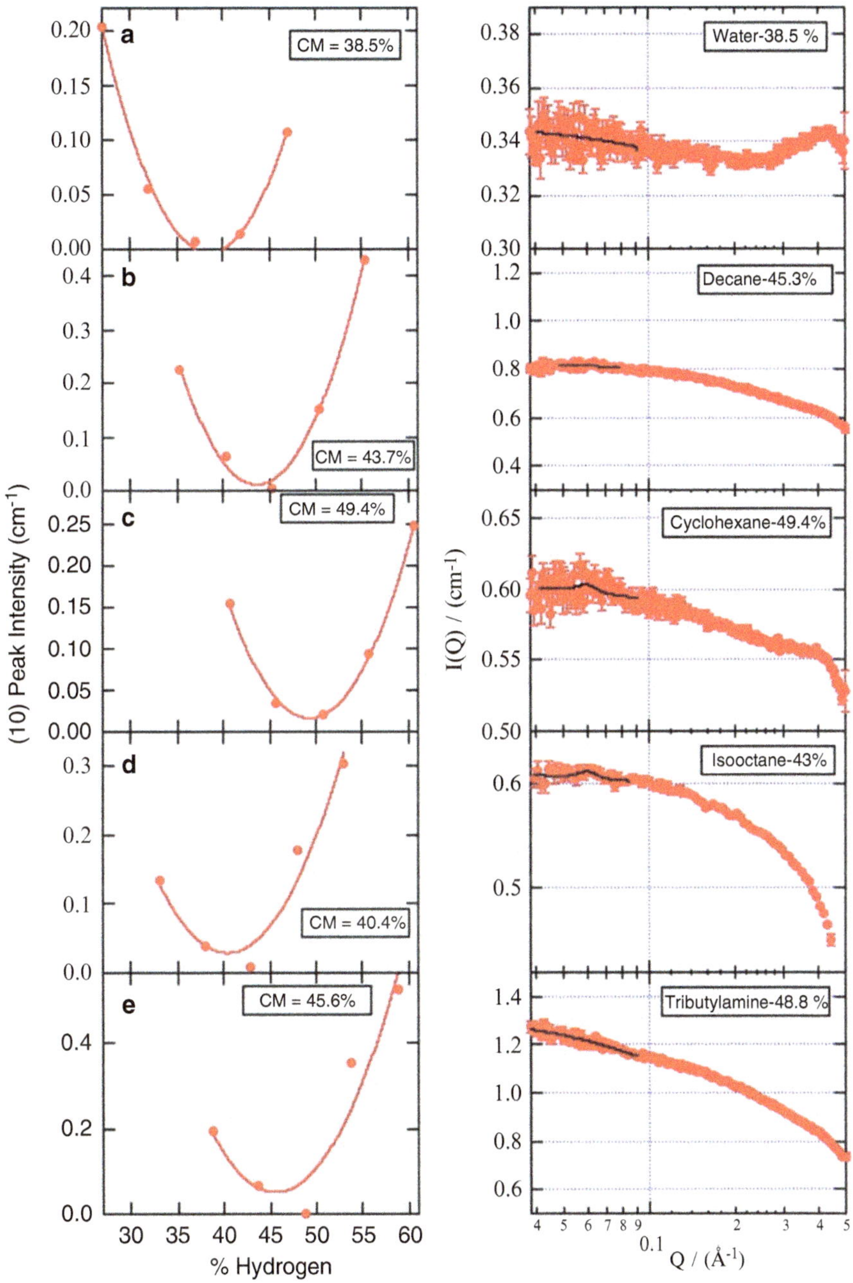

Fig. 5.15 *Left panel*: Diffraction peak intensities plotted as a function of the hydrogen isotopologue mole fraction for (**a**) water, (**b**) decane, (**c**) cyclohexane, (**d**) isooctane, and (**e**) tributylamine mixtures and corresponding contrast match concentration. *Right panel*: scattering curves, $I(Q)$, vs. Q, at the contrast match concentration determined from the parabola minimum Reprinted with permission from [39]. Copyright 2011 American Chemical Society

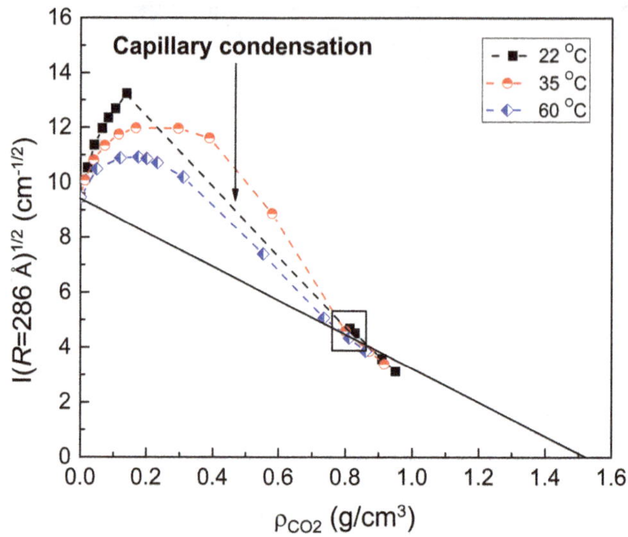

Fig. 5.16 Variation of SANS from 286 Å mesopores in porous silica as a function of CO_2 density and temperature shown in the inset. The straight solid line connects two limits corresponding to full neutron contrast (density of $CO_2 = 0$) and the contrast matching point (density of $CO_2 = 1.52$ g/cm^3). The box highlights the intersection region at which densities of the adsorbed and unadsorbed phases become equal to each other. Reprinted with permission from [40]. Copyright 2010 American Chemical Society

below the linear function, which indicates that fluid density in the pore core exceeded that of the adsorbed layer. The data demonstrate that the density of the adsorbed phase does not show measurable variation as a function of temperature.

5.11 Average Scattering Length Density of Multicomponent Samples

Determination of the total porosity and internal specific surface area of porous media using SAS requires placing scattering data on absolute scale and knowledge of the SLD of the solid matrix (if pores are filled with air) or contrast (neutron or x-ray) if pores are filled with a fluid. Calculation of the SLD for one-component skeletons is fairly straightforward, and examples of such calculations are given in Sect. 1.3. The situation becomes more complicated if the solid matrix consists of multiple components. This is the case for many geological samples (coals, shales, etc.) that consist of numerous compounds, such as minerals and total organic carbon (TOC). The average neutron SLD of such multicomponent samples can be calculated as a volume average over all compounds according to the formula [24]:

$$\overline{\rho_n^*} = \frac{1}{100}\sum\nolimits_1^n \phi_i \rho_{n,i}^*, \tag{5.31}$$

where index i indicates a mineral component (including TOC), n is a total number of compounds, ϕ is the volume concentration of the i-th compound (in percent), and ρ_n^* is the SLD of the i-th compound. Similarly, the average SLD for x-rays is:

$$\overline{\rho_X^*} = \frac{1}{100}\sum\nolimits_1^n \phi_i \rho_{X,i}^*. \tag{5.32}$$

Densities, molecular weights, scattering lengths, and corresponding SLDs for most common naturally abundant inorganic materials are shown in Table 5.2. The typical volume proportion of minerals in a New Albany shale sample 472-1 and the grain densities of individual minerals and their weight percent content are shown in Table 5.3. The corresponding volume fractions, ϕ, were calculated using the conversion algorithm available at http://www.ncnr.nist.gov/resources/fractioncalc.html. Using data from Table 5.3 and Eq. (5.32), the average value of the neutron SLD of the New Albany shale sample was calculated as $\overline{\rho_n^*} = 3.74 \times 10^{10}\ \text{cm}^{-2}$. At the same time, its average x-ray SLD was calculated as $\overline{\rho_X^*} = 26.23 \times 10^{10}\ \text{cm}^{-2}$.

The values of neutron SLDs for coals, hydrocarbons, common minerals, and fluids are shown in Fig. 5.17 [24]. For organic compounds such as whole coals, fractions of crude oil (asphaltenes, polars and aromatics, and saturates), and pure hydrocarbons, the SLD is determined mostly by the atomic ratio of two dominant elements in the organic matter: carbon and hydrogen. The neutron contrast between solid organic and inorganic components of the rock is close to zero; therefore, the dominant scattering contrast in coals originates from the difference between the SLDs of the solid matrix and the pore space. Thus, neutrons "see" rocks as quasi–two-phase systems, independent of the organic matter content [24]. Because of the large difference between the coherent scattering length of deuterium and hydrogen, isotopic mixtures of various solvents can be used to contrast match the SLD of a rock matrix and eliminate the matrix-pore scattering. Contrast matching with mixtures of ($D_2O + H_2O$) has been widely employed for SANS characterization of the structure of various coals, shales, and sandstone [41–46]. Isotopic mixtures of toluene ($C_7D_8 + C_7H_8$) were used to determine closed porosity in coal chars [47], and isotopic mixtures of methanol ($CD_4O + CH_4O$) were used to determine the neutron SLD and structure of kerogen in coal [48]. On the right-hand side of Fig. 5.17 are the experimentally reachable ranges of SLDs for deuterated methane and CO_2 that are tunable as a function of pressure and temperature. Both fluids can be used to contrast match whole coals and shales, and examples of using supercritical fluids to study the structure and closed porosity of coals and shales can be found in [28, 38, 46, 49–53].

X-ray SLDs for coals, hydrocarbons, and common minerals are shown in Fig. 5.18 [24]. For purely organic (e.g., ash-free coal) or purely inorganic (e.g., sandstone)

Table 5.2 Density, molecular weight, neutron ρ_n^*, and x-ray ρ_X^* SLDs for common minerals and TOC

Mineral	Formula	Density (g/cm^3)	Molecular weight (g/mole)	Molecular scattering amplitude $\Sigma s_i b_i$ (10^{-12} cm)	SANS SLD ρ_n^* (10^{10} cm^{-2})	SAXS SLD ρ_X^* (10^{10} cm^{-2})
Quartz	SiO_2	2.2	60	1.58	3.47	18.87
Corundum	Al_2O_3	4.0	102	2.43	5.74	33.32
Hematite	Fe_2O_3	5.2	160	3.65	7.16	42.05
Carbonates	$CaCO_3$	2.8	100	2.23	3.76	24.0
Albite	$NaAlSi_3O_8$	2.61	262	6.59	3.95	22.2
Anatase	TiO_2	3.9	80	0.83	2.44	31.52
Orthoclase	$KAlSi_3O_8$	2.6	278	6.60	3.71	22.16
Pyrite	FeS_2	5.0	120	1.52	3.62	40.67
Clay	$Al_2Si_2O_5(OH)_4$	2.63	263	5.25	3.16	22.72
Periclase	MgO	3.6	40	6.01	6.01	30.34
Lime	CaO	3.25	56	1.07	3.82	7.96
	K_2O	2.32	94	1.32	1.96	19.24
Zircon	$ZrSiO_4$	4.0	183	3.44	4.53	31.95
TOC	$C_{135}H_{96}O_9NS$	1.3	1908	60.32	2.47	11.59

Table 5.3 Mineral composition of the New Albany Shale sample 472-1

Mineral	Quartz	Carbonates	Albite	Orthoclase	Pyrite	Clays	TOC
Weight %	20	51	8	2	2	16	1
Volume %	20.48	49.42	8.32	2.09	1.09	16.51	2.09
Density (g/cm^3)	2.65	2.80	2.61	2.6	5.0	2.63	1.3

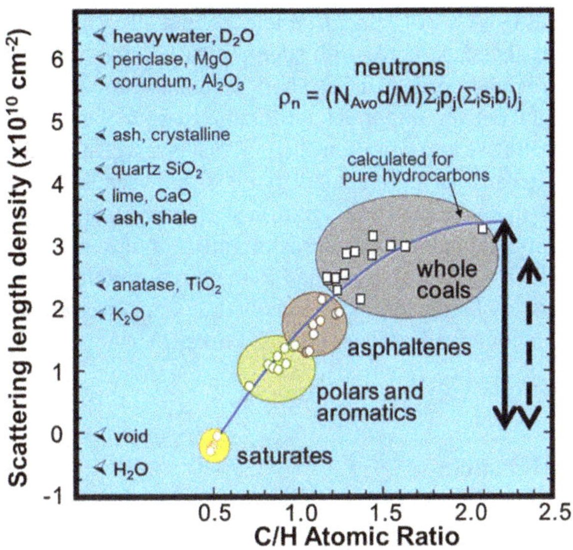

Fig. 5.17 Neutron SLDs for coals, hydrocarbons, and common minerals. Vertical solid and dashed arrows show ranges of variation of the SLD of D-methane and CO_2, respectively as a function of pressure up to 1 Kbar at room temperature [24]

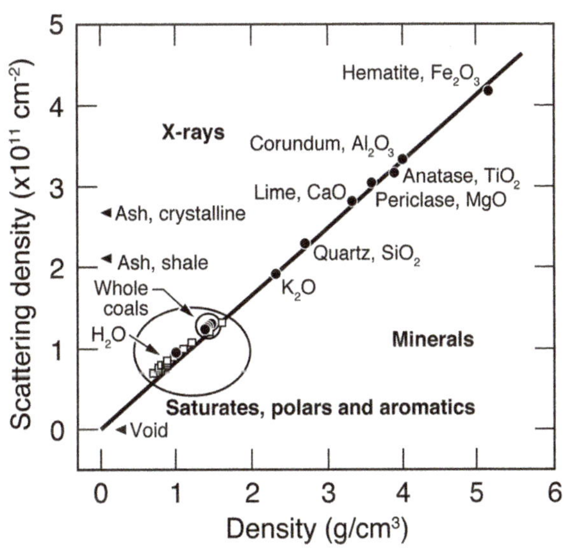

Fig. 5.18 X-ray SLDs (the electron densities) for coals, hydrocarbons, and common minerals [24]

rocks, the system remains quasi-binary from the scattering stand point and the structural information obtained from both SANS and SAXS is identical. If rocks contain both organic and inorganic components, the system remains quasi–two-phase for SANS; however, it could become multiphase for SAXS and the scattering data obtained using the two types of radiation are no longer equivalent [54]. The variation of the x-ray SLD of fluid at pressures below 1 Kbar is limited to ~6–7 × 10^{10} cm^{-2}. This value is significantly smaller than the x-ray SLD of whole coals (~1.2 × 10^{11} cm^{-2}), which does not allow use of x-rays for contrast matching of coal and shale pores with supercritical liquids. On the other hand, because of high fluxes and corresponding short acquisition times, synchrotron SAXS and USAXS might deliver important information about the local site-specific kinetics of gas penetration into porous media [55] that is marginally accessible to SANS and USANS.

Because of significant differences between the neutron and x-ray SLDs of rocks and their components, neutron and x-ray scattering can deliver complementary information about the microstructure of multicomponent systems. The combination of SAXS and SANS applied to the same sample can provide far more information than can be gained by one method alone. Unfortunately, until now, the number of such combined studies remains limited.

References

1. http://neutrons.ornl.gov/users/proposal-tips.shtml
2. Grillo, I.: Small-angle neutron scattering and its applications in soft matter. In: Borsali, R., Pecora, R. (eds.) Soft-Matter Characterization. Springer, New York (2008)
3. Wignall G.D.: Notes on the Collection and Initial Analysis of Cold Neutron Data on the GP SANS instrument. A tutorial for users. ORNL, 2012
4. He, L., Do, C., Qian, S., Wignall, G.D., Heller, W.T., Littrell, K.C., Smith, G.S.: Corrections for the geometric distortion of the tube detectors on SANS instruments at ORNL. Nucl. Instrum. Methods Phys. Res. A. **775**, 63 (2015)
5. Des Cloizeaux, J., Jannink, G.: Polymers in Solution. Their Modelling and Structure. Clarendon Press, Oxford (1990)
6. Lal, J., Wildmaier, J.M., Bastide, J., Boue, F.: Determination of an interpenetrating network structure by small-angle neutron scattering. Macromolecules **27**, 6443 (1994)
7. Schelten, J., Schmatz, W.: Multiple scattering treatment for small-angle scattering problems. J. Appl. Crystallogr. **13**, 385 (1980)
8. Goyal, P.S., King, P.S., Summerfield, G.C.: Multiple-scattering in small-angle neutron scattering measurement of polymers. Polymer **24**, 131 (1983)
9. Berk, N.F., Hardman-Rhyne, K.A.: Analysis of SAS data dominated by multiple scattering. J. Appl. Crystallogr. **21**, 645 (1988)
10. Sabine, T.M., Bertram, W.K.: The use of multiple-scattering data to enhance small-angle neutron scattering experiments. Acta. Crystallogr. **55**, 500 (1999)
11. Cheng, G., Kent, M.S., He, L., Varanasi, P., Dibble, D., Arora, R., Deng, K., Hong, K., Melnichenko, Y.B., Simmons, B., Singh, S.: Effect of ionic liquid treatment on the structures of lignins in solutions: molecular subunits released from lignin. Langmuir **28**, 11850 (2012)
12. Bahadur, J., Radlinski, A., Melnichenko, Y.B., Mastalerz, M., Schimmelmann, A.: SANS/USANS study of the New Albany Shale: a treatise on microporosity. Energy Fuel **29**, 567 (2015)

13. Hayter, J.B., Penfold, J.: Determination of micelle structure and charge by neutron small-angle scattering. Colloid Polym. Sci. **261**, 1022 (1983)
14. Wignall, G.D., Bates, F.S.: Absolute calibration of small-angle neutron scattering data. J. Appl. Crystallogr. **20**, 28 (1087)
15. Russell, T.P., Lin, J.S., Spooner, S., Wignall, G.D.: Intercalibration of small-angle X-ray and neutron scattering data. J. Appl. Crystallogr. **21**, 629 (1988)
16. Schmatz, W., Springer, T., Schelten, J., Ibel, K.: Neutron small-angle scattering—experimental techniques and applications. J. Appl. Crystallogr. **7**, 96 (1974)
17. Wignall, G.D., Christen, D.K., Ramakrishnan, V.: Instrumental resolution effects in small-angle neutron scattering. J. Appl. Crystallogr. **21**, 438 (1988)
18. Mildner, D.F.R.: Design optimization of a small-angle neutron scattering spectrometer—comment. Nucl. Instrum. Methods Phys. Res. A **290**, 259 (1990)
19. Pedersen, J.S., Posselt, D., Mortensen, K.: Analytical treatment of the resolution function for small-angle scattering. J. Appl. Crystallogr. **23**, 321 (1990)
20. Mildner, D.F.R., Carpenter, J.M., Worchester, D.I.: Measurement and calculation of resolution of time-of-flight small-angle scattering. J. Appl. Crystallogr. **19**, 311 (1986)
21. Radlinski, A.P., Claoue-Long, J., Hinde, A.L., Radlinska, E.Z., Lin, J.-S.: Small-angle X-ray scattering measurement of the internal microstructure of natural zircon crystals. Phys. Chem. Miner. **30**, 631 (2003)
22. Melnichenko, Y.B., Wignall, G.D., Compton, R., Bakale, G.: Characterization of fullerenes and fullerene derivatives by small-angle neutron scattering and transmission measurements. J. Chem. Phys. **111**, 4724 (1999)
23. Keiper, J.S., Behles, J.A., Bucholz, T.L., Simhan, R., DeSimone, J.M., Lynn, G., Wignall, G. D., Melnichenko, Y.B., Frielinghaus, H.: Self-assembly of phosphate fluorosurfactants in carbon dioxide. Langmuir **20**, 1065 (2004)
24. Radlinski, A.P.: Small-angle neutron scattering and the microstructure of rocks. Rev. Miner. Geochem. **63**, 363 (2006)
25. Lowell, S., Shield, J.E., Thomas, M.A., Thommes, M. (eds.): Characterization of Porous Solids and Powders: Surface Area, Pore Size and Density. Springer, New York (2006)
26. Spalla, O., Lyonnard, S., Testard, F.: Analysis of the small-angle intensity scattered by a porous and granular medium. J. Appl. Crystallogr. **36**, 338 (2003)
27. Radlinski, A.P., Mastalerz, M., Hinde, A.L., Hainbuchner, M., Rauch, H., Baron, M., Lin, J.S., Fan, L., Thyagarajan, P.: Application of SAXS and SANS in evaluation of porosity, pore size distribution and surface area of coal. Int. J. Coal Geol. **59**, 245 (2004)
28. Mastalerz, M., He, L., Melnichenko, Y.B., Rupp, J.: Porosity of coal and shale: insights from gas adsorption and SANS/USANS techniques. Energy Fuel **26**, 5109 (2012)
29. Vacher, R., Woignier, T., Pelous, J., Courtens, E.: Structure and self-similarity of silica aerogels. Phys. Rev. B **37**, 6500 (1988)
30. Stuhrmann, H.B.: Neutron small-angle scattering of biological macromolecules in solution. J. Appl. Crystallogr. **7**, 173 (1974)
31. Stuhrmann, H.B., Fuess, H.: Neutron small-angle scattering study of hen egg-white lysozyme. Acta Crystallogr. **A32**, 67 (1976)
32. Hirai, M.: Contrast variation. In: Imae, T., Kanaya, T., Furusaka, M., Torikai, N. (eds.) Neutrons in Soft Matter. John Wiley, Hoboken, New Jersey (2011)
33. Allen, A.J.: Characterization of ceramics by X-ray and neutron small-angle scattering. J. Am. Ceram. Soc. **88**, 1367 (2005)
34. Williams, C.E.: Contrast variation in X-ray and neutron scattering. In: Lindner, P., Zemb, T. (eds.) Neutron, X-ray and Light Scattering. Elsevier Science, Amsterdam (1991)
35. Mileeva, Z., Ross, D.K., Wilkinson, D., King, S.M., Ryan, T.A., Sharrock, H.: The use of small angle neutron scattering with contrast matching and variable adsorbate partial pressures in the study of porosity in activated carbons. Carbon **50**, 5062 (2012)

36. Melnichenko, Y.B., Wignall, G.D., Cole, D.R., Frielinghaus, H.: Adsorption of supercritical CO_2 in aerogels as studied by small-angle neutron scattering and neutron transmission techniques. J. Chem. Phys. **124**, 204711 (2006)
37. Wiltzius, P., Bates, F.S., Dierker, S.B., Wignall, G.D.: Structure of porous Vycor glass. Phys. Rev. **A36**, 2991 (1987)
38. Melnichenko, Y.B., He, L., Sakurovs, R., Kholodenko, A.L., Blach, T., Mastalerz, M., Radlinski, A.P., Cheng, G., Mildner, D.F.R.: Accessibility of pores in coal to methane and carbon dioxide. Fuel **91**, 200 (2012)
39. Pollock, R.A., Walsh, B.R., Fry, J., Ghampson, T., Melnichenko, Y.B., Kaiser, H., Pynn, R., DeSisto, W.J., Clayton Wheeler, M., Frederick, B.: Size and spatial distribution of micropores in SBA-15 using CM-SANS. Chem. Mater. **23**, 3828 (2011)
40. Melnichenko, Y.B., Mayama, H., Cheng, G., Blach, T.: Monitoring phase behavior of sub- and supercritical CO_2 confined in porous fractal silica with 85 % porosity. Langmuir **26**, 6374 (2010)
41. Gethner, J.S.: The determination of the void structure of macroporous coals by small-angle neutron scattering: void geometry and structure in Illinois No-6 bituminous coal. J. Appl. Phys. **59**, 1068 (1986)
42. Hall, P.J., Antxustegi, M.M., Ruiz, W.: Contrast-matching small-angle neutron scattering evidence for the absence of a connected pore system in Pittsburgh No. 8 coal. Fuel **77**, 1663 (1998)
43. Antxustegi, M.M., Hall, P.J., Calo, J.: Development of porosity in Pittsburgh No. 8 coal char as investigated by contrast-matching small-angle neutron scattering and gas adsorption techniques. Energy Fuel **12**, 542 (1998)
44. Hall, P.J., Brown, S.D., Calo, J.M.: The pore structure of the Argonne coals as interpreted from contrast matching small angle neutron scattering. Fuel **79**, 1327 (2000)
45. Broseta, D., Barre, L., Vizika, O.: Capillary condensation in a fractal porous medium. Phys. Rev. Lett. **86**, 5313 (2001)
46. Ruppert, L.F., Sakurovs, R., Blach, T.P., He, L., Melnichenko, Y.B., Mildner, D.F.R., Alcantar-Lopez, L.: A USANS/SANS study of the accessibility of pores in the Barnett shale to methane and water. Energy Fuel **27**, 772 (2013)
47. Snook, I., Yarovsky, I., Hanley, H.J.M., Lin, M.Y., Mainwaring, D., Rogers, H., Zulli, P.: Characterization of metallurgical chars by small angle neutron scattering. Energy Fuel **16**, 1009 (2002)
48. Thomas, J.J., Valenza II, J.J., Craddock, P.R., Bake, K.D., Pomerantz, A.E.: The neutron scattering length density of kerogen and coal as determined by CH_3OH/CD_3OH exchange. Fuel **117**, 801 (2014)
49. Melnichenko, Y.B., Radlinski, A.P., Mastalerz, M., Cheng, G., Rupp, J.: Characterization of the CO_2 fluid adsorption in coal as a function of pressure using neutron scattering techniques (SANS and USANS). Int. J. Coal Geol. **77**, 69 (2009)
50. Sakurovs, R., He, L., Melnichenko, Y.B., Radlinski, A.P., Blach, T., Lemmel, H., Mildner, D. F.R.: Pore size distribution and accessible pore size distribution in bituminous coals. Int. J. Coal Geol. **100**, 51 (2012)
51. Clarkson, C.R., Solano, N., Bustin, R.M., Bustin, A.M.M., Chalmers, G.R.L., He, L., Melnichenko, Y.B., Radlinski, A.P., Blach, T.P.: Pore structure characterization of North American shale gas reservoirs using USANS/SANS, gas adsorption, and mercury intrusion. Fuel **103**, 606 (2013)
52. Clarkson, C.R., Freeman, M., He, L., Agamalian, M., Melnichenko, Y.B., Mastalerz, M., Bustin, R.M., Radlinski, A.P., Blach, T.P.: Characterization of tight gas reservoir pore structure using USANS/SANS and gas adsorption analysis. Fuel **95**, 371 (2012)
53. He, L., Melnichenko, Y.B., Mastalerz, M., Sakurovs, R., Radlinski, A.P., Blach, T.: Pore accessibility by methane and carbon dioxide in coal as determined by neutron scattering. Energy Fuel **26**, 1975 (2012)

54. Radlinski, A.P., Boreham, C.J., Wignall, G.D., Lin, J.S.: Microstructural evolution of source rocks during hydrocarbon generation: a small-angle-scattering study. Phys. Rev. B **53**, 14152 (1996)
55. Radlinski, A.P., Busbridge, T.L., Mac, E., Gray, A., Blach, T.P., Cheng, G., Melnichenko, Y. B., Cookson, D.J., Mastalerz, M., Esterle, J.: Dynamic micromapping of CO_2 sorption in coal. Langmuir **25**(2385) (2009)

Further Reading

Guinier, A., Fournet, G.: Small-Angle Scattering of X-rays. John Wiley, New York (1955)
Rennie, R.: Reduction of data from SANS instruments. In: Brumberger, H. (ed.) Modern aspects of Small-Angle Scattering (NATO ASI Series). Kluwer Academic, Dodrecht, Boston, London (1993)
King, S.M.: Small-angle neutron scattering. In: Pethrick, R.A., Dawkins, J.V. (eds.) Modern Techniques for Polymer Characterization. John Wiley, New York (1999)
Higgins, J.S., Benoit, H.C.: Polymers and Neutron Scattering. Clarendon, Oxford (1994)
Cotton, J.P.: Initial data treatment. In: Lindner, P., Zemb, T. (eds.) Neutron, X-ray and Light Scattering. North-Holland, Amsterdam, Oxford, New York, Tokyo (1991)
Mortensen, K.: Small-angle neutron scattering at reactor sources. In: Imae, T., Kanaya, T., Furusaka, M., Torikai, N. (eds.) Neutrons in Soft Matter. John Wiley & Sons, Hoboken, New Jersey (2011)
Grillo, I.: Small angle neutron scattering and applications in Soft matter. In: Borsali, R., Pecora, R. (eds.) Soft Matter Characterization. Springer, New York (2008)
Brulet, D., Lairez, A., Lapp, A., Cotton, J.P.: Improvement of data treatment in small-angle neutron scattering. J. Appl. Crystallogr. **40**, 165 (2007)
Cotton, J.P.: Variations on contrast in SANS: determination of self and distinct correlation functions. Adv. Colloid. Interface Sci. **69**, 1 (1996)
Zaccai, G., Jacrot, B.: Small angle neutron scattering. Annu. Rev. Biophys. Bioeng. **12**, 139 (1983)

Chapter 6
Fundamentals of Data Analysis

The aim of SAS experiments is to extract volume-averaged information on the spatial distribution of scattering length density (neutrons) or the electron density (x-rays) in the sample from the measured differential cross section as a function of scattering vector. Correlation functions can be calculated directly from the SAS data and deliver information on the sample structure. This chapter begins with a brief description of the mathematical form and geometrical meaning of correlation functions, which enter theory of scattering from the two-phase random systems. Asymptotic behavior of scattering in the limit of low and high values of the scattering vector is considered, together with the unified scattering function, which bridges the Guinier and Porod scattering regimes. The rest of the chapter deals with the scattering theory from fractal systems and possible extension of the theory of scattering from the two-phase to multiphase systems.

6.1 Correlation Functions: Mathematical Form and Geometrical Meaning

The amplitude *A(Q)* of neutrons or x-rays scattered from a sample normalized by the amplitude of the incident beam is defined by the generalized scattering length B, which represents either coherent neutron (b_{coh}) or x-ray (b_X) scattering length

$$A(\boldsymbol{Q}) = \sum_j B_j e^{-i\boldsymbol{Q}\cdot\boldsymbol{r}_j}, \tag{6.1}$$

where $\boldsymbol{r}_j$ gives the position of either electron or nucleus. Introducing the density distribution function of the electrons or nuclei $n(\boldsymbol{r})$, Eq. (6.1) can be rewritten as

$$A(\boldsymbol{Q}) = B\int_V n(\boldsymbol{r})e^{-i\boldsymbol{Q}\cdot\boldsymbol{r}}d\boldsymbol{r} = \int_V \rho^*(\boldsymbol{r})e^{-i\boldsymbol{Q}\cdot\boldsymbol{r}}d\boldsymbol{r}, \tag{6.2}$$

Y.B. Melnichenko, *Small-Angle Scattering from Confined and Interfacial Fluids*, DOI 10.1007/978-3-319-01104-2_6

where $\rho^*(\boldsymbol{r})$ is the SLD or the electron density. The differential scattering cross section per unit volume $d\Sigma/d\Omega$ Eq. (1.12) is related to the normalized scattering amplitude as $d\Sigma/d\Omega = \frac{1}{V}|A(\boldsymbol{Q})|^2$; therefore,

$$\frac{d\Sigma}{d\Omega}(\boldsymbol{Q}) = \frac{1}{V}|A(\boldsymbol{Q})|^2 = \frac{1}{V}\left|\int_V \rho^*(\boldsymbol{r})e^{-i\boldsymbol{Q}\cdot\boldsymbol{r}}d\boldsymbol{r}\right|^2, \tag{6.3}$$

where the integration is performed over the illuminated sample volume V. Equation (6.3) implies that the differential cross section measured in a SAS experiment represents the square of the Fourier transform of $\rho^*(r)$. The Fourier transform is essential in the interpretation of the SAS experiments. Properties of the Fourier transform can be found in [1, 2].

Unlike for soft matter materials, the positions r_j of atoms and hence the SLD distribution $\rho^*(\boldsymbol{r})$ in solid porous matrices is fixed in time and the SAS experiment measures an ensemble average of $d\Sigma/d\Omega$. Equation (6.3) can be rewritten as

$$\frac{d\Sigma}{d\Omega}(\boldsymbol{Q}) = \frac{1}{V}\int \Gamma_\eta(\boldsymbol{r})e^{-i\boldsymbol{Q}\cdot\boldsymbol{r}}d\boldsymbol{r}, \tag{6.4}$$

where $\Gamma_\eta(\boldsymbol{r})$ is the autocorrelation function of a new variable $\eta(r) = \rho^*(\boldsymbol{r}) - \langle\rho^*\rangle$ defined as a deviation of $\rho^*(\boldsymbol{r})$ from its mean value $\langle\rho^*\rangle$. The normalized autocorrelation function $\gamma(r)$, defined as

$$\gamma(r) = \frac{\Gamma_\eta(\boldsymbol{r})}{\Gamma_\eta(0)}, \quad \gamma(0) = 1, \tag{6.5}$$

was introduced by Debye and Bueche [1]. For isotropic medium, the correlation function $\gamma(\boldsymbol{r})$ depends on the magnitude of the vector $\boldsymbol{r}$ only and Eq. (6.4) can be rewritten as

$$\frac{d\Sigma}{d\Omega}(Q) = 4\pi(\Delta\rho^*)^2\int_0^\infty r^2\gamma(r)\frac{\sin(Qr)}{Qr}dr. \tag{6.6}$$

The correlation function can be expressed as the inverse Fourier transform of the scattering differential cross section

$$\gamma(r) = \left[2\pi^2(\Delta\rho^*)^2\right]^{-1}\int_0^\infty Q^2\frac{d\Sigma}{d\Omega}(Q)\frac{\sin(Qr)}{Qr}dQ. \tag{6.7}$$

Equations (6.6) and (6.7) are of great importance in SAS data analysis. The correlation function $\gamma(r)$ can be calculated directly from the SAS data and deliver important information on the particle structure. For uniform particles, $\gamma(r)$ and the function

$$p(r) = r^2\gamma(r), \tag{6.8}$$

which is called distance distribution function, allow for direct geometrical interpretation. Consider a volume element ν_i within a particle. The probability of another volume element ν_j being situated at distance r from ν_i within the particle is given by $\gamma(r)$. The number of elements ν_i is proportional to the particle volume V, while the number of elements ν_j is determined by the surface of the sphere with radius r, namely $4\pi r^2$. The product $4\pi r^2 V\gamma(r) = 4\pi p(r)/\rho$, where ρ is the density of the uniform particle, represents the number of distances r joining two arbitrary volume elements within the particle. The same functions for nonuniform particles do not have a straightforward meaning, as each distance is weighed by the product of the (different) SLDs of the corresponding volume elements.

In most cases, calculation of the correlation function $\gamma(r)$ is complicated and requires numerical integration except for a few simple particle shapes, for which analytical expressions are available. For example, in the case of a solid sphere with radius R

$$\gamma(r) = 1 + \frac{3}{4}\frac{r}{R} + \frac{1}{16}\left(\frac{r}{R}\right)^3. \tag{6.9}$$

Figure 6.1 shows functions $\gamma(r)$ and $p(r)$ for this shape. The corresponding form factor is given in Sect. 1.5 (see Eq. (1.24)).

All previous considerations hold true for scattering from single particles (or systems of dilute independently scattering objects without interference effects). For more dense systems, when the correlation in the particle positions or interactions between scattering objects cannot be ignored, the scattering becomes affected by the correlations and this effect is described by the interparticle structure factor $S(Q)$, which is defined as [2]

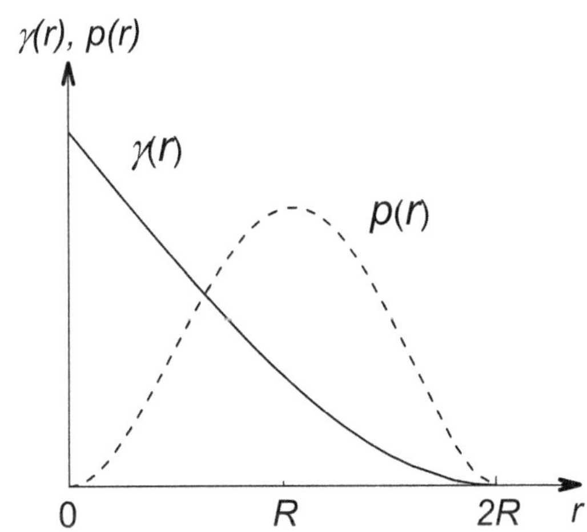

Fig. 6.1 Correlation function $\gamma(r)$ and distance distribution function $p(r)$ for a solid sphere of radius R

$$S(\boldsymbol{Q}) = 1 + \int g(\boldsymbol{r})\exp(-i\boldsymbol{Q}\cdot\boldsymbol{r})d\boldsymbol{r}, \tag{6.10}$$

where $g(\boldsymbol{r})$ is the particle pair distribution function, which describes the spatial distribution of the particles (scattering objects) as a function of the mean separation distance r. For the isotropic system $|\boldsymbol{r}| = r$, and neglecting structure-independent forward scattering, Eq. (6.10) can be rewritten as

$$S(\boldsymbol{Q}) = 1 + \int (g(r) - g_{\infty}(r))\exp(-i\boldsymbol{Q}\cdot\boldsymbol{r})d\boldsymbol{r}, \tag{6.11}$$

where $g_{\infty}(r)$ is the mean number density of the particles in the system and the pair correlation function $g(r)$ is defined as

$$dN(r) = (N_p/V)g(r)4\pi r^2 dr, \tag{6.12}$$

where N is the number of particles N_p in volume V. The number of scattering objects within a sphere of radius r around a given object is

$$N_p(r) = \frac{N_p}{V}\int_0^r g(r)4\pi r^2 dr. \tag{6.13}$$

The quantity $(N_p/V)g(r)4\pi r^2$ is often called the radial distribution function. Using Eq. (6.13), the structure factor $S(Q)$ can be represented as (see also Eq. (1.27))

$$S(Q) = 1 + \frac{N_p}{V}\int (g(r) - 1)\exp(-i\boldsymbol{Q}\cdot\boldsymbol{r})d\boldsymbol{r}. \tag{6.14}$$

Finally, consider the relative contribution of the form and structure factors in the scattering cross section of a set of N_p of monodisperse spheres with the number density N, radius R, and volume V_p. The scattering cross section per unit volume is given by Eq. (1.23) with the form factor of sphere described by Eq. (1.24). For the dilute system, $(NV_p < 0.1)$ $S(Q) \approx 1$. For the semidilute system at low Q, the shape of the scattering curve $d\Sigma/d\Omega$ is defined predominantly by the structure factor $S(Q)$ since $P(Q) \approx 1$ for $QR << 1$. At high Q, $d\Sigma/d\Omega$ depends mainly on the form factor of the individual sphere as $S(Q)$ is ≈ 1 for $QR >> 1$.

As is discussed subsequently, the correlation function $\gamma(r)$ (Eq. (6.5)) can be generalized for isotropic random two-phase systems and used for the calculation of the scattering law from surface fractals [3–5]. The first derivative of $\gamma(r)$ is proportional to the surface-to-volume ratio S/V for both the single particles and random-phase systems, and this property of $\gamma(r)$ allows for the determination of S/V from $d\Sigma/d\Omega$ in the limit of high Q. Equation 6.14 can be used to calculate the scattering law from mass fractals [6–8].

6.2 Scattering from Two-Phase Random Systems: The Porod Invariant

A random two-phase system (also called nonparticulate, stochastic, or inhomogeneous) consists of two different materials (phases) with different SLDs (see Fig. 6.2). A porous medium is an example of a random two-phase system that comprises a solid matrix with a constant SLD (phase 1) and pores that are random in shape, size, and spatial distribution (phase 2), with a sharp interface between the phases. The SLDs of phases 1 and 2 are ρ_1^* and ρ_2^*, respectively, and they are assumed to be constant within the phases. The volume fractions of phase 1 (ϕ_1) and 2 (ϕ_2) are related via

$$\phi_1 + \phi_2 = 1. \tag{6.15}$$

The average values of the SLD $<\rho^*>$ and $<(\rho^*)^2>$ for the two-phase system are

$$<\rho^*> \; = \; \rho_1^*\phi_1 + \rho_2^*\phi_2, \quad \text{and} \tag{6.16}$$

$$<(\rho^*)^2> \; = \; (\rho_1^*)^2\Phi_1 + (\rho_2^*)^2\Phi_2, \tag{6.17}$$

from which follows

$$<(\rho^*)^2> - <(\rho^*)>^2 \; = \; \phi_1\phi_2(\rho_1^* - \rho_2^*)^2. \tag{6.18}$$

The SLD fluctuation introduced in a two-phase system is defined as

$$\eta(r) \; = \; \rho^*(r) - <\rho^*>. \tag{6.19}$$

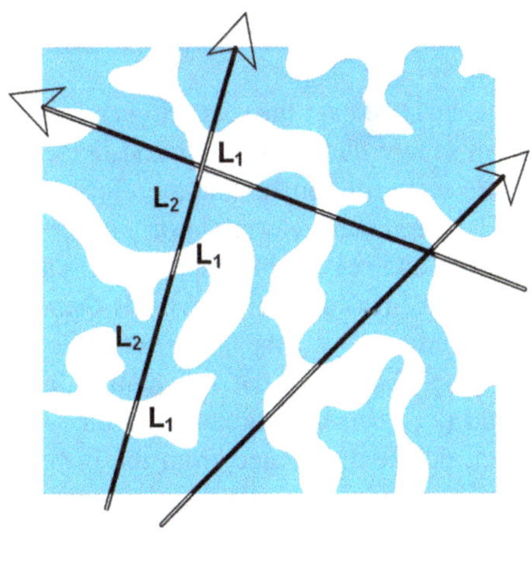

Fig. 6.2 Two-phase random (nonparticulate) system. A line crossing the system in an arbitrary direction cuts out chords of alternating length L_1 and L_2 between the boundaries of the phases. Image courtesy of G. D. Wignall

Using Eq. (6.16) we obtain

$$\eta_1 = \phi_2(\rho_1^* - \rho_2^*), \quad \text{and} \tag{6.20}$$

$$\eta_2 = \phi_1(\rho_2^* - \rho_1^*). \tag{6.21}$$

The mean square average of the SLD fluctuation is

$$<\eta^2> = \phi_1\eta_1^2 + \phi_2\eta_2^2 = \phi_1\phi_2(\rho_1^* - \rho_2^*)^2 = <(\rho^*)^2> - <(\rho^*)>^2. \tag{6.22}$$

It will be shown below that $<\eta^2>$ can be determined from the integral over $Q^2 d\Sigma/d\Omega$.

From Eqs. (6.19) and (6.5) for the two-phase system we get

$$\frac{d\Sigma}{d\Omega}(Q) = 4\pi(\rho_1^* - \rho_2^*)^2\phi_1\phi_2\int_0^\infty r^2\gamma(r)\frac{\sin(Qr)}{Qr}dr. \tag{6.23}$$

Equation (6.23) yields a way of obtaining the correlation function $\gamma(r)$:

$$\gamma(r) = \left[2\pi^2(\rho_1^* - \rho_2^*)^2\phi_1\phi_2\right]^{-1}\int_0^\infty Q^2\frac{d\Sigma}{d\Omega}(Q)\frac{\sin(Qr)}{Qr}dQ. \tag{6.24}$$

The correlation function for random two-phase systems was derived independently by Porod [9] and Debye, Anderson, and Brumberger [10]. Equation (6.23) (Debye-Porod law) is one of the most important equations that can be used for interpretation of SAS data and for extracting the structural parameters of various porous media including total porosity, interphase surface area, the mean domain size, etc. Since $\left[\frac{\sin(Qr)}{Qr}\right]_{r\to 0} = 1$ and by definition $\gamma(r = 0) = 1$, for the two-phase system

$$\int_0^\infty Q^2\frac{d\Sigma}{d\Omega}(Q)dQ \equiv Z = 2\pi^2(\rho_1^* - \rho_2^*)^2\phi_1\phi_2 \equiv 2\pi^2(\Delta\rho^*)^2\phi_1(1 - \phi_1). \tag{6.25}$$

Z, the integral over the scattering cross section weighed by Q^2, is proportional to the mean square fluctuation of the SLD (see Eq. (6.22)). Z is independent of the shape of the scattering object and therefore is called the Porod invariant.

Experimental determination of Z requires $d\Sigma/d\Omega(Q)$ data measured in a wide Q-range and reasonable approximations of the integral for the range from $Q = 0$ to Q_{min} (the first data point taken at the minimum available value of Q) and for the Q-range from Q_{max} (the last measurable data point in the limit of high Q) to $Q \to \infty$. The contribution to the invariant from the (unmeasured) cross section between 0 and $Q_{\min}$ can be estimated via the Guinier approximation, $I(Q) \sim exp(-Q^2 R_g^2)$ (Sect. 6.5), extrapolated from the measured cross section at the lowest Q-values. Similarly, the contribution to the invariant from the Q-range between Q_{max} and ∞

can be estimated via the Porod approximation, $I(Q) \sim Q^{-4}$ (Sect. 6.3). Typically, the low-Q Guinier extrapolation contributes ~1 % of the total invariant integral, as this component is weighted by Q^2 in a region where $Q \Rightarrow 0$. The extrapolation in the high-Q region may contribute a larger fraction (3–20 %) of the invariant, which is why it is important to collect the SAS data in both the low and high Q-limits. The Porod invariant can be used for determining the total porosity $\phi_1 = 1 - \phi_2$ (or volume fraction ϕ_1, ϕ_2 of the phases) in the two-phase system if electron (or scattering length) density contrast is known from the chemical composition of the system. Alternatively, the contrast can be estimated from SAS data using (6.25) if the sample porosity is known from independent measurements.

6.3 Asymptotic Behavior: The Porod Law

Assuming that the correlation function $\gamma(r)$ for a random medium can be approximated by exponential function [1, 9]

$$\gamma(r) = exp(-r/a), \tag{6.26}$$

and combining Eq. (6.26) with Eq. (6.23) we obtain

$$\frac{d\Sigma}{d\Omega}(Q) = 8\pi\phi_1\phi_2\left(\rho_1^* - \rho_2^*\right)^2 \frac{a^3}{\left[1 + a^2Q^2\right]^2}, \tag{6.27}$$

where a is called the Debye correlation length, which can be determined from a Debye-Bueche plot of $(d\Sigma/d\Omega)^{-1/2}$ vs. Q^2.

The small r expansion of $\gamma(r)$ in Eq. (6.26) in case of smooth (sharp) interfaces between the phases can be represented as [9, 10]:

$$\gamma(r \to 0) = 1 - r/a. \tag{6.28}$$

The Debye correlation length is related to the specific surface area of the interface (S/V, total surface area S per sample volume V) as

$$a = 4\left(\frac{S}{V}\right)^{-1} \phi_1\phi_2. \tag{6.29}$$

Insertion of Eq. (6.29) into Eq. (6.23) gives in the limit of large $Q >> 1/a$:

$$\frac{d\Sigma}{d\Omega}(Q) = 2\pi\left(\rho_1^* - \rho_2^*\right)^2 \frac{S}{V} Q^{-4}. \tag{6.30}$$

Equation (6.30) represents the Porod law, which states that the scattering cross section $(d\Sigma/d\Omega)$ from a two-phase system with a smooth interface between the phases decays as Q^{-4} in the limit of high values of Q. A plot of $Q^4(d\Sigma/d\Omega)$ vs. Q is called a Porod plot. After subtraction of the background, it shows a straight horizontal line in the large Q limit (Porod constant P) that can be used to evaluate specific surface area, provided the scattering length densities (or electron densities) of both phases are known. The value of (S/V) can also be determined from Debye's correlation length (Eq. 6.29) if volume fractions of phases are known. The Porod law is valid for dilute sets of monodisperse scattering objects with well-defined shape and sharp density discontinuity at the surface.

For random two-phase systems, deviations from the Q^{-4} behavior can be related to the variation of the SLD (the electron density) in the transition layer, as well to the diffuse character of the interface boundary (Sect. 6.6). Another source of deviation can be strong interface curvature with the curvature radius $R >> 1/Q$, as the Porod law is satisfied in the limit of a flat surface. For small radii of curvature, which is, for instance, anticipated for nanopores in many porous materials, Kirste and Porod introduced a correction term to the Porod law that accounts for the curvature effects [12, 13]:

$$\begin{aligned} \frac{d\Sigma}{d\Omega}(Q) &= 2\pi(\rho_1^* - \rho_2^*)^2 \frac{S}{V} Q^{-4} \\ &\times \left\{1 + Q^{-2}\left[\frac{1}{4} < (C_1 + C_2)^2 > + \frac{1}{8} < (C_1 - C_2)^2 >\right]\right\}. \end{aligned} \tag{6.31}$$

where C_1 and C_2 are the local principal curvatures of the surface, and the average is taken over all the surface. The curvature correction to Porod's law varies as Q^{-6}, which indicates that the Porod plot should exhibit at least one "hump" before reaching the constant value (Fig. 6.3). The hump position can be used for estimating the characteristic length scale of a two-phase system independently of any specific structure model. The width of the hump is related to the curvature distribution function in the medium [13].

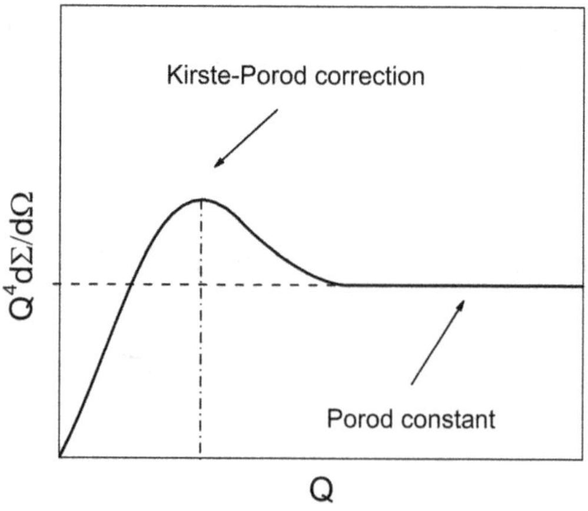

Fig. 6.3 Porod plot with Kirste-Porod correction for finite surface curvature

6.4 Radius of Gyration

The size of an object of arbitrary shape can be characterized in terms of its radius of gyration R_g, which is defined as the root-mean-square distance between its center of mass and all points within the object. In case of a constant scattering length density, R_g is defined as:

$$R_g^2 = \frac{1}{V_p} \iiint r^2 d^3 \boldsymbol{r}, \tag{6.32}$$

where the origin of $\boldsymbol{r}$ is in the center of the mass of the object with the volume V_p. Examples of the radius of gyration expressed by the dimensions of simple uniform geometrical bodies are given in Table 6.1.

6.5 Asymptotic Behavior: The Guinier Approximation

For noninteracting scattering objects of arbitrary shape the exponential factor in Eq. (6.2) can be expanded in terms of the radius of gyration in the limit of small Q-values. This leads to the Guinier law (Guinier approximation):

$$\frac{d\Sigma}{d\Omega}(Q) = NV_p^2 (\Delta\rho^*)^2 \exp\left(\frac{-Q^2 R_g^2}{3}\right), \tag{6.33}$$

or, after expansion of the exponential in the $QR_g << 1$ limit,

Table 6.1 Radii of gyration of some homogeneous bodies

Shape of geometrical bodies	Radius of gyration
Sphere of radius R	$R_g^2 = \frac{3}{5} R^2$
Spherical shell with radii $R_1 > R_2$	$R_g^2 = \frac{3}{5} \frac{R_1^5 - R_2^5}{R_1^3 - R_2^3}$
Solid rod with length L and cross section of radius R	$R_g^2 = \frac{L^2}{12} + \frac{R^2}{2}$
Thin rod $(R \rightarrow 0)$	$R_g^2 = \frac{L^2}{12}$
Thin disk $(L \rightarrow 0)$	$R_g^2 = \frac{R^2}{2}$
Elliptical cylinder with semiaxes a and b and height h	$R_g^2 = \frac{a^2 + b^2}{4} + \frac{h^2}{12}$
Hollow cylinder with radii $R_1 > R_2$ and height h	$R_g^2 = \frac{R_1^2 + R_2^2}{2} + \frac{h^2}{12}$

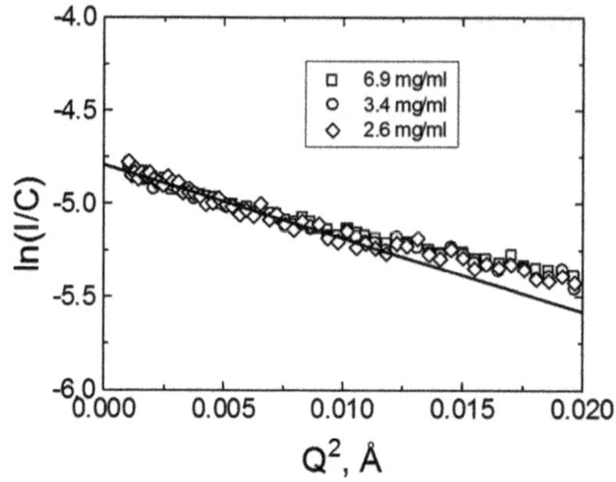

Fig. 6.4 Scattering intensity (in cm^{-1}) of $C_{60}/(C_6H_{12})_X$ complexes in D-toluene normalized to the concentration C of the complex shown in the inset. A slope of the solid line gives $R_g = 11.7$ Å. The theoretical limit $QR_g = 1$ of the Guinier law validity is $Q^2 \cong 0.0073$ Å. Reprinted with permission from [15]. Copyright [1999], AIP Publishing LLC

$$\frac{d\Sigma}{d\Omega}(Q) = NV_p^2(\Delta\rho^*)^2\left[1 - \frac{Q^2R_g^2}{3}\right]. \tag{6.34}$$

The Guinier approximation is valid only in the limit of small Q values ($QR_g < 1$) for a system of dilute, independently scattering and randomly oriented objects. Equations (6.33) and (6.34) allow for determination of the radius of gyration R_g of a scattering object of unknown shape and size from SAS experiments. The value of $R_g^3/3$ is determined directly from the initial slope of the "Guinier plot" (see Fig. 6.4):

$$ln\frac{d\Sigma}{d\Omega}(Q) \;\; vs. Q^2. \tag{6.35}$$

Number density (concentration) of the scattering objects N in the sample, as well as the neutron (or x-ray) contrast $(\Delta\rho^*)^2$ between the particles and surrounding medium, is usually known *a priori*. Extrapolation of the measured SAS data (in absolute units) to $Q \rightarrow 0$ allows for evaluating the particle volume V_p from thus determined value of $\frac{d\Sigma}{d\Omega}(0) = NV_p^2(\Delta\rho^*)^2$.

The Guinier plot can be modified to describe scattering from elongated quasi 1D and flat 2D objects in the "intermediate" Q-range [14]. For a cylinder (rod) with length L and radius R, the low-Q region is still described by Eq. (6.33) $\left(ln\frac{d\Sigma}{d\Omega}(Q) \; vs. \, Q^2\right)$ with $R_g^2 = \frac{L^2}{12} + \frac{R^2}{2}$ (Table 6.1). However, in the intermediate Q-range the variation of scattering is different:

$$\frac{d\Sigma}{d\Omega}(Q) = \frac{d\Sigma/d\Omega(0)}{Q}\exp\left(\frac{-Q^2R_g^2}{2}\right), \tag{6.36}$$

where $R_g^2 = R^2/2$ and thus the "intermediate" Guinier plot is described by (see Fig. 1.6):

$$ln\, Q\frac{d\Sigma}{d\Omega}(Q) \; vs.\, Q^2. \tag{6.37}$$

Similarly, for flat objects with thickness *L*,

$$\frac{d\Sigma}{d\Omega}(Q) = \frac{d\Sigma/d\Omega(0)}{Q^2}\exp\left(\frac{-Q^2R_g^2}{1}\right), \tag{6.38}$$

where $R_g^2 = L^2/12$ and therefore the modified Guinier plot in the intermediate Q-range becomes (see Fig. 1.6):

$$ln\, Q^2\frac{d\Sigma}{d\Omega}(Q) \; \text{vs.}\, Q^2. \tag{6.39}$$

Guinier law is still applicable if scattering objects are polydisperse in their in size; however, in this case the radius of gyration and the particle volume represent certain average numbers. Scattering from polydisperse in size and independently scattering objects characterized by the same value of the scattering length (or the electron) density (Eq. (6.33)) can be represented as [16]:

$$\frac{d\Sigma}{d\Omega}(Q) = N\left\langle V_p^2\right\rangle_n (\Delta\rho^*)^2 \exp\left(-\frac{Q^2}{3}\left\langle R_g^2\right\rangle_z\right), \tag{6.40}$$

where $\langle V_p^2\rangle_n$ is the number average of V_p^2 and $\langle R_g^2\rangle_z$ is the number average of R_g^2.

6.6 Structural Parameters of the Two-Phase Porous Medium

The inverse of the specific surface area (*S*/*V*) in the Porod law (Eq. (6.30)) has dimension of length and can be regarded as the length scale that characterizes the structure of the two-phase medium. Straight lines drawn in random directions through such medium will cut out "chords" of alternating length (L_1, L_2; see Fig. 6.2) with the average lengths $< L_1 >$ and $< L_2 >$ that are related to *V*/*S* via [17]:

$$< L_1 > = 4\frac{V}{S}\phi_1. \tag{6.41}$$

$$< L_2 > = 4\frac{V}{S}\phi_2. \tag{6.42}$$

The Debye length a (sometimes called the length of inhomogeneity) is defined by Eq. (6.29), and therefore for statistically random medium:

$$a^{-1} = < L_1 >^{-1} + \ < L_2 >^{-1}.$$

Combining this equation with Eqs. (6.29), (6.41), and (6.42) gives:

$$< L_1 > = \frac{a}{\phi_1}, \quad \text{and} \quad < L_2 > = \frac{a}{\phi_2} = \frac{a}{(1-\phi_1)}. \tag{6.43}$$

Equation (6.30) can be represented as

$$\frac{d\Sigma}{d\Omega}(Q) = 2\pi(\rho_1^* - \rho_2^*)^2 \frac{4\phi_1(1-\phi_1)}{a} Q^{-4}. \tag{6.44}$$

The pore volume fraction (sample porosity) ϕ_1 can be determined from the Porod invariant Z (6.25), and the Debye correlation length a is related to Z in the limit $Q \to \infty$ as

$$a = \frac{4Z}{\pi \lim_{Q\to\infty}\left(\frac{d\Sigma}{d\Omega}\right)Q^4}. \tag{6.45}$$

With ϕ_1 and a thus determined, using Eq. (6.43) it becomes possible to calculate two characteristic structural parameters of the porous medium: the average pore size $< L_1 >$ and the average size of the solid phase $< L_2 >$.

6.7 Bridging the Asymptotic Behavior: The Unified Scattering Function

Many real systems represent a complex mix of multilevel morphologies, all of which contribute to scattering on length scales from subnanometers to millimeters. USAS and SAS patterns from such systems are nearly featureless, displaying broad shoulders or humps on monotonically changing scattering functions. As discussed previously, in general, such scattering patterns are described by the asymptotic exponential (Guinier) regime in the limit of low Q and by power-law scaling regimes in the limit of high Q. Beaucage [18–20] developed the unified Guinier power-law mathematical model that can help to describe a wide range of USAS/SAS data with multiple levels of structures. In the simplest case of scattering from sharp interfaces $\left(\frac{d\Sigma}{d\Omega} \sim Q^{-4}\right)$, the scattering is described by a sum of two components

$$\frac{d\Sigma}{d\Omega}(Q) = Gexp\left(-\frac{Q^2R_g^2}{3}\right) + B\left\{\frac{\left[\text{erf}\left(QR_g/\sqrt{6}\right]^3}{Q}\right\}^{D=4}, \tag{6.46}$$

where $G = NV_p^2(\Delta\rho^*)^2$ and $B = 2\pi(\Delta\rho^*)^2\frac{S}{V}$—the Guinier and the Porod law prefactors, respectively. In the more general case of fractal morphologies (Sect. 6.8.1) $\frac{d\Sigma}{d\Omega} \sim Q^{-D}$ where the range of D-values $3 \leq D < 4$ corresponds to surface fractals, $D < 3$ to mass fractals and $D > 4$ to nonfractal diffuse interfaces. The error function (*erf*) is available in numerous fitting programs and/or downloadable macros [21, 22]. If combined, USAS and SAS data are available over several decades in Q, and Eq. (6.46) can be extended to describe an arbitrary number of individual structural levels. Taking into account the exponential cutoff to a power-law scattering at high Q (Eq. (6.46)), Beaucage demonstrated that in general [19]

$$\frac{d\Sigma}{d\Omega}(Q) = \sum_{i=1}^{n}\left\{G_i\exp\left(-Q^2R_{g,i}^2/3\right) + B_i\exp\left(Q^2R_{g,(i+1)}^2/3\right)\left[erf\left(QR_{g,i}/\sqrt{6}\right)^3/Q\right]^D i\right\}, \tag{6.47}$$

where $i = 1$ refers to the largest-size structural level. Importantly, Eq. (6.46) and its extensions do not introduce new fitting parameters over local Guinier and power-law fits and are widely used for adequate fits of the data from various materials with multilevel morphology (see Fig. 6.5).

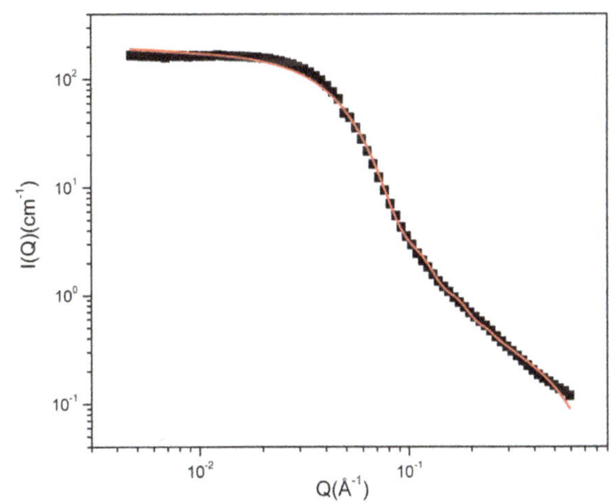

Fig. 6.5 SANS profile from carbon aerogel. The solid line shows fit to the Beaucage unified model (Eq. (6.46)) with two structural levels ($i = 2$). The values of fitting parameters are shown in Table 6.1. The average sizes (i.e., radii of gyration) corresponding to the two structural levels (i.e., two populations of pores of different sizes) in the aerogel matrix are $R_{g1} \sim 41.3$ Å and $R_{g2} \sim 12.6$ Å. Assuming the spherical shape of pores and using $R = R_g\sqrt{5/3}$ (Table 4.1), the corresponding average pore radii are $R_1 \sim 53$ Å and $R_2 \sim 16$ Å. Reprinted from [23]. Copyright (2012), with permission from Elsevier

6.8 Scattering from Fractal Systems and the Polydisperse Spherical Model

6.8.1 Scattering from Mass, Surface, and Pore Fractals

As is discussed in Sect. 6.3, the scattering cross section $d\Sigma/d\Omega(Q)$ decays as Q^D with different power-law exponents ($D=-4$ for a sphere, $D=-2$ for a thin disk, and $D=-1$ for a thin rod), which indicates that the power-law exponent D is related to the dimensionality of the scattering object. In many instances, the value of D in the limit of large Q is fractional and differs from the aforementioned values for three-, two-, and one-dimensional objects. Such nonintegral values of D can be understood based on the concept of fractality [24]. A fractal is an object or quantity that displays *self-similarity* under scale transformations. In other words, if a fractal is examined through a microscope, each of its enlarged fragments looks just like the original. Artificial (mathematically constructed) fractals such as Gosper island, Koch snowflake, Menger sponge, and Sierpiński sieve represent unlimited sequences of self-similar structural details. At the same time, the self-similarity of real physical objects is usually observed within a relatively narrow space length interval restricted by the lower (L_l) and upper (L_u) cutoff length scales that usually differ by 1.5–2 orders of magnitude [25]. A notable exception is sedimentary rocks, which can show fractal behavior of more than 3 orders of magnitude of the length scale [26, 27]. The lower cutoff is typically defined by the size of the basic building block unit of the system (such as atom, molecule, microcrystal, or small aggregate). The upper cutoff is usually much smaller than the size of the object because of limited mechanical strength, or the depletion of building resources during formation of the object structure.

The most important property of a fractal is its *fractal dimension*, D, the concept of which provides a new description of the structure of random systems. The mass fractal dimension quantifies the manner in which the mass, M, of an object increases with length r:

$$M \sim r^{D_m}, \tag{6.48}$$

where the mass fractal dimension, D_m, is a number between 1 and 3 and could be fractional. Objects generated by different growth processes (e.g., diffusion-limited aggregation, cluster-cluster aggregation, and percolation) form structures that are characterized by different mass fractal dimensions. Therefore, the fractal dimension can deliver important information about the growth process [28]. In general, the smaller values of D_m correspond to more open structures.

Surface fractals are bulk objects with a rough surface, the roughness of which displays a property of self-similarity within a certain length scale. The fractal dimension D_s of a 3D surface fractal can be defined similar to Eq. (6.48). If the surface area of an object S is measured with the resolution r^2, then:

$$S \sim r^{2-D_s}, \tag{6.49}$$

where the surface fractal dimension D_s in a 3D space ranges from 2 to 3. D_s can be taken as a measure of the space-filling capability of the surface. It is equal to 2 for perfectly smooth surfaces and approaches 3 when the surface is so folded that it almost completely fills the space, like a tightly crumpled napkin.

Some porous solids (e.g., activated carbons) host an extended network of channels that reveal yet another type of fractality, the so-called pore fractals described by a fractal dimension D_p. Here, unlike in mass fractals, the volume of the void space, rather than the sample mass, obeys scaling law (Eq. (6.48)). In other words, a mass fractal becomes a pore fractal on exchanging the roles of the solid matrix and pores [3]. All three types of fractals are illustrated schematically in Fig. 6.6.

SAS technique represents an important tool for studying the fractality of scattering objects. The interparticle structure factor is related to the particle pair distribution function $g(r)$ via Eq. (1.27). For a mass fractal aggregate of particles with radius $R_p \equiv L_l$, the deviation of the number density of particles N from its mean value at a distance r from the center of the aggregate can be represented as [6, 8]:

$$N[g(r) - 1] = \frac{D_m}{4R_p^{D_m}} r^{D_m - 3} \exp\left(-\frac{r}{\xi}\right), \tag{6.50}$$

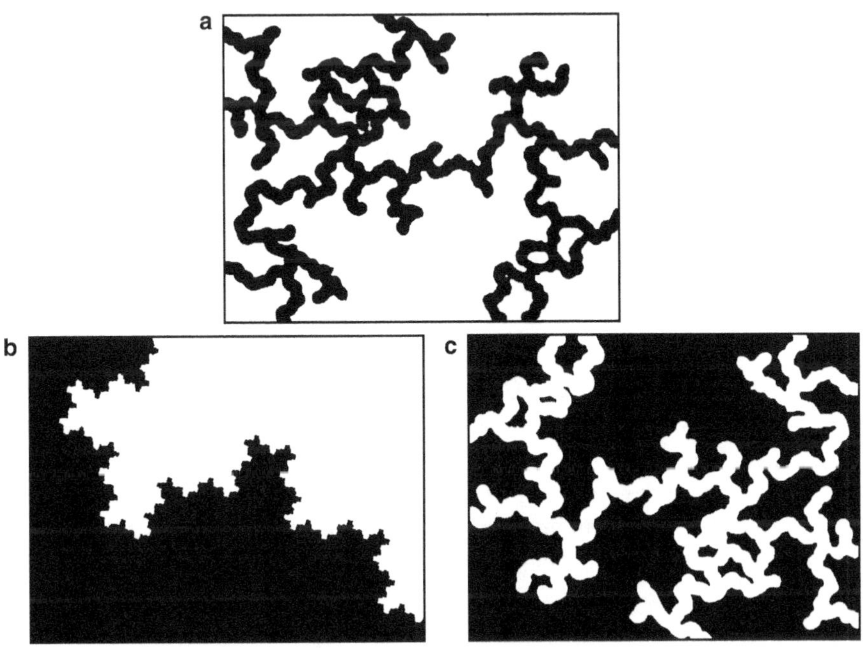

Fig. 6.6 Different types of fractals: (a) mass fractal, (b) surface fractal, and (c) pore fractal [29]

where $\xi \equiv L_u$ represents the upper cutoff length scale of the mass fractal [31], that is, the distance above which the mass distribution is no longer described by Eq. (6.48). Performing the Fourier transform, we obtain:

$$S(Q) = 1 + \frac{D_m \Gamma(D_m - 1)}{(QR_p)^{D_m}\left[1 + (Q\xi)^{-2}\right]^{(D_m-1)/2}} \sin\left[(D_m - 1)\arctan(Q\xi)\right], \quad (6.51)$$

where $\Gamma(x)$ is the gamma function of argument x. In the limit of low Q, Eq. (6.51) becomes:

$$S(Q \to 0) \cong 1 + \Gamma(D_m + 1)\ (\xi/R_p)^{D_m}\left[1 - (Q\xi)^2 D_m(D_m - 1)/6\right]. \quad (6.52)$$

This gives an estimate of the radius of gyration for a mass fractal:

$$R_g^2 = \frac{D_m(D_m + 1)\xi^2}{2}. \quad (6.53)$$

In the high-Q limit, Eq. (6.51) is reduced to

$$S(Q \to \infty) \cong 1 + \frac{D_m \Gamma(D_m - 1)\sin\left[(D_m - 1)(\pi/2)\right]}{(QR_p)^{D_m}}. \quad (6.54)$$

For $QR_p >> 1$, $S(Q) \to 1$ and $d\Sigma/d\Omega(Q) \cong NV_p^2(\Delta\rho^*)^2 P(Q)$ (see Eq. (1.23)), which implies that the scattering from primary particles is observed predominantely at large Q. The fractal character of scattering dominates in a certain intermediate rangy only, namely $1/\xi << Q << 1/R_p$, when $P(Q) \cong 1$ and $d\Sigma/d\Omega(Q) \cong NV_p^2\$\$(\Delta\rho^*)^2 S(Q)$. In this range, Eq. (6.54) becomes:

$$\frac{d\Sigma}{d\Omega}(Q) \cong NV_p^2(\Delta\rho^*)^2 D_m \Gamma(D_m - 1)\sin\left[(D_m - 1)(\pi/2)\right](QR_p)^{-D_m}, \quad (6.55)$$

and hence

$$\frac{d\Sigma}{d\Omega}(Q) \sim Q^{-D_m}, \quad \xi^{-1} \ll Q \ll R_p^{-1}. \quad (6.56)$$

The mass fractal dimension of a scattering object can be obtained from the slope of a log-log plot of the scattering cross section vs. Q. Deviations from the straight line at low and high Q (with background properly subtracted) provide estimates of the upper and lower cutoff length scales ξ (or L_l) and R_p (or L_u).

The scaling law for fractal surfaces is based on the correlation function:

$$\gamma(r \to 0) = 1 - \frac{1}{4\phi(1-\phi)} \frac{\nu_b(r)}{V}, \tag{6.57}$$

where ϕ is the volume fraction of pores, V is the illuminated sample volume, and $\nu_b(r)$ is the volume of a (rough) boundary layer. Equation (6.57) holds if r is much less than the correlation length ξ_s of the surface fractal behavior. Using Eq. (6.49), the boundary volume can be set as [3, 6]:

$$\nu_b(r) = Sr^{(3-D_s)}, \tag{6.58}$$

where S is total sample area. Combining Eqs. (6.23), (6.57), and (6.58), we obtain in the limit of $\xi Q >> 1$:

$$\frac{d\Sigma}{d\Omega}(Q) \cong KQ^{-(6-D_s)}, \quad \text{and} \quad \frac{1}{Q_{max}} < \xi_S < \frac{\pi}{Q_{min}}, \tag{6.59}$$

where Q_{min} and Q_{max} are the end points of the Q-range in which Eq. (6.49) is satisfied and

$$K = \pi(\rho_1^* - \rho_2^*)^2 \frac{S}{V} \Gamma(5 - D_s) \sin[(3 - D_s)\pi/2]. \tag{6.60}$$

For a random two-phase system with a constant SLD in either phase, and interface smooth on the scale $1/Q$ ($D_s = 2$), Eq. (6.59) reduces to the Porod law (6.30).

In general, for a surface/mass/pore fractal system, the cross section can be represented as [30]:

$$\frac{d\Sigma}{d\Omega}(Q) \sim Q^{D_s - 2(D_m + D_p) + 6}. \tag{6.61}$$

For a surface fractal, $D_m = D_p = 3$ and Eq. (6.61) reduces to Eq. (6.59). For a mass fractal, $D_s = D_m$ and $D_p = 3$ and it reduces to Eq. (6.56). For a pore space fractal, $D_s = D_p$ and $D_m = 3$ and Eq. (6.61) becomes

$$\frac{d\Sigma}{d\Omega}(Q) \sim Q^{-D_P}. \tag{6.62}$$

In some random two-phase systems, a transition layer of finite width between the phases can have certain distribution of the SLD (the electron density). For a Gaussian profile of the SLD at the interface, the cross section in the limit of high Q can be represented as

$$\frac{d\Sigma}{d\Omega}(Q \to \infty) = 2\pi(\rho_1^* - \rho_2^*)^2 \frac{S}{V} Q^{-4}(1 - \sigma^2 Q^2), \tag{6.63}$$

where σ is the standard deviation of the density distribution function. For such systems, a plot $Q^4 d\Sigma/d\Omega(Q)$ *vs.* Q^2 will reveal a straight line with negative slope $-\alpha$ for large Q, which is known as *negative* deviation from the Porod law [32–34] ($\alpha > 4$ in $d\Sigma/d\Omega(Q \sim Q^{-\alpha})$. Such deviation was observed experimentally in SAXS studies of porous silica with the surface tethered with short hydrocarbon chains C_nH_m (derivatized silicas) [35] and organo-modified mesoporous molecular sieves with the organic groups covalently linked to the surface [36]. *Positive* deviations from the Porod law with α slightly less than 4 were documented in experiments on the structure of carbons and explained by short-range density fluctuations in these materials [17].

6.8.2 Polydisperse Spherical Model

SAS from sedimentary rocks is usually linear on a log-log scale and is well described by Eq. (6.59), with the values of D_S generally close to 3 [37] (see Fig. 6.7). It was demonstrated [38] that scattering from a fractal surface can be considered to be equivalent to the scattering from a polydisperse system of spheres, with a certain number distribution of sizes (the number of spheres with radii between R and $R + dR$) given by

$$f(R)dR \sim R^{-(1+D_s)}dR, \tag{6.64}$$

where D_s is the surface fractal dimension determined from the slope of the power-law scattering. In practice, the distribution described by Eq. (6.64) is both upper and lower bound, that is, extends only in the range $R_{min} \leq R \leq R_{max}$. This is similar to

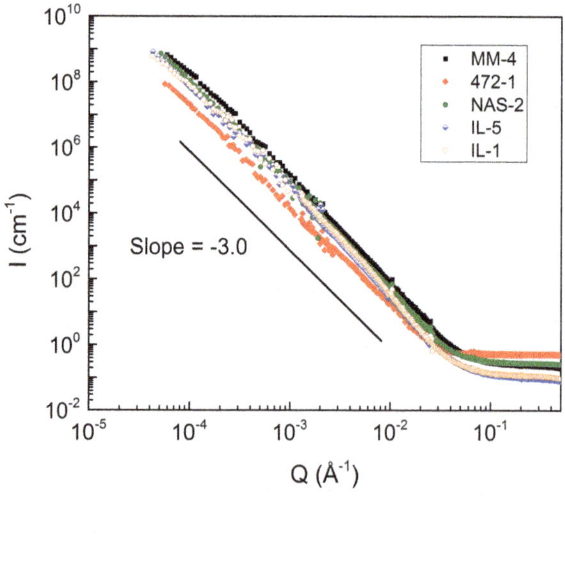

Fig. 6.7 Combined SANS and USANS absolute scattering neutron cross section vs. the scattering vector acquired from five samples of New Albany shales. The straight line has a slope of −3 for comparison. Reprinted with permission from [57]. Copyright 2015 American Chemical Society

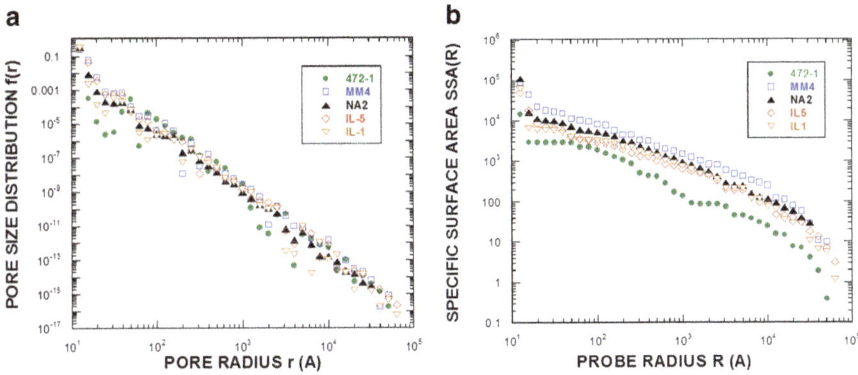

Fig. 6.8 Pore size distribution (**a**) and specific surface area (**b**) of New Albany shales calculated using the polydisperse spherical model. Reprinted with permission from [57]. Copyright 2015 American Chemical Society

the upper and lower cutoff parameters that limit fractal behavior to a certain range of the length scales. Scattering from a sample composed of a system of polydisperse spheres also has a linear region with a similar slope $-(1+D_s)$ and is described by [39, 40]:

$$\frac{d\Sigma}{d\Omega}(Q) = \left(\rho_1^* - \rho_2^*\right)^2 \frac{\phi}{\overline{V}} \int_{R_{min}}^{R_{max}} V^2 f(r) P(Q) dr, \tag{6.65}$$

where $\overline{V}$ is the average pore volume and $P(Q)$ is a form factor of a sphere described by Eq. (1.24). Such a representation is often referred to as the polydisperse spherical pore (PDSP) model. This model turned out to be remarkably useful for analyzing scattering data from sedimentary rocks as it allows for determining such important structural parameters as pore size distribution function, surface area, and the average number of pores per unit volume (see Fig. 6.8). The PDSP model is not restricted to fractal microstructures and can be considered as a general representation of the microarchitecture of the rock-pore interface [41]. Windows-based *PRINSAS* software is available for the processing and interpretation of SAS data tailored to the analysis of scattering from sedimentary rocks [39].

6.9 Beyond the Two-Phase Model

6.9.1 *Partial Scattering Functions of Multiphase Systems*

Multicomponent (or multiphase) systems are much more common in real life and are used in many technological applications. Of particular interest to the scope of this book are three-phase systems consisting of a porous matrix with films deposited on the pore surface resulting from coating, wetting/condensation, and adsorption of fluids. For macroscopic samples, the density distribution function and the thickness of the deposited layer can be determined directly by measuring the force between two macroscopic solid plane walls separated by a very thin film of fluid [42]. For obvious reasons, such measurements, however, cannot be carried out in the case of fluid-saturated porous solids. In this situation SAS provides a useful and noninvasive way to structurally characterize deposited layers and films in pores of different sizes at various thermodynamic conditions.

For a multiphase incompressible system with p different phases and phase 1 arbitrarily chosen as a "reference" phase, the coherent cross section per unit volume can be split into partial scattering functions $S(Q)_{i,j}$ [44]:

$$\frac{d\Sigma}{d\Omega}(Q) = \sum_{i=1}^{p} \left(\rho_i^* - \rho_1^*\right)^2 S_{ii}(Q) + 2\sum_{i<j}^{p} \left(\rho_i^* - \rho_1^*\right)\left(\rho_j^* - \rho_1^*\right) S_{ij}(Q)\,, \quad (6.66)$$

where ρ_i^*, ρ_j^* is the SLD (or the electron density) of *i-th* and *j-th* phase. The partial scattering functions $S_{ij}(Q)$ are represented by [45]:

$$S_{ij}(Q) = \frac{1}{V}\int_V \left\langle \phi_i(\boldsymbol{r})\phi_j\left(\boldsymbol{r}'\right)\right\rangle \exp\left[i\boldsymbol{Q}\left(\boldsymbol{r}-\boldsymbol{r}'\right)\right] d^3r\, d^3r', \quad (6.67)$$

where the integration is performed over the illuminated sample volume V and $\phi_i(\boldsymbol{r})$ describes the volume fraction of the phase i at position $\boldsymbol{r}$. In the case of ternary systems, there are six partial scattering functions interrelated via:

$$\begin{aligned} 2S_{12}(Q) &= -S_{11}(Q) - S_{22}(Q) + S_{33}(Q)\,. \\ 2S_{23}(Q) &= S_{11}(Q) - S_{22}(Q) - S_{33}(Q)\,. \\ 2S_{13}(Q) &= -S_{11}(Q) + S_{22}(Q) - S_{33}(Q)\,. \end{aligned} \quad (6.68)$$

Hence, all cross terms can be substituted by the squared terms, so that the cross section can be described only by the partial scattering functions S_{ii} with $1 \le i \le 3$ [46]:

$$\begin{aligned} \frac{d\Sigma}{d\Omega}(Q) = {} & \left(\rho_1^* - \rho_2^*\right)\left(\rho_1^* - \rho_3^*\right)S_{11}(Q) + \left(\rho_2^* - \rho_1^*\right)\left(\rho_2^* - \rho_3^*\right)S_{22}(Q) \\ & + \left(\rho_3^* - \rho_1^*\right)\left(\rho_3^* - \rho_2^*\right)S_{33}(Q). \end{aligned} \quad (6.69)$$

Equation (6.69) is useful for neutron contrast variation techniques as contrast matching any pair of the SLDs allows for experimental investigation of individual structure functions (S_{11}, S_{22}, or S_{33}).

Porous material with a homogeneous adsorbed layer of thickness δ on the pore walls is a special case of the three-phase system (see Fig. 6.9). If all pores are accessible (i.e. connected to the sample surface), the scattering cross section of the

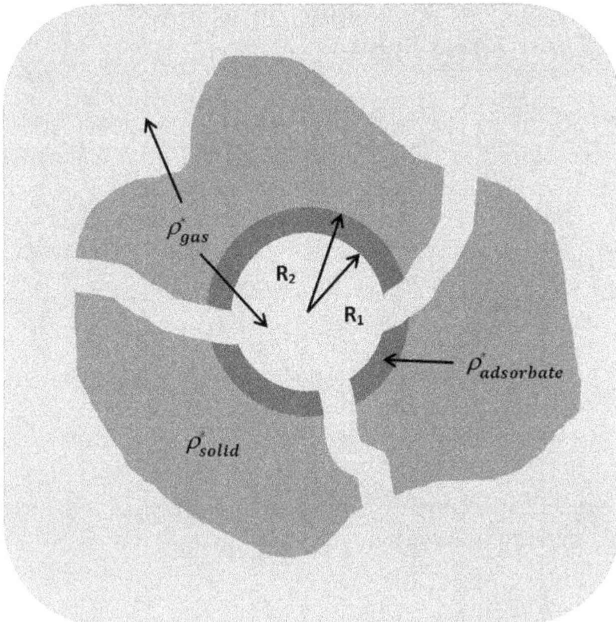

Fig. 6.9 Schematic diagram of the model of porous material, which is used for the calculation of $d\Sigma/d\Omega(Q)$ in Eq. (6.70). The ρ^* variables are the coherent SLDs (or the electron densities) that define corresponding neutron (x-ray) contrasts in Eqs. (6.71) and (6.72). The thickness of the adsorbed layer is $\delta = R_2 - R_1$

porous medium with gaseous fluid in the pore core and a layer of fluid adsorbed on the pore wall is given by [46]:

$$\begin{aligned} \frac{d\Sigma}{d\Omega}(Q) = \phi_1 \Bigg[& \left(\Delta\rho_a^*\right)^2 \frac{\left(v_1^2\right)^2}{v_2} S_{11}(Q,R_1) + \left(\Delta\rho_b^*\right)^2 v_2 S_{22}(Q,R_2) \\ & + 2\left(\Delta\rho_a^*\right)\left(\Delta\rho_b^*\right) v_1 S_1(Q,R_1) S_2(Q,R_2) \Bigg], \end{aligned} \tag{6.70}$$

where ϕ_1 is the volume fraction of pores, v_1 is the pore volume with radius R_1 (adsorbed layer deposited on the pore wall), and v_2 is the pore volume with radius R_2 (no adsorbed layer present, see Fig. 6.9). For spherical pore scattering functions, S_{11} and S_{22} are described by Eq. (1.24) with $R \equiv R_1$ and $R \equiv R_2$, respectively, and contrast factors $\Delta\rho_a^*$, $\Delta\rho_b^*$ are defined as:

$$\Delta\rho_a^* = \rho_{gas}^* - \rho_{adsorbate}^*, \quad \text{and} \tag{6.71}$$

$$\Delta\rho_b^* = \rho_{solid}^* - \rho_{adsorbate}^*. \tag{6.72}$$

Fitting the experimental SAS data to Eq. (6.70) can be used for estimating the thickness of the adsorbed layer as a function of temperature and pressure.

6.9.2 Scattering Contrast and the Invariant of a Three-Phase System

The equations describing scattering contrast and the invariant from three-phase systems were developed in [47] to model scattering from microvoids in binary composite materials. Consider a system with the total volume V_t, which consists of three "phases": solid porous material (phase 1), unadsorbed fluid (phase 2), and adsorbed fluid (phase 3) with the volumes V_1, V_2, and V_3, respectively:

$$V_t = V_1 + V_2 + V_3. \tag{6.73}$$

The volume fraction of phase i is defined as $\phi_i = V_i/V_t$, and therefore

$$\phi_1 + \phi_2 + \phi_3 = 1. \tag{6.74}$$

Based on the equations for correlation functions of three-phase systems [47, 48], the corresponding invariant is given by

$$\begin{aligned} \Xi &= \int_0^\infty Q^2 I(Q) dQ = 2\pi^2 \langle \Delta\rho^2 \rangle \\ &= 2\pi^2 \left[\left(\rho_1^* - \rho_2^*\right)^2 \phi_1\phi_2 + \left(\rho_2^* - \rho_3^*\right)^2 \phi_2\phi_3 + \left(\rho_1^* - \rho_3^*\right)^2 \phi_1\phi_3 \right], \end{aligned} \tag{6.75}$$

where ρ_i^* is the SLD of phase i. As mentioned previously, this expression was originally developed to model the scattering from microvoids in binary composite materials but can also be used to analyze the scattering from fluids adsorbed in porous media, where the adsorbed phase takes the place of the microvoids in the original formulation. In the case of zero adsorption ($\phi_3 = 0$), Eq. (6.75) reduces to Eq. (6.25), describing the invariant of a two-phase system (see Sect. 6.2). Combining Eqs. (6.75) and (6.25), we obtain

$$\phi_3 = \frac{\Xi - Z}{2\pi^2 \left[\phi_2 \left(\rho_2^* - \rho_3^*\right)^2 + \phi_1 \left(\rho_1^* - \rho_3^*\right)^2 - \phi_1 \left(\rho_1^* - \rho_2^*\right)^2 \right]}. \tag{6.76}$$

Equation (6.76) can be applied for determining the properties of the adsorbed phase in porous solids saturated with fluids under pressure P at a fixed temperature T [49, 50], which requires calculating the invariant Z (Eq. (6.24)) of the hypothetical two-phase system (with no adsorption) at each P,T. This procedure can be avoided by eliminating the volume fraction of the adsorbed phase ϕ_3 in the right-hand side of Eq. (6.76) using Eq. (6.74), which leads to [51]

$$\Xi = \int_0^\infty Q^2 I(Q) dQ = 2\pi^2 \langle \Delta\rho^2 \rangle = 2\pi^2 \big[(\rho_1^* - \rho_3^*)^2 \phi_1 (1 - \phi_1) + (\rho_2^* - \rho_3^*)^2 \phi_2 (1 - \phi_2) - 2(\rho_1^* - \rho_3^*)\ (\rho_2^* - \rho_3^*)\ \phi_1 \phi_2 \big]. \tag{6.77}$$

The quadratic Eq. (6.77) can be solved against the volume fraction of the adsorbed phase

$$\phi_3 = 1 - \phi_1 - \phi_2 = \frac{-b - \sqrt{b^2 - 4ac}}{2a}, \tag{6.78}$$

where

$$\begin{aligned} a &\equiv (\rho_2^* - \rho_3^*)^2, \\ b &\equiv 2(\rho_1^* - \rho_3^*)\ (\rho_2^* - \rho_3^*)\phi_1 - (\rho_2^* - \rho_3^*)^2, \quad \text{and} \\ c &\equiv \Xi/2\pi^2 - (\rho_1^* - \rho_3^*)^2 \phi_1 (1 - \phi_1). \end{aligned} \tag{6.79}$$

Most of the parameters in Eq. (6.78) are usually known *a priori* or can be calculated based on the known chemical composition of the solid matrix or equation of state of bulk (unadsorbed) fluids. Two unknown parameters of interest in Eq. (6.78) are the volume fraction ϕ_3 and the SLD ρ_3^* of the adsorbed phase. To determine the unknowns, additional information is required from independent measurements. Gravimetry or neutron/x-ray transmission measurements can deliver information on the average density of the adsorbed phase, which allows for calculating ρ_3^* and, consequently, ϕ_3 from Eq. (6.78). Examples of SAS experiments combined with other methods that give information on both ϕ_3 and ρ_3^* as a function of pressure and temperature are given in the subsequent chapters.

6.9.3 Oscillatory Deviations from the Porod Law

The analysis of SAS data from fluid-saturated solids with an adsorbed layer is usually conducted based on a few simplifying assumptions. Firstly, it is assumed that the density profile of the adsorbed phase is constant in a direction perpendicular to the pore surface (box model). Secondly, it is supposed that the sample has only two interfaces separated by distance δ, which represents the thickness of the adsorbed layer (or phase 3). In other words, unadsorbed fluid that occupies the pore core is never in contact with pore walls and is in contact only with the adsorbed layer. For large enough pore curvature radii, the two interfaces can be considered as parallel to each other. As shown in [52], the presence of parallel interfaces is responsible for a finite discontinuity in the second derivative of the auto-correlation function $\gamma''(r)$ at $r = \delta$, and the discontinuity gives rise to an oscillatory contribution in the Porod plot of the scattering intensity. The equation that takes into account the oscillatory contribution reads [52, 53]:

$$\frac{d\Sigma}{d\Omega}(Q) = \frac{A}{Q^4}[1 + R(\nu)\cos\,(Q\delta)] + B\,, \tag{6.80}$$

where

$$A \equiv \frac{2\pi\left(\rho^*_{1,2}\right)^2 S\,\alpha(\nu)}{V}, \tag{6.81}$$

$$R(\nu) \equiv (1-\nu^2)/(1+\nu^2), \quad \alpha(\nu) \equiv (1+\nu^2)/2\,, \tag{6.82}$$

$$\nu \equiv \frac{\rho^*_{1,3} - \rho^*_{3,2}}{\rho^*_{1,2}}, \tag{6.83}$$

$$\rho^*_{i,j} \equiv \rho^*_i - \rho^*_j, \quad \text{and} \quad i,j = 1,2,3\,. \tag{6.84}$$

Here V is the illuminated sample volume, ρ^*_i is the SLD (the electron density) of the *i-th* phase (as before, phase 1 is the solid matrix, phase 2 is the unadsorbed fluid, and phase 3 is the adsorbed phase), and B is the background term. Derivation of Eq. (6.80) is based on the assumption that the surface area of the interface between phase 1 and 2 is approximately equal to that between phase 2 and 3, and thus the volume of phase 3 (the volume of the adsorbed film) is $V_3 \cong \delta S$. Parameters A, R, δ, and B can be determined by fitting the measured cross section to Eq. (6.80).

The Porod invariant for a three-phase system given by Eq. (6.75) can be rewritten as:

$$\Xi \equiv \int_0^\infty Q^2 \frac{d\Sigma}{d\Omega}(Q)dQ = 2\pi^2\left[\left(\rho^*_{1,2}\right)^2 \phi_1\phi_2 + \left(\rho^*_{1,3}\right)^2\phi_1\phi_3 + \left(\rho^*_{2,3}\right)^2\phi_2\phi_3\right], \tag{6.85}$$

and the value of Ξ can be numerically evaluated from the cross section values $\frac{d\Sigma}{d\Omega}(Q)$ using the expression:

$$\Xi = \left[\int_0^\infty Q^2\left(\frac{d\Sigma}{d\Omega}(Q) - B\right)Q^2 dQ\right] + \frac{A}{Q_{\max}}. \tag{6.86}$$

In most cases the scattering length density of the porous matrix ρ^*_1 and the sample porosity ϕ_1 are known *a priori* or can be measured in a separate SAS experiment. Hence it is convenient to introduce the dimensionless scattering length densities $\widehat{\rho_\iota^*}$ as:

$$\widehat{\rho_2^*} \equiv \frac{\rho_2^*}{\rho_1^*}, \quad \text{and} \quad \widehat{\rho_3^*} \equiv \frac{\rho_3^*}{\rho_1^*}. \tag{6.87}$$

Thus, the parameter ν becomes:

$$\nu = \left(1 + \widehat{\rho_2^*} - 2\widehat{\rho_3^*}\right) / \left(1 - \widehat{\rho_2^*}\right). \tag{6.88}$$

Using the condition $\phi_1 + \phi_2 + \phi_3 = 1$, Eq. (6.85) can be rewritten as:

$$\begin{aligned} \Xi = 2\pi^2 \left(\rho_1^*\right)^2 \Big[& \left(1 - \widehat{\rho_2^*}\right)^2 \phi_1(1 - \phi_1 - \phi_3) + \left(1 - \widehat{\rho_3^*}\right)^2 \phi_1\phi_3 \\ & + \left(\widehat{\rho_2^*} - \widehat{\rho_3^*}\right)^2 \phi_3(1 - \phi_1 - \phi_3) \Big]. \end{aligned} \tag{6.89}$$

The parameters A and R can now be represented as follows:

$$A = 2\pi\left(\rho_1^*\right)^2 (S/V) \left[1 - 2\widehat{\rho_3^*} - 2\widehat{\rho_2^*}\widehat{\rho_3^*} + \left(\widehat{\rho_2^*}\right)^2 + 2\left(\widehat{\rho_3^*}\right)^2\right], \tag{6.90}$$

and

$$R = \frac{2\left(\widehat{\rho_2^*} - \widehat{\rho_3^*}\right)\left(\widehat{\rho_3^*} - 1\right)}{1 - 2\widehat{\rho_3^*} - 2\widehat{\rho_2^*}\widehat{\rho_3^*} + \left(\widehat{\rho_2^*}\right)^2 + 2\left(\widehat{\rho_3^*}\right)^2}. \tag{6.91}$$

These equations can be solved to express $\widehat{\rho_2^*}$ and $\widehat{\rho_3^*}$ in terms of A and R. The physically meaningful solution reads

$$\widehat{\rho_2^*} = 1 - \frac{1}{\rho_1^*}\sqrt{\frac{A(1+R)}{2\pi\widehat{S}}}, \quad \text{and} \tag{6.92}$$

$$\widehat{\rho_3^*} = 1 - \frac{1}{2\rho_1^* R}\sqrt{\frac{A\left(1 - \sqrt{(1 - R^2)}\right)}{\pi\widehat{S}}}. \tag{6.93}$$

The value of (S/V) can be determined independently by fitting the scattering cross section from the porous matrix in vacuum to the Porod law (Eq. (6.30)). Fitting of the scattering cross section to Eq. (6.80) provides the values of parameters A, R and δ. Thus, the dimensionless SLDs $\widehat{\rho_2^*}$ and $\widehat{\rho_3^*}$ can be determined from

Eqs. (6.92) and (6.93), and physical densities of the adsorbed and unadsorbed phases calculated using Eqs. (1.15) and (1.17). Volume fractions of the unadsorbed ϕ_2 and adsorbed ϕ_3 phase are readily obtained using $\phi_3 = (S/V)\delta$ and $\phi_2 = 1 - \phi_1 - (S/V)\delta$. The outlined scheme allows for determining the average densities of the adsorbed and unadsorbed phases, as well as their volume fractions, at different thermodynamic conditions. For the known densities and volume fractions of the adsorbed and unadsorbed phases, the mean density of the adsorbed fluid can be calculated using the relation

$$\bar{\rho} = \frac{\rho_2\phi_2 + \rho_3\phi_3}{\phi_2 + \phi_3}. \tag{6.94}$$

Example applications of the described methodology are given in the following chapters [54–56].

6.10 Interrelation Between the Reciprocal and Real Space

Many SANS and SAXS spectra are characterized by a broad peak emerging from the scattering from partially ordered structures (e.g., pores) in a sample. The characteristic distance (spacing) (D) corresponding to the position of the "correlation" peak (Q_{peak}) can be calculated using Bragg's law Eq. (1.14):

$$D = 2\pi/Q_{peak}. \tag{6.95}$$

For example, the position of the peak from Vycor porous glass ($Q_{peak} = 0.023$ Å^{-1}; (see Fig. 5.13a) because of the correlation in position of pores corresponds to $D \cong 273$ Å. Such interrelation between the reciprocal space (Q) and real space (D) is straightforward for periodic systems. Scattering curves obtained from random systems are usually featureless, which requires use of specific geometrical models to estimate the characteristic sizes of scattering objects corresponding to a certain value of the scattering vector Q.

As discussed in Sect. 6.8, scattering from a fractal surface is equivalent to scattering from a polydisperse system of pores, with the distribution of pore sizes (the number of pores with radii between R and $R + dR$) given by Eq. (6.64). Generally, this equation applies to polydisperse pores of arbitrary shapes. For specific pore geometry, the scattering can be modeled numerically and the contribution of a selected range of pore sizes to the overall scattering at a specific value of the scattering vector can be assessed quantitatively.

Such calculations for a polydisperse system of spheres was conducted by Radlinski et al. [41] using a form factor of spheres given by Eq. (1.24). The results of consecutive calculations are shown in Fig. 6.10. Figure 6.10a shows the Q-dependence of the cross section for a monodisperse sphere of radius $R = 1000$ Å.

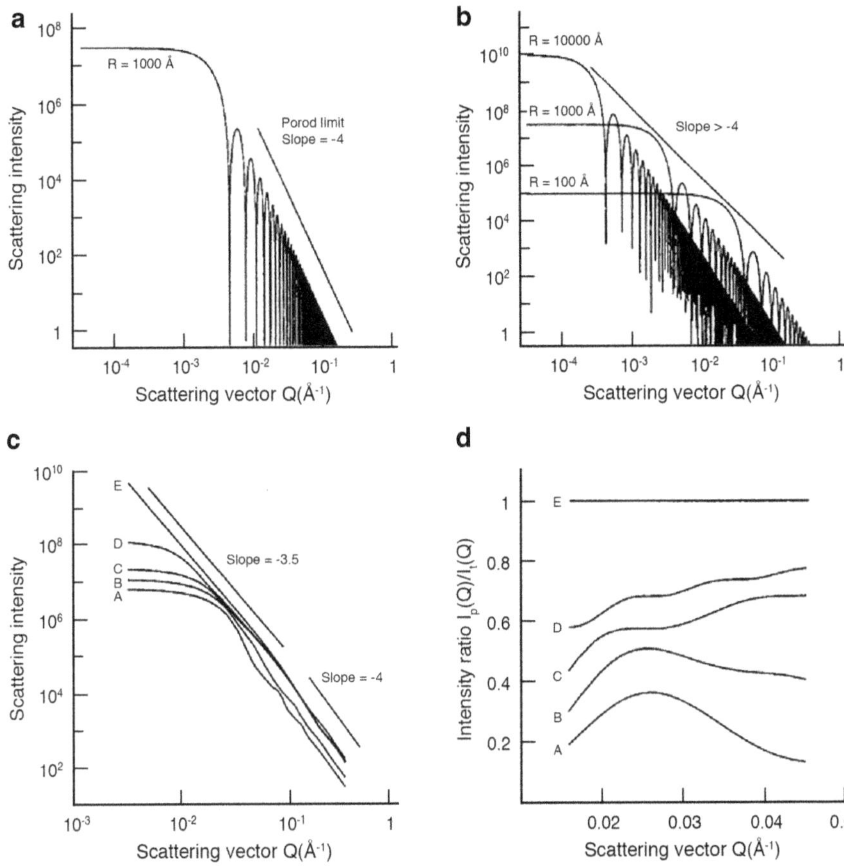

Fig. 6.10 Results of numerical simulations of SAS for spherical pores: (**a**) Q-dependence of the scattering intensity for a monodisperse spheres; (**b**) scattering curves for spheres with radii of 10,000, 1,000, and 100 Å with the pore size distribution given by Eq. (4.64); (**c**) the summed scattering curve for the spherical pore size distribution (Eq. (4.64)), calculated for $D_s = 2.5$; (**d**) the relative contribution of various R-ranges of pore sizes to the total scattering intensity. See explanations in the text. Reprinted with permission from Radlinski et al. Reprinted from [41]. Copyright (2000), with permission from Elsevier

As discussed in Sect. 6.5.7, in real experiments the deep minima simulated for the monodisperse spheres are smeared out by the polydispersity effects and instrument resolution and only the envelope with the slope of −4 is observed experimentally.

Figure 6.10b shows calculated scattering curves for spheres of 10,000; 1,000; and 100 Å radii, weighed by the number distribution given by Eq. (6.64). As shown in the figure, the slope in the limit of large Q of the summed (total) scattering is less steep than the Porod value of −4 seen in the large Q region for monodisperse spheres. This is the expected result since a surface fractal slope is D_s-6 ($2 \leq D_s \leq 3$, Sect. 6.8.1). Figure 6.10d gives the relative contribution of various R-ranges of pore sizes to the total scattering intensity for the full pore size distribution. Figure 6.10c shows the summed scattering curve for the spherical pore size distribution (Eq. (6.64)) calculated using $D_s = 2.5$ with the range of $(R_{min} - R_{max})$ changing from 50–150 Å (curve A) to 25–40,000 Å (curve E). The widest range of the pore sizes corresponds to the range of fractal scattering common for sedimentary rocks [26], and curve E shows the anticipated crossover of the slope value from −3.5 to −4 in the large Q-limit. As the distribution of pore sizes narrows down at both ends of the R-range, the simulated scattering curves start to deviate from the truly fractal behavior both in the large and low-Q limits. This deviation is least pronounced for the Q-range around $Q' \cong 2.5 \times 10^{-2}$ Å^{-1} and is related to the average value of $\overline{R} = (R_{max} - R_{min})/2$ by the empirical relationship $\overline{R} \cong 2.5/Q'$. Further analysis shows that although at every Q-value the total scattering intensity contains a contribution of scattering from the entire R-distribution of pore sizes, the *dominant* contribution comes from the pores with average radius $\overline{R} \pm 50\%$ and is independent of D_s. Therefore, the quantity

$$\overline{R} \cong 2.5/Q \tag{6.96}$$

can be considered as the average value of pore radii contributing most to fractal scattering at a given value of Q. We note that the numerical factor in Eq. (6.96) can be quite different for 2D and 1D scattering objects such as discs or rods.

References

1. Debye, P., Bueche, A.M.: Scattering by an inhomogeneous fluid. J. Appl. Phys. **20**, 518 (1949)
2. Feigin, L.A., Svergun, D.I.: Structure Analysis by Small-Angle X-Ray and Neutron Scattering. Plenum, New York, London (1987)
3. Bale, H.D., Schmidt, P.W.: Small-angle X-ray scattering investigation of submicroscopic porosity with fractal properties. Phys. Rev. Lett. **53**, 596 (1984)
4. Kjems, J.K., Schofield, P.: Neutron and X-ray studies of interfaces. In: Pynn, R., Skjeltorp, A. (eds.) Scaling Phenomena in Disordered Systems (NATO ASI Series B: Physics). Springer, New Year (1991)
5. Wong, P.Z., Bray, A.J.: Fractal dimension of rough surfaces in the solid-on-solid model. Phys. Rev. Lett. **59**, 1057 (1987)
6. Kjems, J.K., Freltoft, T.: Neutron and X-ray scattering from aggregates. In: Pynn, R., Skjeltorp, A. (eds.) Scaling Phenomena in Disordered Systems (NATO ASI Series B: Physics). Springer, New York (1991)
7. Teixeira, J.: Experimental methods for studying fractal aggregates. In: Stanley, H.E., Ostrowski, N. (eds.) On the Growth and Form: Fractal and Non-fractal Patterns in Physics (NATO Science series E: Applied Sciences). Springer, New York (1986)
8. Teixeira, J.: Small-angle scattering by fractal systems. J. Appl. Crystallogr. **21**, 781 (1988)
9. Porod, G.: General theory. In: Glatter, O., Kratky, O. (eds.) Small Angle X-Ray Scattering. Academic, London (1982)
10. Debye, P., Anderson, H.R., Brumberger, H.: Scattering by an inhomogeneous solid. 2. The correlation function and its application. J. Appl. Phys. **28**, 679 (1957)

11. Melnichenko, Y.B., Wignall, G.D., Cole, D.R., Frielinghaus, H.: Adsorption of supercritical CO_2 in aerogels as studied by small-angle neutron scattering and neutron transmission techniques. J. Chem. Phys. **124**, 204711 (2006)
12. Kirste, V.R., Pord, G.: Rontgenkleinwinkwlstreum an kolloiden systemen—asymptotisches verhalten der streukurven. Colloid Polym. Sci. **184**, 1 (1962)
13. Auvray, L., Auroy, P.: Scattering at interfaces: variations on Porod's Law. In: Lindner, P., Zemb, T. (eds.) Neutron, X-Ray and Light Scattering: Introduction to an Investigative Tool for Colloidal and Polymeric Systems. North-Holland, Amsterdam (1991)
14. Glatter, O., Kratky, O. (eds.): Small Angle X-Ray Scattering. Academic, London, New York (1982). A PDF file with this book is available at: http://physchem.kfunigraz.ac.at/sm/Software.htm
15. Melnichenko, Y.B., Wignall, G.D., Compton, R.N., Bakale, G.: Characterization of fullerenes and fullerene derivatives by small-angle neutron scattering and transmission measurements. J. Chem. Phys. **111**, 4724 (1999)
16. Roe, R.J.: Methods of X-Ray and Neutron Scattering in Polymer Science. Oxford University Press, New York, Oxford (2000)
17. Hoinkis, E.: Small-angle scattering of neutrons and X-rays from carbons and graphites. In: Thrower, P.A. (ed.) Chemistry and Physics of Carbon. Marcel Dekker, New York (1997)
18. Beaucage, G., Schaefer, D.W.: Structural studies of complex systems using small-angle scattering—a unified Guinier power—law approach. J. Non-Cryst. Solids **172–174**, 797 (1994)
19. Beaucage, G.: Approximations leading to a unified exponential power-law approach to small-angle scattering. J. Appl. Crystallogr. **28**, 717 (1995)
20. Beaucage, G.: Small-angle scattering from polymeric mass fractals of arbitrary mass-fractal dimension. J. Appl. Crystallogr. **29**(134) (1996)
21. http://usaxs.xray.aps.anl.gov/staff/ilavsky/irena.html
22. http://www.ncnr.nist.gov/programs/sans/data/red_anal.html
23. Chathoth, S.M., He, L., Mamontov, E., Melnichenko, Y.B.: Effect of carbon dioxide and nitrogen on the diffusivity of methane confined in nano-porous carbon aerogel. Micropor. Mesopor. Mat. **148**, 101 (2012)
24. Mandelbrot, B.B.: The Fractal Geometry of Nature. Freeman, San Francisco (1983)
25. Avnir, D., Biham, O., Lidar, D., Malcai, O.: Is the geometry of nature fractal? Science **279**, 39 (1997)
26. Radlinski, A.P., Radlinska, E.Z., Agamalian, M., Wignall, G.D., Lindner, P., Randl, O.G.: Fractal geometry of rocks. Phys. Rev. Lett. **82**, 3078 (1999)
27. Ruppert, L.F., Sakurovs, R., Blach, T.P., He, L., Melnichenko, Y.B., Mildner, D.F.R., Alcantar-Lopez, L.: A USANS/SANS study of the accessibility of pores in the Barnett shale to methane and water. Energy Fuel **27**, 772 (2013)
28. Schaefer, D.W., Keefer, K.D.: Fractal geometry of silica condensation polymers. Phys. Rev. Lett. **53**, 1383 (1984)
29. Pfeifer, P.: Mapping a Pore Fractal. Physics News Update No. 578 (American Institute of Physics, 27 Feb 2002). Reproduced with permission. The original image may be viewed at http://www.fractal.org/Life-Science-Technology/Publications/Pore-fractal.htm
30. Pfeifer, P., Ehrburger-Dolle, F., Rieker, T.P., Gonzalez, M.T., Hoffman, W.P., Molina-Sabio, M., Rodriguez-Reinoso, F., Schmidt, P.W., Voss, D.J.: Nearly space-filling fractal networks of carbon nanopores. Phys. Rev. Lett. **88**, 115502 (2002)
31. Freltoft, T., Kjems, J.K., Sinha, S.K.: Power-law correlations and finite size effects in silica particle aggregates studied by small-angle neutron scattering. Phys. Rev. B **33**, 269 (1986)
32. Vonk, C.G.: Investigation of non-ideal 2-phase polymer structures by small-angle X-ray scattering. J. Appl. Crystallogr. **6**, 81 (1973)
33. Ruland, W.: Small-angle scattering of 2-phase systems—determination and significance of systematic deviations from Porod law. J. Appl. Crystallogr. **4**, 70 (1971)
34. Koberstein, J.T., Morra, B., Stein, R.S.: Determination of diffuse-boundary thicknesses of polymers by small-angle X-ray scattering. J. Appl. Crystallogr. **13**, 34 (1980)

35. Schmidt, P.W., Avnir, D., Levy, D., Hohr, A., Steiner, M., Roll, A.: Small-angle X-ray scattering from the surfaces of reversed-phase silicas—power-law scattering exponents of magnitudes greater than 4. J. Chem. Phys. **94**(1474) (1991)
36. Li, Z.H., Gong, Y.J., Wu, D., Sun, Y.H., Wang, J., Liu, Y., Dong, B.Z.: A negative deviation from Porod's law in SAXS of organo-MSU-X. Micropor. Mesopor. Mat. **46**, 75 (2001)
37. Radlinski, A.P.: Small-angle neutron scattering and the microstructure of rocks. Rev. Miner. Geochem. **63**, 363 (2006)
38. Schmidt, P.W.: Interpretation of small-angle scattering curves proportional to a negative power of the scattering vector. J. Appl. Crystallogr. **15**, 567 (1982)
39. Hinde, A.L.: PRINSAS—a windows-based computer program for the processing and interpretation of small-angle scattering data tailored to the analysis of sedimentary rocks. J. Appl. Crystallogr. **37**, 1020 (2004)
40. Radlinski, A.P., Ioannidis, M.A., Hinde, A.L., Hainbuchner, M., Baron, M., Rauch, H., Kline, S.R.: Angstrom-to-millimeter characterization of sedimentary rock microstructure. J. Colloid Interface Sci. **274**, 607 (2004)
41. Radlinski, A.P., Boreham, C.J., Lindner, P., Randl, O., Wignall, G.D., Hinde, A., Hope, J.M.: Small angle neutron scattering signature of oil generation in artificially and naturally matured hydrocarbon source rocks. Org. Geochem. **31**, 1 (2000)
42. Israelachvili, J.N.: Forces between surfaces in liquids. Adv. Colloid Interface Sci. **16**, 31 (1982)
43. Higgins, J.S., Benoit, H.C.: Polymers and Neutron Scattering. Clarendon, Oxford (1994)
44. Bacon, G.E.: Neutron Diffraction. Clarendon, Oxford (1975)
45. Endo, H., Schwahn, D., Colfen, H.: On the role of block copolymer additives for calcium carbonate crystallization: small angle neutron scattering investigation by applying contrast variation. J. Chem. Phys. **120**, 9410 (2004)
46. Hoinkis, E., Allen, A.J.: A small-angle neutron scattering study of porous graphitic materials before and after adsorption and condensation of C_6D_6 within the accessible pores. J. Colloid Interface Sci. **145**, 540 (1991)
47. Wu, W.L.: Small-angle X-ray study of particulate reinforced composites. Polymer **23**, 1907 (1982)
48. Peterlin, A.: Small-angle scattering by a 3 component system. Macromol. Chem. Phys. **87**, 152 (1965)
49. Melnichenko, Y.B., Wignall, G.D.: Small-angle neutron scattering in materials science: recent practical applications. J. Appl. Phys. **102**, 021101 (2007)
50. Rother, G., Melnichenko, Y.B., Wignall, G.D., Cole, D.R., Frielinghaus, H.: Microstructural characterization of adsorption and depletion regimes of supercritical fluids in nanopores. J. Phys. Chem. C **111**, 15736 (2007)
51. Melnichenko, Y.B., Wignall, G.D.: Density and volume fraction of supercritical CO_2 in pores of native and oxidized aerogels. Int. J. Thermophys. **30**, 1578 (2009)
52. Ciccariello, S.: The best two-phase idealization of a sample in small-angle scattering. Acta Crystallogr. A **58**, 460 (2002)
53. Pikus, S., Kobylas, E., Ciccariello, S.: Small-angle scattering characterization of n-aliphatic alcohol films adsorbed on hydroxyled porous silicas. J. Appl. Crystallogr. **36**, 744 (2003)
54. Ciccariello, S., Melnichenko, Y.B., He, L.: Supercritical carbon dioxide behavior in porous silica aerogel. J. Appl. Crystallogr. **44**, 43 (2011)
55. Ciccariello, S., Melnichenko, Y.B., He, L.: Phase behavior of carbon dioxide confined in silica aerogel in the vicinity of the bulk critical point. J. Phys. Chem. C **115**, 22336 (2011)
56. Melnichenko, Y.B., Ciccariello, S.: Small-angle neutron scattering study of deuterated propane adsorption in silica aerogel. J. Phys. Chem. C **116**, 24661 (2012)
57. Bahadur, J., Radlinski, A., Melnichenko, Y.B., Mastalerz, M., Schimmelmann, A.: SANS/USANS study of the New Albany Shale: a treatise on microporisity. Energy Fuel **29**, 576 (2015)

Chapter 7
Structural Characterization of Porous Materials Using SAS

This chapter begins with an overview of various engineered and naturally occurring porous solids and their applications in modern technologies. Different modes of confinement, such as topological, chemical, biological, and interfacial, can affect structural and kinetic aspects of the fluid phase equilibria, mechanical and viscoelastic properties, diffusion, and flow in confined spaces. Such dramatic modification of the fluid properties under confinement can be efficiently exploited in a variety of technologies, provided there are ways of controlling the internal structures and surface chemistry of the target porous material at the macroscopic, mesoscopic, microscopic, and molecular levels. SAS represents a nonintrusive technique that can provide information on the structure of porous media that is inaccessible to other methods of structural characterization. The rest of the chapter is devoted to the description of SAS structural analysis of porous silicas, carbons, membranes, polymer monoliths, sedimentary rocks and other common materials.

7.1 Porous Media for Energy, Environmental, and Biomedical Applications

Phase behavior of bulk fluids is now well understood, and their properties can be predicted accurately using equations of state over a wide range of pressures and temperatures [1]. The behavior of fluids changes dramatically when they are injected into small pores because of the increasing importance of the boundary conditions and molecule-surface interactions. Various modes of confinement, such as topological, chemical, biological, and interfacial, can affect structural and kinetic aspects of fluid behavior, such as phase equilibria, mechanical and viscoelastic properties, diffusion, fluid flow, and, therefore, the matrix penetrability [2]. Such dramatic modification of the fluid properties under confinement can be efficiently

Y.B. Melnichenko, *Small-Angle Scattering from Confined and Interfacial Fluids*, DOI 10.1007/978-3-319-01104-2_7

exploited in a variety of technologies, provided there are ways of controlling the surface chemistry and internal structures of the target porous material at the macroscopic (pores with sizes $R > 500$ Å), mesoscopic ($20 \leq R \leq 500$ Å), microscopic ($R < 20$ Å), and molecular levels [3]. Here we present a brief description of some of the engineered and naturally occurring porous materials and their major applications.

Porous carbons are widely used in many energy- and environmental-related applications. Their use in diverse technologies is directly related to the unique combination of physical and chemical properties, such as electric and thermal conductivity, chemical stability, and low density (high porosity). In addition to the rapidly growing field of electrochemical energy storage and conversion (batteries, fuel cells, supercapacitors), high surface area and relative chemical stability makes porous carbons attractive as catalyst supports, efficient matrices for gas capture, separation, and storage, as well as and chemically inert hosts for biomolecule immobilization.

The existing methods of synthesizing amorphous porous carbons can be divided in two broad categories: activation processes and template methods. Traditional activation schemes include chemical and physical activation: catalytic activation of carbon precursors using metal salts or organometallic compounds, as well as carbonization of polymer blends and polymer aerogels synthesized and dried under supercritical fluid conditions. In general, amorphous porous carbons produced using activation methods are characterized by relatively nonuniform pore size distributions and could have significant amounts of isolated pores (see Fig. 7.1). The local structure of porous carbons consists of nanometer-sized domains composed of imperfectly stacked, curled, and structurally defective graphenes. The natural curvature of defective graphenes and their disordered arrangement generates micro- and mesopores, which contribute to the internal surface area of porous carbons. Values as large as the theoretical surface area of isolated graphenes (2600 m^2/g) have been reported for the specific surface area of porous carbons obtained by chemical activation.

Over the past 10 years, new template methods of synthesis were developed using various inorganic templates such as MCM-41 (Mobil Composition of Matter #41), SBA-15 (Santa Barbara University #15), and zeolites. Template-based porous carbons exhibit more uniform pore sizes, higher surface areas, and larger pore volumes [4, 5]. Carbide-derived carbons with controlled pore sizes have been obtained by leaching metallic components from chemical carbides.

Other types of porous carbonaceous materials of practical importance include cokes, graphite, pyrolytic and glassy carbons, chars, charcoals, and coals. *Coke* is a highly carbonaceous product of pyrolysis of organic material, some parts of which have passed through a liquid or liquid-crystalline state during the carbonization process. *Graphite* consists of layers of hexagonally arranged carbon atoms in a planar condensed ring system. The layers are stacked parallel to each other, which provides the closest packing of the layer planes with the interlayer spacing ~3.35 Å. *Pyrolytic carbons* are made by pyrolyzing hydrocarbons (e.g., methane or propane)

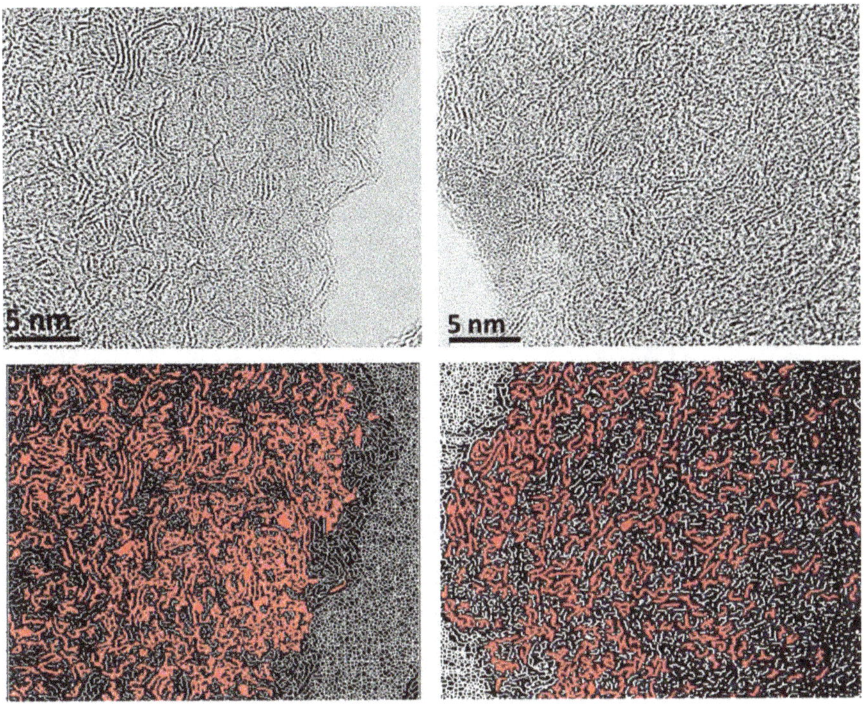

Fig. 7.1 *Top row*: bright field STEM images of polyfurfuryl alcohol-derived activated carbon (PFAC, *left*) and ultramicroporous carbon (UMC, *right*) obtained from secondary activation of wood. *Bottom row*: image-processed areas showing fringes associated with graphene layers viewed edge-on. Reprinted from [6]. Copyright (2014), with permission from Elsevier

on a hot substrate at high temperatures >1200 °C. Their microstructure depends on the substrate temperature, gas composition, and pressure. The layer plane spacing is ~3.44 Å, the total porosity varies between 7 and 42 vol.%, and the accessible porosity varies between 0 and 20 vol.%. *Glassy carbons* are made by heating polymeric carbons (such as blends of poly(vinylidenechloride) and a phenolic resin) to temperatures >700 °C at a rate of <50 ° C/min. During this process the accessible pores in the matrix are sealed off as a result of the ~10 % matrix contraction. A *char* is a carbonization product of natural or synthetic organic materials that have not passed through a fluid stage during carbonization, and *charcoal* is a traditional term for a char obtained from wood.

Sedimentary rocks such as *coal* and *shale* are naturally occurring porous materials characterized by extremely wide pore size distribution from subnanometer to micron sizes (Fig. 7.2). In general, sedimentary rocks have both organic and inorganic component. The limiting cases are represented by sandstones and lean shales, which are completely inorganic, and the entirely organic ash-free rocks. Pure sandstones or carbonates are often found in hydrocarbon (oil and gas) reservoirs. Organic-rich shales are hydrocarbon source rocks, and in some cases also

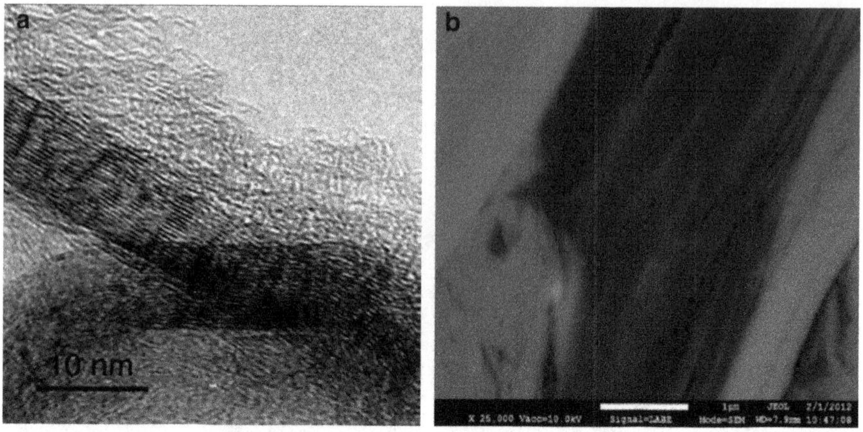

Fig. 7.2 (**a**) SEM of the lamellar micropores in a sample of anthracite (microphotograph courtesy of M. Mastalerz). (**b**) SEM of mesopores in organic matter (dark material) of the Barnett shale sample. The light phase is clay. Reprinted with permission from [7]. Copyright 2013 American Chemical Society

reservoir rocks (in case of tight gas, tight oil, shale gas and shale oil). Coals represent energy resource on their own, but can also be source rocks for oil and gas (coalbed methane).

Coal is composed primarily of carbon along with variable quantities of other elements, chiefly hydrogen, sulfur, oxygen, and nitrogen. Shale is a fine-grained laminar sedimentary rock composed of clay minerals and grains of other minerals, especially quartz and calcite, as well as a variable amount of organic matter. Carbon capture and sequestration in underground geological formations is one of the proposed interim measures for arresting the rising concentration of atmospheric carbon dioxide. Noneconomic (unmineable) coal seams are attractive sequestration reservoirs because of their potential for long-term containment of very large amounts of carbon dioxide (CO_2) through its adsorption into the coal microstructure, and their proximity to coal-fired power stations. The injection of CO_2 in coal or shale under pressure can be used to replace and expel natural gas or oil from the underground formation, contributing to enhanced oil and methane recovery.

Coals are often classified by the parameter called rank, which is related to the degree of transformation (or maturation) that the coal has undergone after deposition of the source organic material. It is a general rule that the higher the percentage of carbon in coal, the higher its rank. Mineral matter in coal includes silicates, pyrite, and metal oxides. Porosity (both accessible and total) and hydrocarbon composition vary significantly with rank. Low-rank coals (lignites and subbituminous coals) with carbon content below 80 wt% consist largely of randomly oriented layers, each containing two or three fused benzene rings connected by aliphatic links. The carbonaceous matrix of low-rank coals is usually highly porous. The carbon content of *bituminous* coals is between 80 and 91 wt%, and their porosity is significantly lower. High-rank coals (anthracites) have carbon contents above 91 wt%.

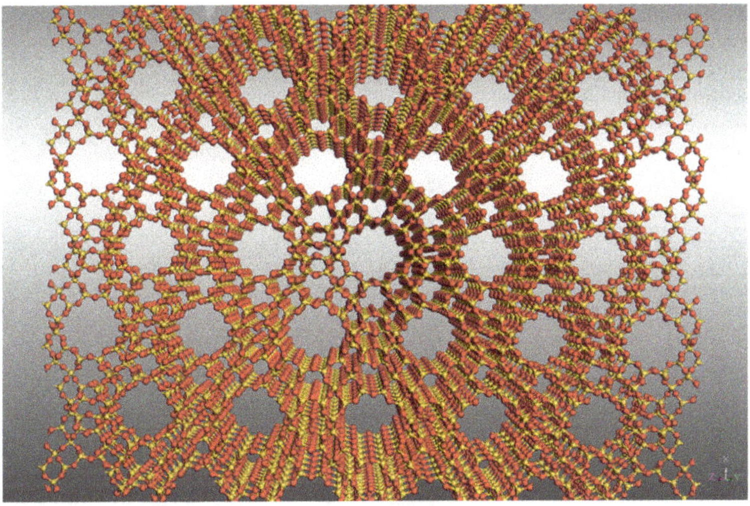

Fig. 7.3 The channels of Y zeolite with the diameter of 7.4 Å. Adapted with permission from [9]. Copyright (2013) American Chemical Society

The local structure of both bituminous coals and anthracites is characterized by moderately and highly ordered graphitic layers, respectively. A comprehensive summary of the SAS structural characterization of various carbonaceous materials can be found in [8].

Silica and silicates are important components of many materials applied in different technological fields, such as catalyst supports, adsorbents, enzyme immobilizers, liquid and gas separating media, drug delivery, etc. Zeolites are crystalline microporous silicates composed of tetrahedral $[SiO_4]^+$ building blocks (Fig. 7.3). A certain number of silicon atoms can be isomorphously substituted by other elements capable of tetrahedral coordination with oxygen atoms [9]. The most frequent substituent is aluminum. The charge imbalance caused by aluminum substitution is compensated by cations (Na^+, K^+, H^+, NH_4^+), which can be readily ion-exchanged from aqueous solutions. The pore space in zeolites consists of regular networks of cavities and connecting channels that can selectively "sieve" gas molecules based on their size.

Amorphous silica xerogels and aerogels are produced by the sol–gel process in monolithic form or as a powder and have a disordered pore structure with broad pore size distribution (Fig. 7.4). Gelsils (Geltech Inc., Florida, USA) are commercially available monolithic mesoporous xerogels with relatively narrow pore size distribution. Controlled porous glasses (CPGs, Electro-Nucleonics Inc., New Jersey, USA) and Vycor glass (Corning Glass Works, New York, USA) are produced by leaching the borate-rich phase of a phase-separated silica/alkali borate mixture. CPGs are available in granular form only, whereas Vycor glass is fabricated in sheet and rod forms [11]. Fabrication methods of ultrathin porous nanocrystalline silicon (pnc-Si) membranes were developed recently [12].

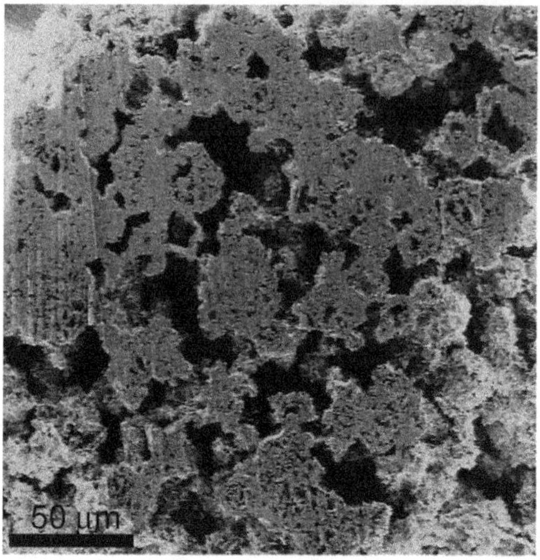

Fig. 7.4 SEM image of the cross section of porous fractal silica xerogel. Reprinted with permission from [10]. Copyright 2010 American Chemical Society

Fig. 7.5 MCM-41 is a "mesoporous molecular sieve" silica nanomaterial with a hexagonal or "honeycomb" arrangement of its practically monodisperse pores, as shown in the TEM image on the *left*. TEM image on the *right* shows the straight cylindrical pores of MCM-41 as they lie perpendicular to the viewing axis. ®2015 ACS Material, LLC by permission only

Surfactant-templated mesoporous materials, such as MCM-41 or SBA-15, have highly uniform periodic mesoporous structures that consist of a two-dimensional hexagonal arrangement of well-defined (in terms of both shape and size) cylindrical pores (see Fig. 7.5). These inorganic porous materials are synthesized in the presence of surfactants as templates for the polycondensation of various silica species [13]. These materials are of interest for liquid chromatography, as supports for immobilizing biomolecules, as catalysts, as agents for polymer reinforcement, and as hard templates for the preparation of mesoscopic carbons.

Hybrid porous solids represent a new class of crystallized porous materials, commonly named metal-organic frameworks (MOFs) or Porous Coordination Polymers (PCPs) [14]. MOFs are formed from metal complexes, known as secondary building units, joined by organic linkers [15]. Their considerable advantage, compared with zeolites, is the ability to accept in their skeleton almost all the cations of the periodic table. In combination with the virtually unlimited number of possible organic linkers, this offers almost infinite possibilities of creating crystal structures and chemical compositions, which can be tuned by changing both the nature and coordination of the metal centers, as well as the length and or functionalization of the organic linker. Nanoporous or mesoporous MOFs have a low framework density (0.2–0.6 g/cm^3). They are often seen as a potential technological breakthrough with respect to the conventional materials (zeolites, carbons) used traditionally for separation, capture, and storage of greenhouse gases. The different type of porous solids suitable for capture and storage of gases is represented by microporous organic polymers (MOPs). MOPs are categorized into four classes according to the types of organic reactions and the structure of the resulting material: noncrystalline polymers of intrinsic microporosity, hypercrosslinked polymers, conjugate microporous amorphous polymers (see Fig. 7.6), and mesoporous crystalline covalent organic frameworks (COFs). COFs are organic analogues of MOFs, formed from light elements such as H, N, C, O, and Si [16].

It is virtually impossible to even mention here all available porous materials, and new types of porous solids are being synthesized in various research groups and industrial companies on a regular basis [18]. One of the key conditions of the successful application of porous solids is the availability of accurate information on their structure. Some of the conventional characterization methods are given in

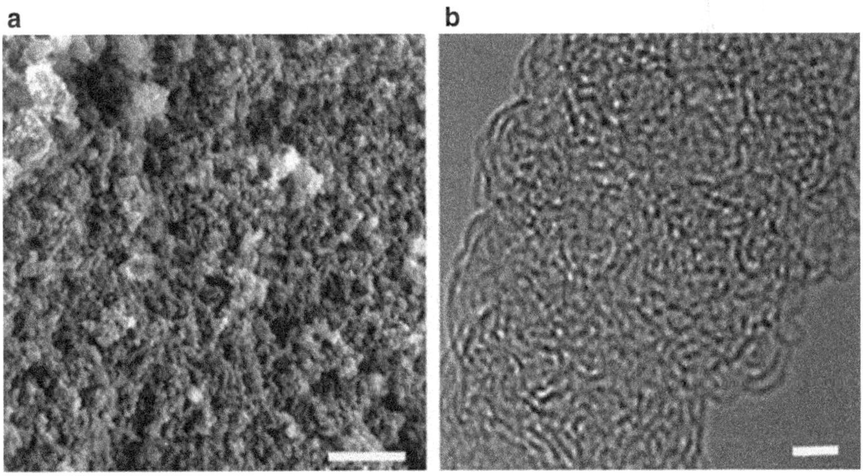

Fig. 7.6 (**a**) SEM image of the microporous conjugated polymer Co-CMP (scale bar 500 nm). (**b**) High-resolution TEM of Co-CMP (scale bar 20 nm). Reprinted by permission from Macmillan Publishers Ltd.: Nature Communications [17], copyright (2013)

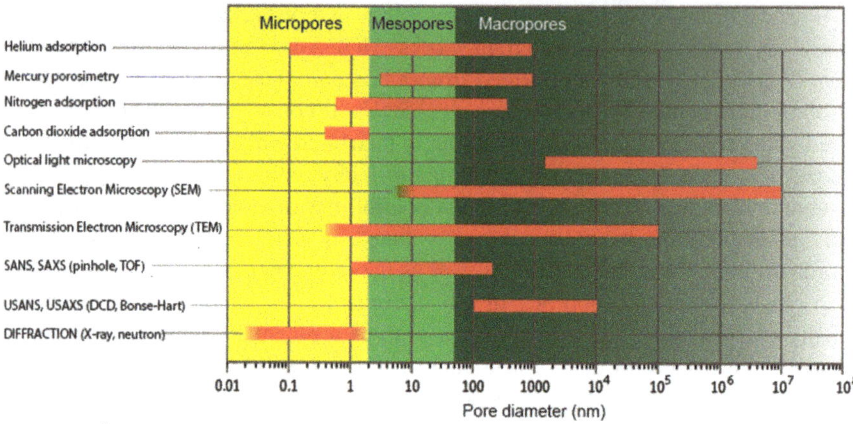

Fig. 7.7 Comparison of the ranges of pore sizes accessible to various characterization methods (image courtesy of M. Mastalerz)

Fig. 7.7. The most common techniques include helium porosimetry, low-pressure nitrogen, CO_2 adsorption, and mercury injection porosimetry, along with various microscopy methods [19]. New techniques based on the evaluation of the shift of the solid–liquid phase transition temperatures under confinement using Nuclear Magnetic Resonance (NMR), cryoporometry, and thermoporometry are being actively developed [20]. As shown in Fig. 7.7, no single fluid-probe-based porosimetry technique can provide information about the whole range of the pores present in the sample. Usually a combination of methods must be applied to characterize porosity and pore size distribution, especially in fractal materials, such as sedimentary rock, porous polymer monoliths, or porous fractal silica. A combination of nondestructive USAS and SAS can deliver structural information from the same sample in the range of pores with diameters from nearly molecular to submilimeter (5 to ~200,000 Å).

Improving the existing and developing new technologies requires fundamental understanding of the influence of confinement on the thermodynamic and kinetic properties of the invasive fluids. The traditional methods for evaluating the properties of the adsorbed phase provide information averaged over the whole sample volume. As such, they cannot help explain how pores of different sizes and chemical composition contribute to the total adsorption capacity of the material. In contrast, USAS and SAS techniques allow the investigator to "look inside pores" and monitor changes in the fluid behaviour as a function of pore size. Application of the neutron contrast variation methods opens new ways of studying confined multicomponent mixtures of gases (e.g., $CO_2 + CH_4$) as scattering from the adsorbed phase of each gas can be highlighted by matching out scattering from the other fluid. Examples of SAS applications for the structural characterization of the select host porous matrices, as well as the assessment of the physical properties of confined fluids in pores of different size, shape, and tortuosity, are given in the subsequent chapters.

7.2 Porous Silica

7.2.1 Porous Vycor Glass

Porous Vycor glass No. 7930 (PVG-7930) is manufactured from a borosilicate glass-forming melt with a high concentration of B_2O_3. The melt is quenched below demixing temperature forcing it to spinodally decompose into SiO_2-rich and B_2O_3-rich phases. The latter is acid soluble and can be leached out leaving a fully penetrable mesoporous matrix that consists predominantly of SiO_2. Vycor is widely used for different applications as filters, semipermeable membranes, catalyst supports, and thermal insulators. It is also considered as a model mesoporous system for investigation of adsorption, flow, and diffusion under confinement. Vycor has a bicontinuous structure (all pores are connected to each other), with a total pore volume of 32–36 % and an internal pore surface of 90–125 m^2/g. SANS reveals a characteristic maximum ("correlation" or "Vycor" peak) at the value of scattering vector $Q_{peak} \cong 0.023$ Å^{-1} (Fig. 7.8, [21]; see also Fig. 5.13a), which indicates certain correlation between the position of pores in the sample with the characteristic correlation length $2\pi/Q_{peak} \cong 270$ Å and the radius of gyration of pores $\cong 70$ Å [22]. This scattering pattern is reminiscent of the structure factors measured from two-component systems that underwent spinodal decomposition. For such systems, the Q-dependence of the scattering cross section is a function of dimensionless time $\overline{\tau}$ [23]:

$$\frac{d\Sigma}{d\Omega}\left(\overline{Q},\overline{\tau}\right) = \frac{\frac{d\Sigma}{d\Omega}(0,0)}{\overline{Q}^2+1}\exp\left[-2\overline{Q}^2\overline{\tau}\left(\overline{Q}^2-1\right)\right], \tag{7.1}$$

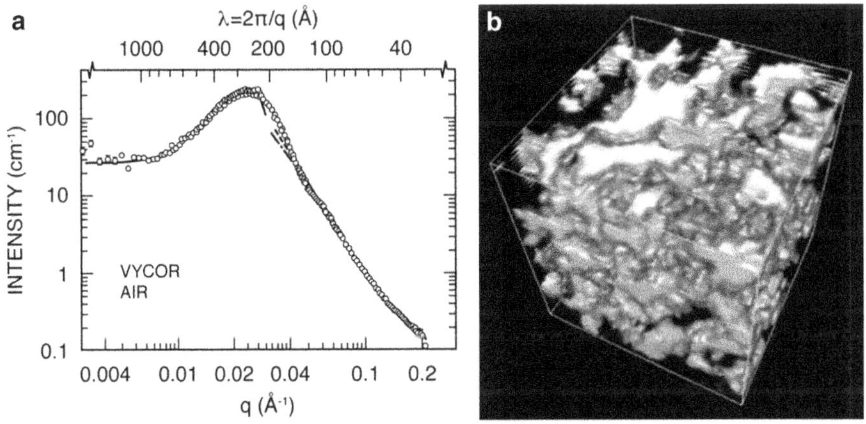

Fig. 7.8 (**a**) Intensity of neutron scattering (in units of cm^{-1}) from dry Vycor as a function of the scattering vector (q). Reprinted figure with permission from [35]. Copyright (1987) by the American Physical Society. (**b**) Reconstructed 3D image of porous Vycor glass. Pores are shown in *white* or *grey*, while solid matrix is *transparent*. Image courtesy of E. Garboszi and D. Bentz (NIST, USA)

where $\overline{Q} = Q\Lambda/2\pi$ is a dimensionless scattering vector and Λ is the wavelength of the concentration fluctuations. Fitting SANS data to Eq. (7.1) gives $\Lambda = 192$ Å and $\overline{\tau} = 5.7$.

Early x-ray studies of the Vycor structure in the limit of high Q revealed deviation from the Porod law, which was interpreted as scattering from the surface fractal with the dimension $D_s = 2.4$ [24]. However, later it was noticed that in SANS experiments [21] saturation of Vycor pores with contrast matched liquid did not completely eliminate coherent scattering from pores [25], which might indicate the presence of the third, unknown phase. Further SANS studies demonstrated that the apparent "surface roughness" revealed in [24] is caused by the presence of chemisorbed water in the form of compact molecular clusters of 40–60 molecules. The chemisorbed water could be completely removed by heating the sample to 650 °C. After such treatment, the coherent SANS signal was completely eliminated with contrast matched solutions and the scattering in the limit of high Q changed from $Q^{-3.5}$ to Q^{-4} before and after treatment, respectively, indicating transition to a smooth internal surface [25].

USANS study of the Vycor structure in the Q-range $3 \times 10^{-5} \leq Q \leq 0.001$ Å^{-1} demonstrated the existence of the strong USANS signal that could not be eliminated by contrast matching. The origin of this scattering signal remains unknown. There are reasons to believe that it might be coming from residual clusters of B_2O_3 that remain in the glass because of incomplete leaching [26].

7.2.2 Silica Aerogels

The sol–gel process is one of the most practical and energy-saving preparation methods for synthesis of amorphous porous materials. Slow drying of synthesized wet gels at ambient conditions results in a gradual shrinking (sintering) of the polymer network and is used to produce xerogels. Shrinking can be prevented and the structural features of the original gel network preserved if a gel-saturating solvent is removed from the matrix under supercritical conditions (i.e., at pressure and temperature that exceed its liquid–gas critical pressure and temperature). Supercritical drying of polymer gels synthesized from silanes (tetramethylsilane, tetraethyloxisilane, etc.) produces silica aerogels, which are composed of thin silica strands with mesh sizes of 60–70 Å and a surface area of ~700 m^2/g. The microstructure of aerogel can be controlled by varying the gelation methods, aging time, and drying conditions.

It is generally accepted that silica gels form fractal structures [27, 28]. The fractal dimension of silica aerogels prepared from precursor gels synthesized in neutral solutions at different densities was studied using SANS in [29]. The results obtained are shown in Fig. 7.9. In the intermediate Q-region, a typical power-law decrease characteristic of fractal structures is observed (Sect. 6.8.1). The scattering signal levels off at small Qs as a result of reaching the upper cutoff length scale ξ of

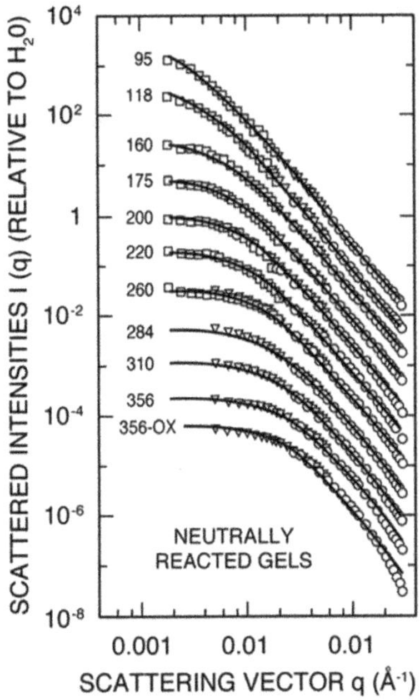

Fig. 7.9 Scattered intensities (in units of cm^{-1}) for 11 samples of silica aerogel with different matrix densities in kg/m^3 shown as labels to corresponding curves. For the sake of clarity, the intensity of scattering from each sample was divided by four compared with the previous one starting from the top. The *solid lines* are fits to Eq. (6.51), with fitting parameters discussed in the text. Reprinted figure with permission from [29]. Copyright (1988) by the American Physical Society

the mass fractal behavior. The data for $Q < 0.15$ Å^{-1} can be fitted to Eq. (6.51), which describes the Guinier and fractal regimes, as well as the crossover between them. The results show that the mass fractal dimension of the studied samples is $D_m = 2.40 \pm 0.03$. This value is close to $D_m \cong 2.5$ of the infinite cluster of a percolating network, and it does not change with the matrix density. On the contrary, the upper cut-off length scale ξ decreases linearly with the density between ~400 Å and ~30 Å. At high *Qs*, the deviation from the power law is observed. This deviation increases with the matrix density and indicates the existence of a gradual crossover from the mass-to-surface fractal behavior with the surface fractal dimension close to $D_s \cong 2.9$. SAXS data obtained from similar aerogels confirm the existence of such crossover [30].

7.2.3 Porous Fractal Silica

As can be seen in Fig. 7.9, the fractality (or self-similarity) of silica aerogels is observed within a relatively narrow space length interval that does not exceed one and one-half orders of magnitude in terms of Q. Extension of the material fractality range might add new useful functional properties, such as localization of electromagnetic waves [31] or super hydrophobicity (super hydrophilicity) [32]. A new method of synthesizing porous silica with broad pore size distribution and extended range of fractality was developed in [33] based on sol–gel transition using flake-shaped crystals of alkylketene dimers as a template. Structural characterization of the resulting samples with extremely broad pore size distribution requires application of both SANS and USANS. The results of such experiments for two samples with different volume fractions of silica in the matrix, $\phi_{SiO2} = 0.15$ and $\phi_{SiO2} = 0.04$, are shown in Fig. 7.10 [34].

Close examination allows for distinguishing between three different ranges of scattering behavior as a function of scattering vector. Region 1 ($Q \leq 0.02$ Å^{-1}) represents scattering from primary SiO_2 particles. Region 2 ($0.001 \leq Q \leq 0.02$ Å^{-1}) and Region 3 ($Q < 0.001$ Å^{-1}) represent scattering from the surface and mass fractal structure, respectively. The combined USANS and SANS data were fitted to a modified Ornstein-Zernike formula [35]:

$$I(Q) = I(0)\frac{1}{\left(1 + \beta^2 Q^2\right)^{\alpha}}\frac{1}{\left(1 + \delta^2 Q^2\right)^{\eta}}, \tag{7.2}$$

where $I(0)$ is extrapolated scattering at zero scattering angle, $Q = 0$; β and δ are related to the upper and lower cutoff length scale limits of the fractal behavior, respectively; and α and η are related to the mass and surface fractal dimensions: $D_m = 2\alpha$; $D_s = 6-2(\alpha + \eta)$. The results demonstrate that the mass fractality reduces from $D_m = 2.68$ for a sample with $\phi_{SiO2} = 0.15$ to $D_m = 2.28$ for a sample with $\phi_{SiO2} = 0.04$. The surface of both samples is rough and is described by the surface fractal dimension $D_s = 2.48$ ($\phi_{SiO2} = 0.15$) and $D_s = 2.62$ ($\phi_{SiO2} = 0.04$). The results indicate that the reduction of the silica concentration during formation of the porous structure leads to a more open porous structure with a smoother wall surface. The obtained USANS and SANS data are consistent with the structural model shown in Fig. 7.10.

7.2.4 *Ordered Mesoporous Silica*

Amphiphillic triblock copolymers can be used as templates for preparing highly ordered, 2D hexagonal mesoporous silica structures (SBA-15) with uniform pore sizes. Calcination at 500 °C results in a formation of porous media with large interlattice *d*-spacing of 75–320 Å between the (100) planes, pore sizes from ~40 to ~300 Å, pore volume fractions up to 0.85, and silica wall thicknesses of ~30 to ~60 Å [36, 37].

To the first approximation, the structure of SBA-15 can be modeled as an array of ideal cylinders (see Fig. 7.5). However, closer examination shows that it actually contains a network of much smaller interconnected pores that extend from the mesopores into the walls of the bulk silica matrix. The spatial distribution of the secondary pore network can be viewed as a "corona" surrounding the main mesopores [38, 39].

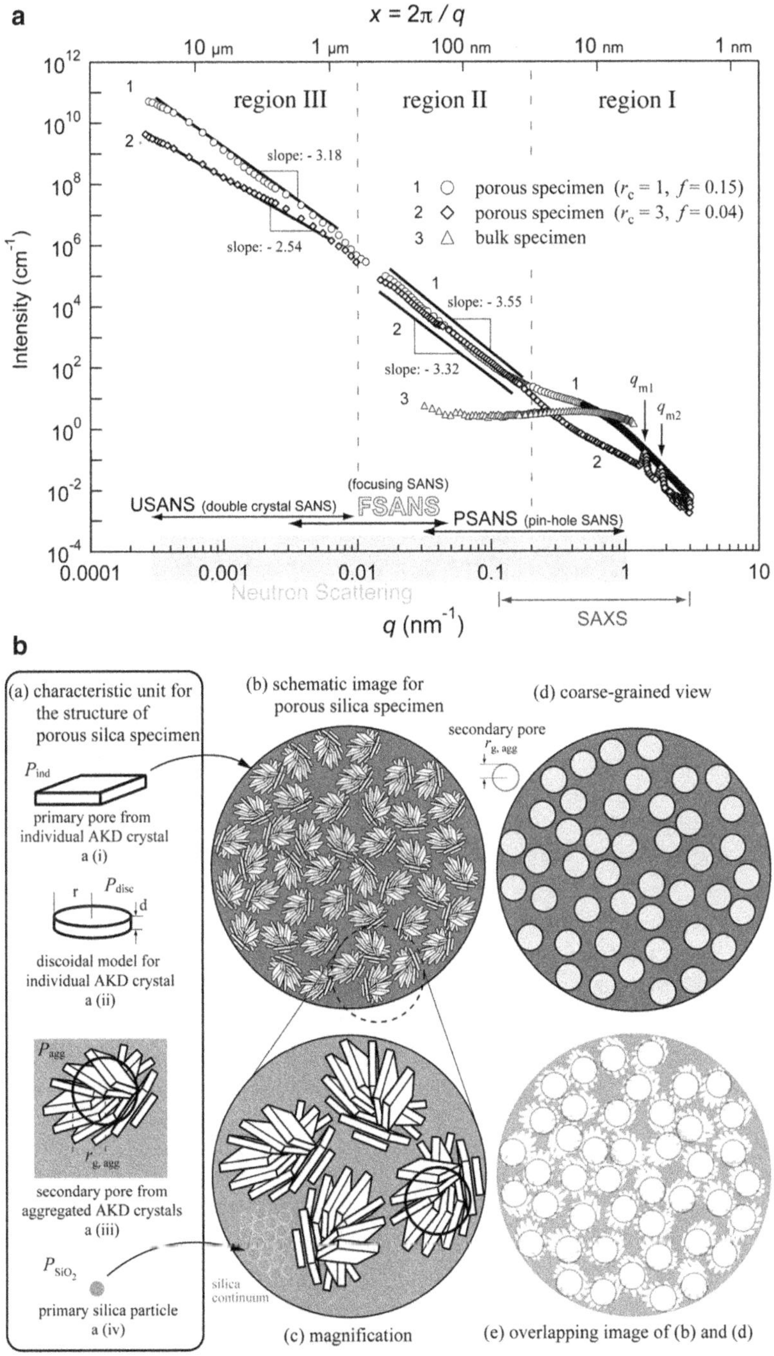

Fig. 7.10 (**a**) SANS scattering intensity (in units of cm^{-1}) from porous fractal silica with volume fraction $\phi_{SiO2} = 0.15$ and $\phi_{SiO2} = 0.04$. For the sake of clarity, data for $\phi_{SiO2} = 0.15$ are shifted upwards by a factor of 100. Solid lines are fits to Eq. (7.2). (**b**) Structural model consistent with the neutron scattering data. Reprinted with permission of D. Yamaguchi [34]

Understanding the microporous structure and the hydrothermal stability of SBA-15 is highly important for production of biofuels, where pyrolysis and liquefaction produce “bio-oils” with high water content [40]. Upgrading these oils to fungible fuels is conducted using catalysts, such as SBA-15, and it is essential that the used catalyst is stable at high reaction temperatures in the presence of water.

The structural modifications of SBA-15 during heat treatment in steam and liquid water were studied using SANS, x-ray diffraction, and nitrogen porosimetry in [41]. Figure 7.11 shows the SANS scattering curves for as-prepared SBA-15, as well as evolution of the scattering curves from the samples that were hydrothermally treated in steam and in water at different temperatures. Coherent scattering from the hexagonal array of mesopores results in the (10), (11), and (20) Bragg diffraction peaks. The position of the (10) peaks for as-prepared samples is observed at $Q_{10} = 0.06$ Å^{-1}. Corresponding center-to-center distance between the pores can be calculated using $d = 4\pi/\left(Q_{10}\sqrt{3}\right) \cong 120$Å [41]. The mesopore radius from SANS and SAXS (44 Å) agrees closely with the nitrogen adsorption measurements, and the onset of the secondary pore size distribution (7 Å) is confirmed by contrast matching experiments using probe molecules of varying kinetic diameter (5.7–6.2 Å) [40] (see also Sect. 5.4 and Fig. 5.15). The specific surface area, *S/V*, was estimated using Eq. (6.29) to be $S/V = 4.5 \times 10^5$ m^2/m^3 using the SLD of silica $\rho^*_{SiO2} = 3.47 \times 10^{10}\text{cm}^{-2}$ (Table 1.2).

Steam treatment results in a decrease in the lattice spacing, which is evidenced by a shift of the (10) peak toward higher *Qs*. At the same time, the lattice spacing for water-treated samples remains relatively constant for all treatment temperatures. Similar trends are observed in the x-ray diffraction patterns. Schematic graphical representation of the SBA-15 structures is shown in Fig. 7.11. Heat treatment of wet SBA-15 significantly reduces the number of small pores initially present in the in the mesopore corona and results in a ~10 % increase in the radius of mesopores [41].

The structure of MCM-41 and its range of applications are similar to that of SBA-15. This material is synthesized by a liquid-crystal templating mechanism in which silicate material forms inorganic walls between ordered surfactant micelles. Calcination produces mesoporous material that consists of a dense silica matrix with 2D array of hexagonally packed cylindrical pores. The dimension of pores can be tailored in the range of ~15 to ~100 Å through the choice of surfactants, auxiliary chemicals, and reaction conditions [42].

The structure of MCM-41 and its modification from exposure to water was studied in [43] using SAXS, transmission electron microscopy (TEM), and nitrogen adsorption. Figure 7.12a shows the SAXS patterns of dry MCM-41 and the same sample immersed in water at 60 °C at different exposure times. The scattering pattern from MCM-41 is qualitatively similar to that from SBA-15 (Fig. 7.11a). The SAXS profile from the dry sample shows the Bragg peaks corresponding to the (10), (11), (20), and (21) reflections. From the position of the main peak, the center-to-center distance between the pores is estimated to be $d \cong 490$ Å.

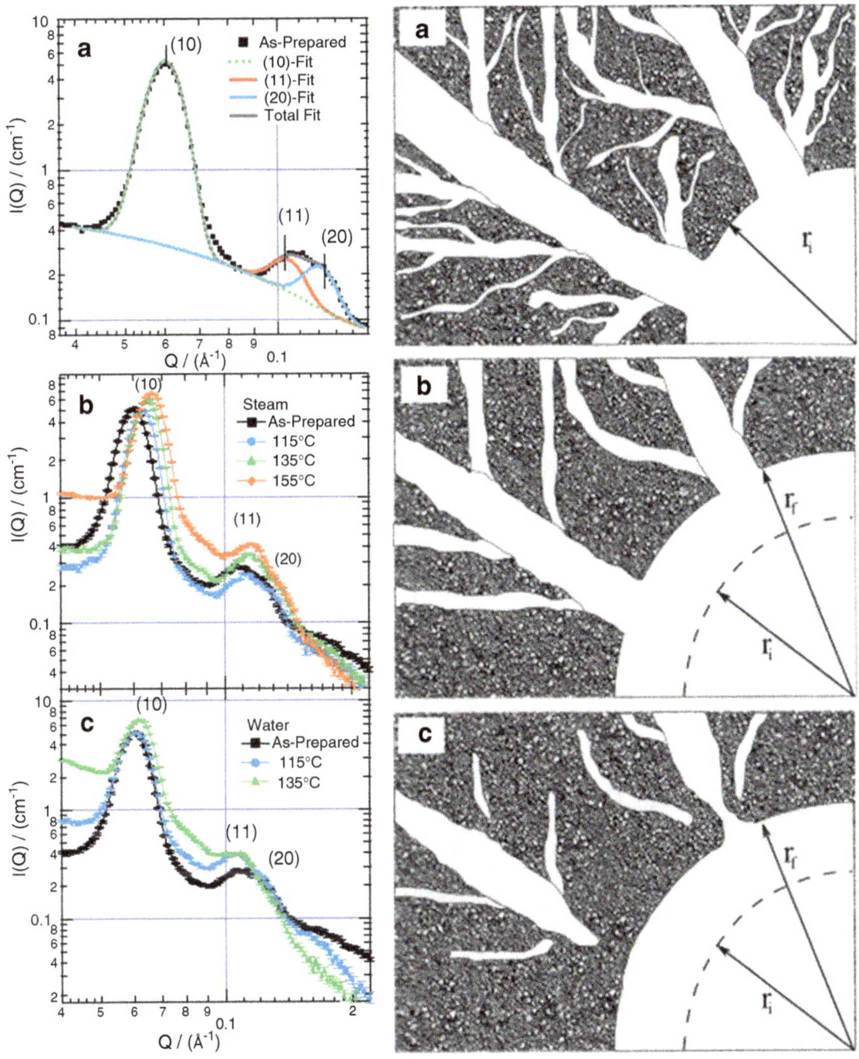

Fig. 7.11 *Left*: SANS scattering curves measured from (**a**) as-prepared, (**b**) steam treated, and (**c**) water treated SBA-15. Changes in the relative intensity of the higher order diffraction peaks indicate changes in the long-range order and the radial density profile around the periodic mesopores. *Right*: structural model consistent with the SANS data. (**a**) For as-prepared SBA-15, the mesopores are surrounded by a "corona" of micropores. (**b**) For steam-treated SBA-15, there is a decrease in the pores less than 30 Å in diameter and a significant reduction of pores in the first corona shell. (**c**) For water-treated SBA-15, micropores in the first corona shell are almost completely eliminated. The mesopore radius in both vapor and water treated samples increases from $r_i \cong 88$ Å to $r_j \cong 98$ Å. Reprinted with permission from [41]. Copyright 2012 American Chemical Society

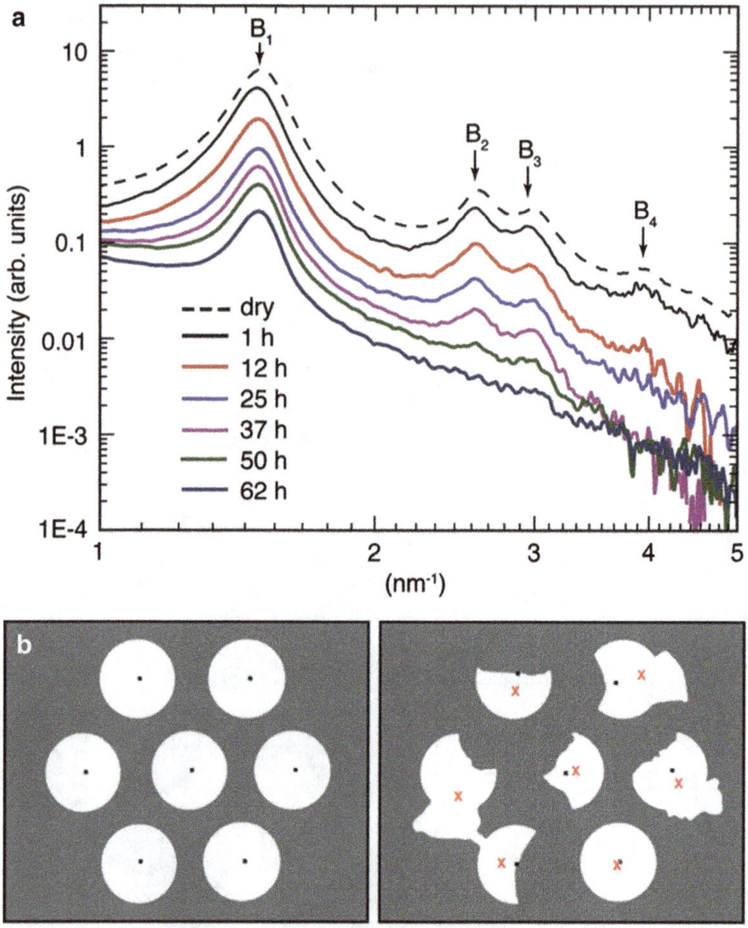

Fig. 7.12 (**a**) Experimental SAXS pattern of dry MCM-41 sample and its evolution with time after exposure to water. The B_1, B_2, B_3, and B_4 peaks denote the (10), (11), (20), and (21) crystallographic planes of the mesoporous 2D hexagonal lattice, respectively. (**b**) Schematic representation of the evolution of the porous structure of the MCM-41 sample before (*left*) and after (*right*) its exposure to water. *Black dots* show the initial position of the pore centers on the hexagonal lattice. These positions marked as crosses become shifted after exposure to water. Reprinted from [43]. Copyright (2014), with permission from Elsevier

SAXS data demonstrate that short-time exposure to water has little effect on the amplitude of the Bragg peaks. Longer exposure time decreases peak amplitudes, and after ~50 h, (11), (20), and (21) peaks completely disappear. The observed reduction of the peak amplitudes can be explained by progressive loss of the hexagonal order caused by randomization of pore positions, as well as distortion and deformation of pore walls (see Fig. 7.12b). The latter is mainly driven by the dissolution and recondensation of hydrolyzed silica, which results in changes of the wall shape. The pore centers are consequently shifted from the original positions on the hexagonal lattice, which results in decreasing the amplitude of the Bragg peaks. The proposed mechanism was confirmed by the TEM images of the dry and water-treated MCM-41.

7.3 Porous Carbons

7.3.1 Activated Carbons

The performance of activated carbons and derived devices is strongly influenced by the carbon structure, as well as the pore morphology [44]. Carbide-derived carbon (CDC) has attracted particular interested because of the possibility of tuning its pore structure and controlling the average pore size with subangstrom accuracy [45].

SANS structural characterization of microporous titanium CDC (TiC-CDC) with a narrow pore size distribution was conducted in [46] as a function of the synthesis (chlorination) temperature (300, 400, 600, and 800 °C). Figure 7.13 shows the background-corrected scattering data from the carbons studied. Contrast matching experiments have shown that the strong SANS signal at small values of Q is the result of scattering from the external surface of porous carbon particles. With this signal subtracted, the remaining scattering originates from the pores in the carbon matrix.

Pore radii in the activated carbons are expected to be on the order of a few Angströms and therefore should have very small radii of curvature. For such scattering objects, Kirste and Porod derived a correction term to the Porod law, which accounts for the curvature effects (see Sect. 6.3). Combining Eqs. (6.27) and (6.30), and taking into account that the radius of gyration of pores R_g and the Debye correlation length a in Eq. (6.27) are related *via* $R_g = a\sqrt{6}$, we get [47, 48]:

$$I(Q) = \frac{A}{Q^a} + \frac{B}{\left(\frac{6}{R_g^2} + Q^2\right)^2}\left(1 + \frac{1}{Q^2}C\right), \tag{7.3}$$

where $C = [1/4 < (C_1 + C_2)^2 > + 1/8 < (C_1 - C_2)^2 >]$, and C_1 and C_2 are the local principal curvatures of the surface [49, 50]. In addition to the Kirste-Porod (KP) correction term, Eq. (7.3) has a power-law term A/Q^α added to account for the scattering from the particle surface. The radii of gyration were obtained by fitting the SANS data to Eq. (7.3). The fits were conducted either taking into account or ignoring the KP correction. The results shown in Fig. 7.14 demonstrate that the values of R_g increase gradually from ~1.5 to ~5.0 Å with the chlorination temperature. The exponential factor α in $A/Q^{-\alpha}$ $(3.6 \leq \alpha \leq 3.9)$ agrees with the corresponding parameter determined for similar carbons by SAXS $(2.8 \leq \alpha \leq 3.8)$.

At the same time, the measured radii of gyration are systematically smaller than previously reported (5 Å $\leq R_g \leq$ 12 Å) [51, 52]. The inclusion of the KP correction in Eq. (7.3) does not affect the values of R_g within the experimental error.

As discussed in Chapter 6, the radius of gyration of a scattering object of arbitrary shape can be determined in the limit of $Q \to 0$ using the Guinier law (Eq. (6.33)). The modified Guinier formulae for 1D and 2D objects are also available. It is generally assumed that the shape of pores in microporous carbons can be approximated by elongated slits or cylinders. For cylindrical scattering

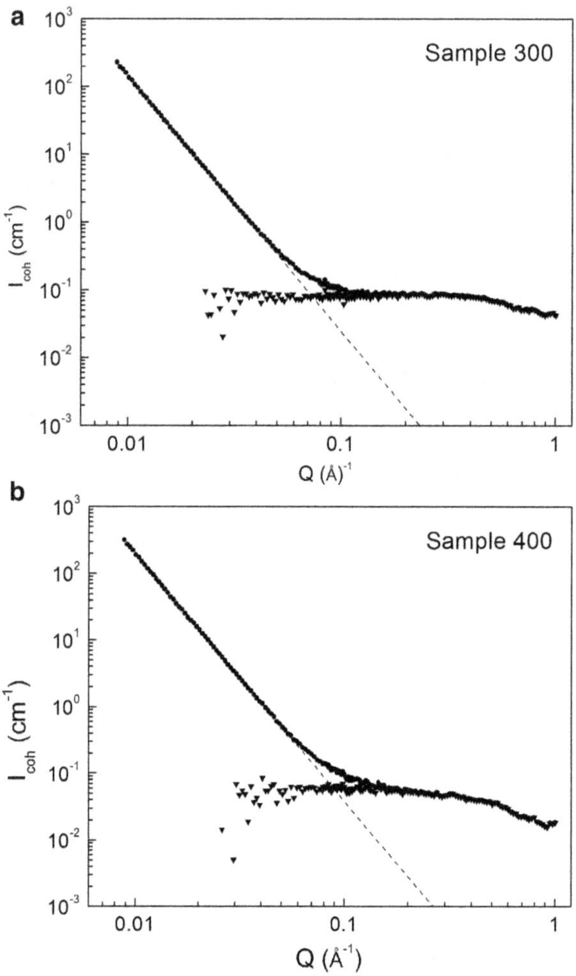

Fig. 7.13 Background-subtracted SANS cross section per unit volume *I(Q)* from the TiC-CDC carbons. Samples are labeled according to the chlorination temperature, which is shown in the insets. The dashed lines represent the power-law contribution of the first term in the right-hand side of Eq. (7.3). Triangles show the SANS data after subtraction of the power-law scattering. Reprinted from [46]. Copyright (2012), with permission from Elsevier

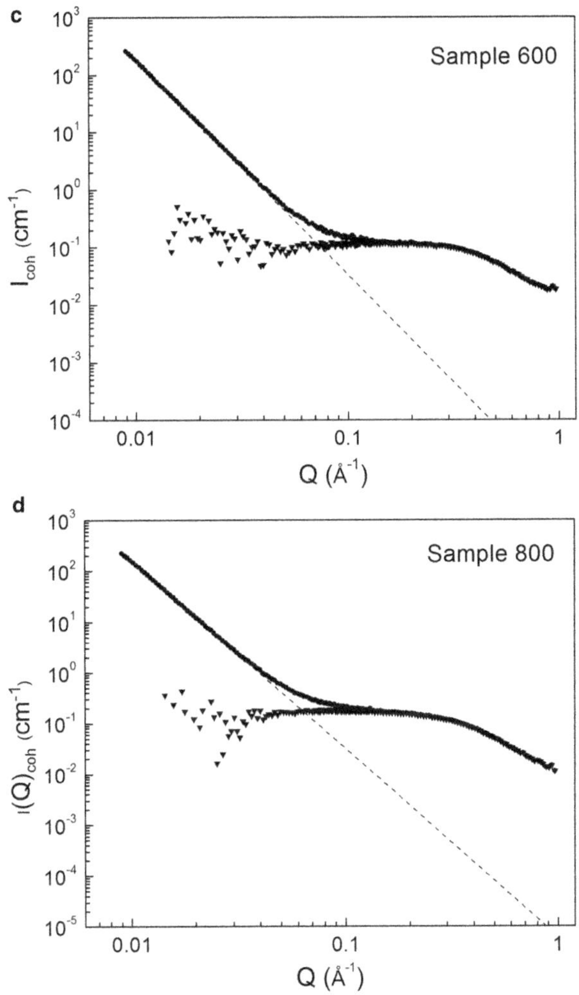

Fig. 7.13 (continued)

objects of length L and the radius of gyration of the cylinder cross section $R_{g,cyl}$ ($R_{g,cyl} << 1/Q << L$), the modified Guinier formula is given by Eq. (6.38), or more precisely:

$$\ln[Q\,I(Q)] = Const - Q^2 R_{g,cyl}^2/2, \tag{7.4}$$

where $R_{g,cyl}$ is related to the radius of the cylinder R_{cyl} as $R_{g,cyl} = R_{cyl}/\sqrt{2}$. Guinier plots are shown in Fig. 7.15, and the corresponding fitted values of R_g are shown in Fig. 7.14. Modified Guinier plots for cylindrical objects are presented in the insets in Fig. 7.15. The modified Guinier plot for sample 300 with the smallest pores (R_g ~1.8 Å) has a positive slope, which indicates that there is no significant

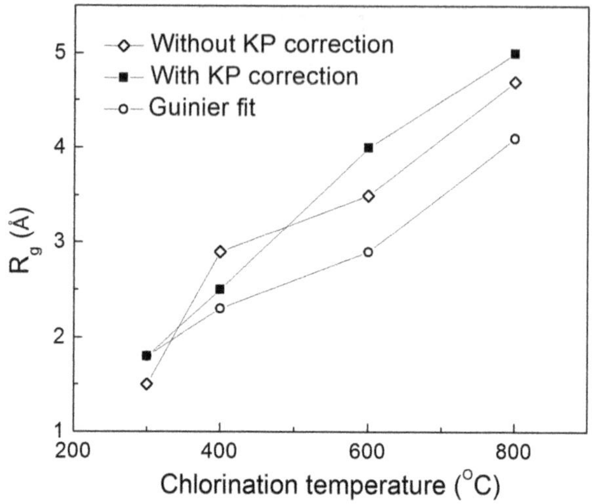

Fig. 7.14 Variation of the radii of gyration of pores in the TiC-CDC carbons obtained by fitting SANS to Eq. (7.3), both taking into account and ignoring the KP correction. R_g obtained from the Guinier fits are also shown. Reprinted from [46]. Copyright (2012), with permission from Elsevier

elongation of these pores with respect to the cross-sectional dimension. Assuming that the shape of these pores is close to spherical, the radius and the "length" (i.e., the sphere diameter D_{sph}) can be calculated using $R_{sph} = R_g\sqrt{5/3} = 2.3$ Å, and thus $L = D_{sph} = 2R_{sph} = 4.6$ Å. All other samples reveal negative slopes; therefore, modified Guinier plots can be used for extracting cross-sectional radii and the length of the pores in the studied samples using the previously determined Guinier radii of gyration, as well as the relation $R_g^2 = R_{cyl}^2/2 + L^2/12$ (Table 6.1).

The analysis shows that the radii of cylindrical pores are practically independent of the chlorination temperature, whereas the pore length increases almost by a factor of three—from $L = 4.6$ Å for sample 300 to $L \cong 12$ Å for sample 800. This indicates that micropores in TiC-CDC become more elongated with the increased synthesis temperature. Calculated pore sizes from SANS data are in good agreement with independent gas sorption measurements [46].

7.3.2 *Glassy Carbon*

Glassy carbon consists of stacks of graphene sheets with lateral extensions between 25 and several 100 Å, stack numbers between 2 and 3, and up to several dozen sheets, depending on the heat treatment conditions during preparation [53, 54]. As prepared, glassy carbon can be formally considered as a "nonporous" material. The pores in such matrices are predominantly closed (not connected to the surface), as evidenced by the He and H_2 permeability experiments. The originally closed pores can be opened by oxidation (activation) that can be carried out with controlled kinetics. Activation opens and connects closed pores and results in a creation of carbon material with a sponge-like structure and the developed internal surface area up to 1000 m^2/g. Understanding the impact of the activation process on glassy carbon morphology is a key to optimizing its properties for specific applications [55].

Evolution of the structural changes of glassy carbon during oxidation was studied using SAXS and nitrogen adsorption in [55, 56]. Figure 7.16 shows SAXS curves measured from the nonactivated carbon and two samples activated in an air-filled muffle furnace, one for 210 min and the other one for 260 min. Scanning electron microscopy (SEM) images showed that the sample activated for 260 min was fully activated and had no unreacted core, whereas the other sample had an unreacted core. Scattering from micropores in both samples was represented

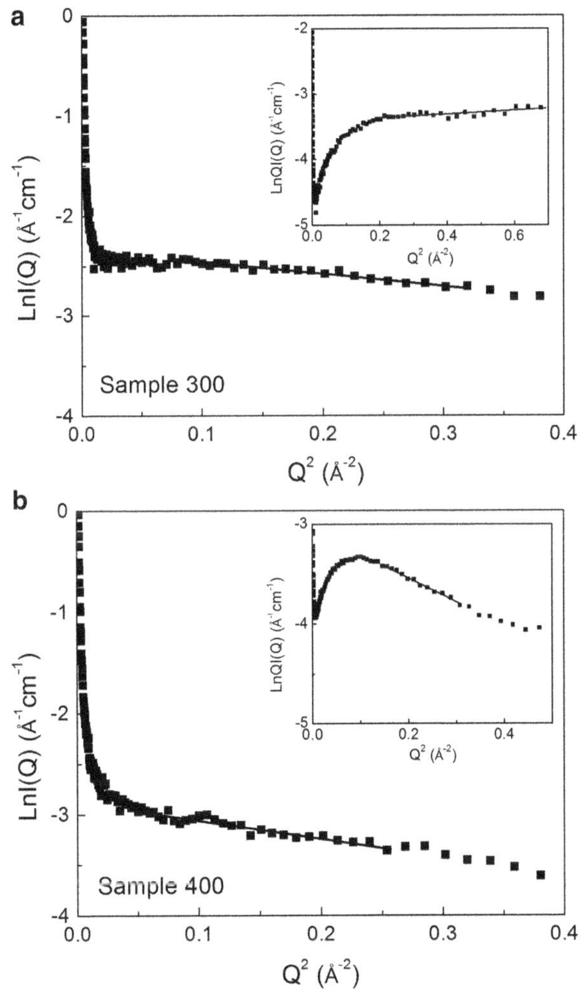

Fig. 7.15 Guinier plots and modified Guinier plots (*cylindrical pores*, *insets*) for the activated carbon. Reprinted from [46]. Copyright (2012), with permission from Elsevier

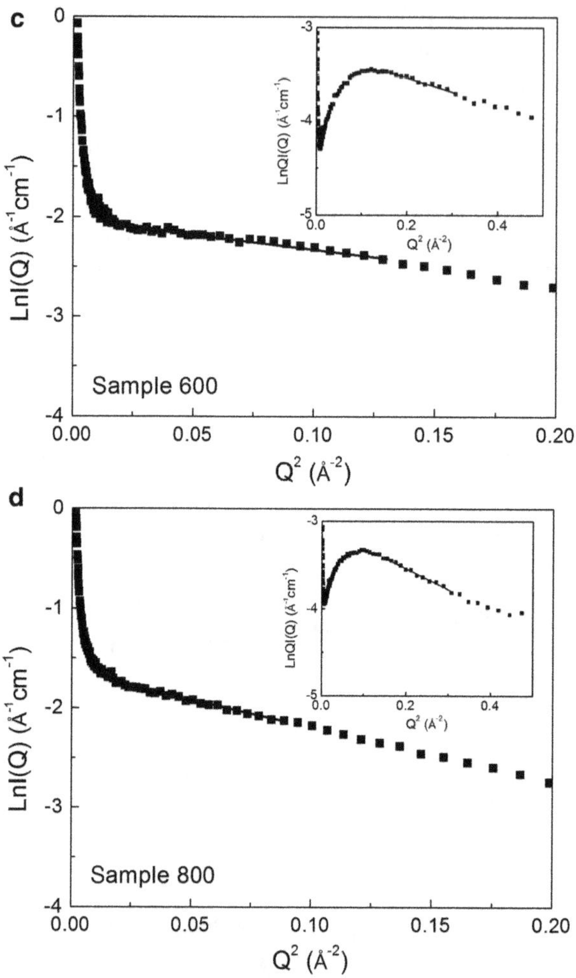

Fig. 7.15 (continued)

by a shoulder at ~0.03 $Å^{-1}$. For the fully activated sample, this shoulder is shifted to smaller Q-values, which indicates an increase in pore size.

Quantitative analysis of the SAXS curves was performed using the unified scattering function (Sect. 6.7) and the Irena software package [57] assuming that the scattering occured from a dilute system of noninteracting spherical pores with a log-normal distribution of sizes. The pore size distribution obtained for both samples is shown in Fig. 7.16. The fully activated sample (activated for 260 min) has pores twice as large as those in the sample activated for 210 min.

Additional structural information can be assessed by analyzing the variation of the power-law (Porod) exponent α in $I(Q) \sim Q^{-\alpha}$ describing scattering in the limit of high Q. The value of α for most carbon materials is typically $\alpha \cong 2.5$. Figure 7.17a

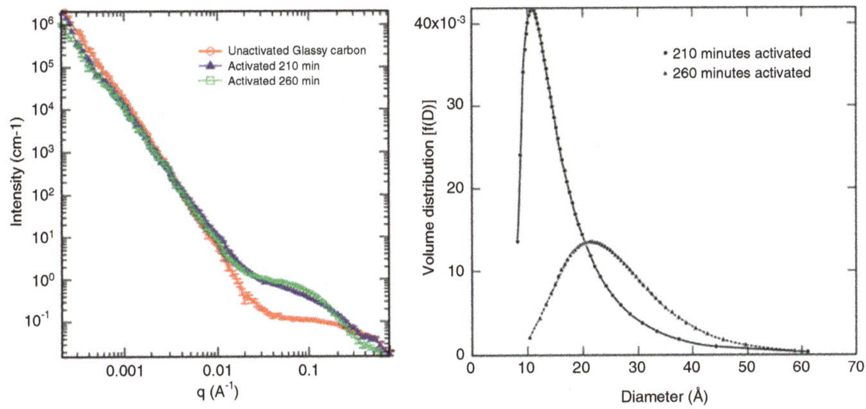

Fig. 7.16 *Left*: SAXS curves from nonactivated and activated glassy carbon. *Right*: pore size distribution in two glassy carbon samples activated for 210 and 260 min. Reproduced with permission from [56]

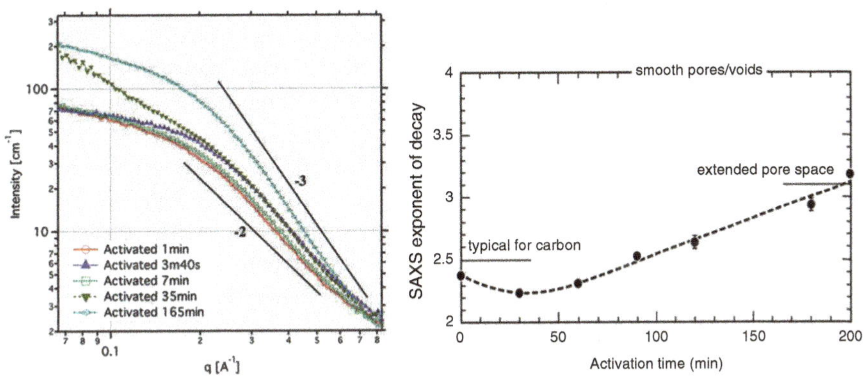

Fig. 7.17 *Left*: SAXS curves of glassy carbon activated for (*left* to *right*) 1 min, 3 min, 7 min, 35 min, and 165 min showing different exponential decay in the limit of high Q. *Right*: SAXS exponent of $I(Q) \sim Q^{-\alpha}$ decay in the limit of high Q for glassy carbon activated at 450 °C in air as a function of activation time. The exponent is approaching three for fully activated carbon. Reproduced with permission from [56]

shows the variation of the Porod exponent as a function of the activation time. For materials with fully open pore surface uniformly extended throughout the entire pore space, the value of the slope should be $\cong 3$ [58]. As can be seen in Fig. 7.17(b), the Porod exponent changes between $\alpha \cong 2.2$ for nonactivated and $\alpha \cong 3$ for fully activated glassy carbon, which provides indirect evidence of the gradual formation of a network of interconnected pores with the activation time.

7.3.3 Carbon Aerogel

Carbon aerogels are characterized by well-developed micro- and mesoporosity and a relatively large surface area. Their structure represents a network of interconnected nanoscopic primary particles. The population of micropores is associated with the intraparticle structure, whereas mesopores represent the interparticle voids. This difference of origin allows for independent control of the micro- and mesoporosity of carbon aerogels by varying the concentration of the original ingredients, curing and drying conditions, as well as the carbonization protocol [58].

The structure of carbon aerogel obtained by pyrolysis of an organic phenolic-like aerogel was studied using SANS in [59]. The SANS profile and an SEM image of the studied sample are shown in Fig. 7.18. This figure also shows a fit to the PDSP model, which is based on an assumption that the pore space in the sample can be approximated as an ensemble of spheres with a broad distribution of sizes randomly distributed over the sample volume (see Sect. 6.8.2). The fitting procedure was conducted using the PRINSAS software [60], which allows for determining the pore size distribution and the specific surface area of the sample. The results of the PDSP fits are shown in Fig. 7.19. As expected, the pore size distribution of the studied sample is quasi-bimodal, with two main maxima corresponding to the intraparticle micropores with radius $R \cong 12$ Å and the intreparticle mesopores with $R \cong 50$ Å. In addition, the data were fitted to the unified scattering function (see Eq. (6.46), Sect. 6.7), which describes the scattering from a multihierarchical structure, with each structure level described by a Guinier regime in the limit of low Q followed by a power-law scattering. For the studied sample it is sufficient to account for two structural levels with two populations of pores in the aerogel matrix with $R_{g,1} \cong 41$ Å and $R_{g,2} \cong 13$ Å. Assuming the spherical shape of pores and using $R = R_g\sqrt{5}/3$, the calculated pore radii $R_1 \cong 53$ Å and $R_2 \cong 16$ Å agree with the results of the PDSP model analysis.

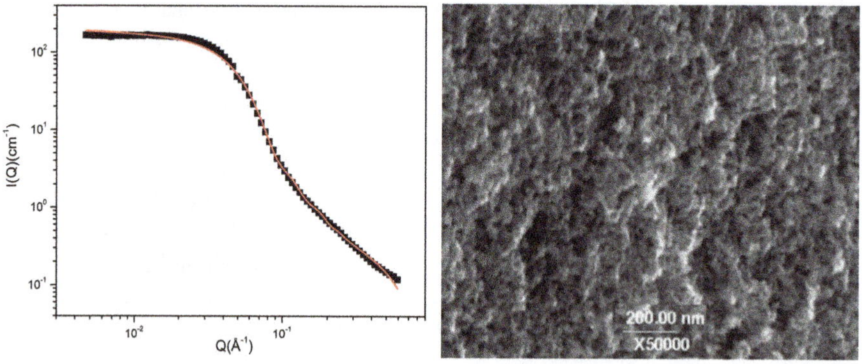

Fig. 7.18 *Left*: SANS profile for carbon aerogel and fit to the PDSP model (*solid line*). *Right*: SEM image of the cross section of the studied aerogel. Reprinted from [59]. Copyright (2012), with permission from Elsevier

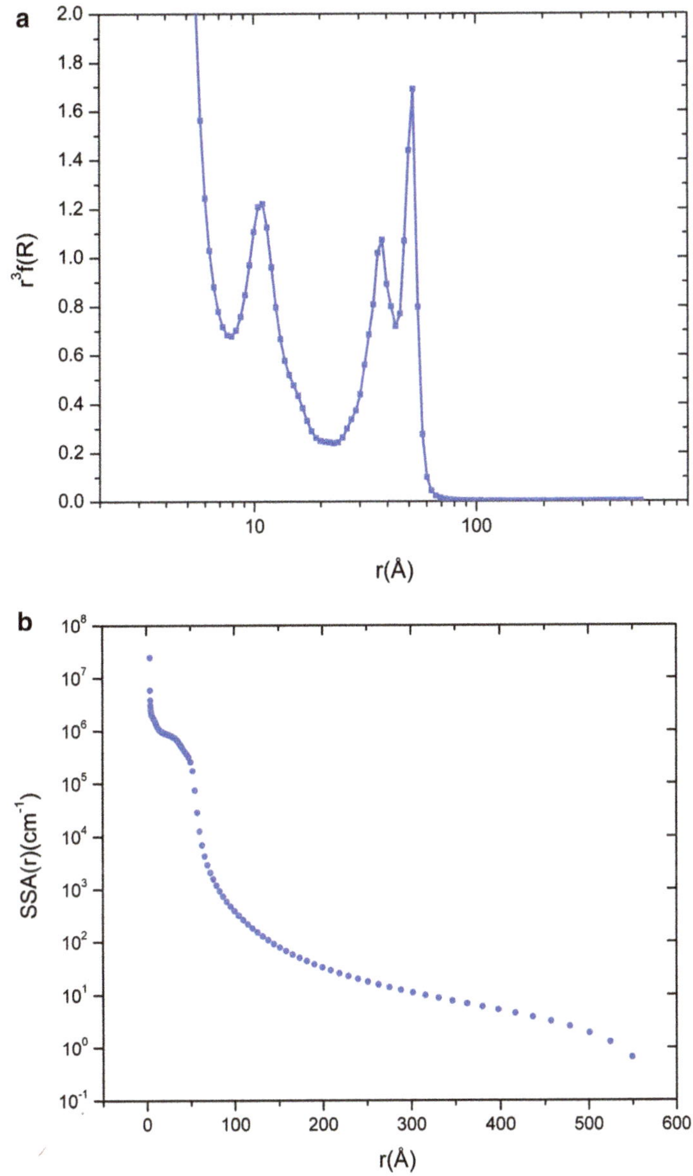

Fig. 7.19 Volume-weighed pore size distribution (**a**) and the specific surface area (**b**) of the studied carbon aerogel obtained using the PDSP model. Reprinted from [59]. Copyright (2012), with permission from Elsevier

SANS data presented in Fig. 7.18 can be used for evaluation of the porosity ϕ_1 and the specific sample surface area S/V described by Eqs. (6.25) and (6.30), respectively. SANS data in the $Q^2I(Q)$ vs. Q coordinates are shown in Fig. 7.20a. The value of the scattering invariant corresponds to the area under the curve in

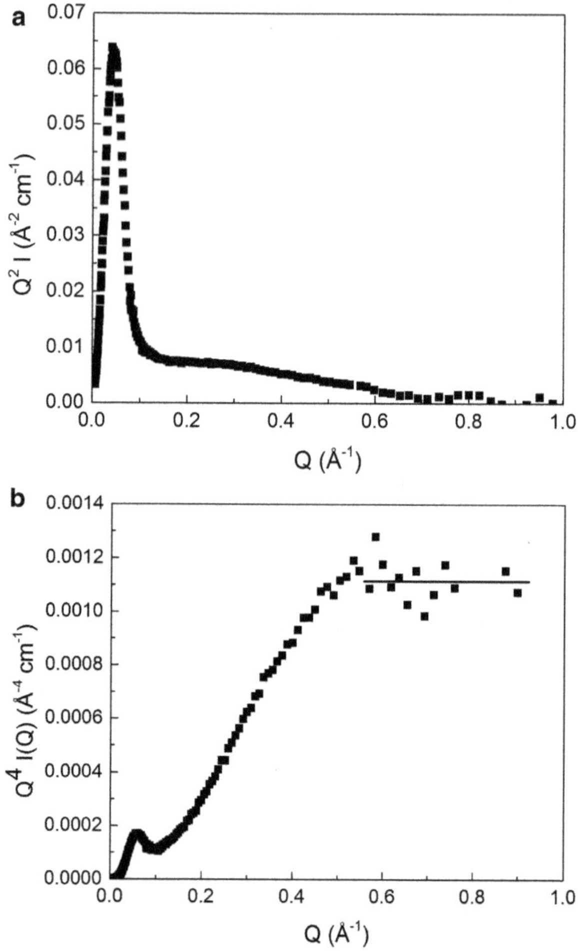

Fig. 7.20 SANS cross section per unit volume of the carbon aerogel in the (**a**) $Q^2I(Q)$ *vs.* Q and (**b**) $Q^4 I(Q)$ *vs.* Q coordinates. The *solid line* in (**b**) shows the Porod plateau used to calculate the specific surface area of the sample studied

Fig. 7.20a. The numerical integration was performed in three parts, as the SANS data were measured only between the minimum and maximum values of the scattering vector, Q_{min} and Q_{max} [61]. The contribution to the invariant from the (unmeasured) cross section between 0 and Q_{min} was estimated via the Guinier approximation $\left[I(Q)\tilde{}exp\left(-Q^2R_g^2\right)\right]$, which was extrapolated from the measured cross section at the lowest Q-values. Similarly, the contribution to the invariant from the Q-range between Q_{max} and ∞ was estimated via the Porod approximation $[I(Q) \sim Q^{-4}]$, which was used to extrapolate from the measured cross section at the highest Q-values. Thus determined area was calculated to be 0.00636 $(\text{Å}^{-1})^3\,\text{cm}^{-1} = 6.36 \times 10^{21}\ \text{cm}^{-4}$. Assuming that the SLD of amorphous carbon is $6 \times 10^{10}\ \text{cm}^{-2}$ (Table 1.2), the volume fraction of pores in the sample studied was calculated from Eq. (6.25) to be $\phi_1 \cong 0.66$.

The Porod plot of the SANS data in the $Q^4\ I(Q)$ *vs.* Q coordinates is shown in Fig. 7.20b. The plot flattens out in the limit of high Q, which allows for estimating the "Porod constant" $P = Q^4\ I(Q) = 0.0011$ ($Å^{-4}$ cm^{-1}) $= 1.1 \times 10^{29}$ cm^{-5}. From Eq. (6.30) one calculates $S/V = 487$ m^2/cm^3 or, using the density of the studied aerogel 1.06 g/cm^3, $S/V = 459$ m^2/g. Both $\phi_I \cong 0.66$ and $S/V = 459$ m^2/g, determined using SANS, are in a good agreement with the Brunauer-Emmett-Teller (BET) data supplied by the aerogel manufacturer ($\phi_I \cong 0.60$ and $S/V = 480$ m^2/g). Other examples of the comparative SANS, SAXS, and gas adsorption studies of carbon aerogels prepared using different conditions can be found in the literature (see, for example, [62] and references therein).

7.4 Alumina Membranes

Mesoporous aluminum membranes are produced by two-stage anode oxidation of aluminum, which allows the production of Al_2O_3 films having uniform cylindrical pores with diameters in the range 50–2500 Å and pore densities up to 10^{11} cm^{-2} [63]. The pores form regular hexagonal arrays of parallel, nonintersecting channels aligned perpendicularly to the film surface, which allows them to be used as components of model hybrid bilayer membranes for biomedical research [64, 65]. In addition, their high aspect ratio enables the fabrication of long, monodisperse metal and semiconductor nanowires, metal clusters, colloids, luminescent siloxanes, and carbon nanotubes (see, for example, [64] and references therein).

The structure of the aluminum membranes is qualitatively similar to that of MCM-41 and SBA-15 porous materials (Sect. 7.2.4, Fig. 7.5). Thus, the SAS data treatment and interpretation is similar to that used for structural characterization of the ordered mesoporous silica. Examples of SAXS and SANS studies of the morphology of the aluminum membranes formed at different stages of the anodizing process are available in [64, 66–69].

7.5 Porous Polymer Monoliths

Macroporous polymer monolithic materials have been introduced recently as a new generation of porous media that can be used as stationary phases for different types of chromatography, high-throughput bioreactors, microfluidic chip applications, as well as separation of large molecules (e.g., proteins and DNA). These materials can be synthesized *in situ* in a mold and can therefore be made in different shapes and sizes, such as large rods suitable for high-performance liquid chromatography columns. The monoliths consist of the aggregates of the interconnected polymer microglobules, and their structural rigidity is defined by the high degree of cross linking between polymers. Macroporous size and skeleton thickness can be independently controlled by changing the composition of the precursor mixtures. The main morphological feature of polymer monoliths is the presence of large pores, which permits their use in liquid chromatography at high flow rates and low backpressures [70, 71].

As for any porous medium, optimizing performance requires detailed information on the variation of the morphology of the polymer monoliths as a function of the preparation conditions. In addition, it is highly desirable to be able to monitor possible structural changes under conditions at which monoliths are normally operated. In particular, of great interest is information on the monolith morphology in the solvated environment that might significantly impact the material's properties via nanoscale solvation and swelling. Currently, the structural characterization of polymer monoliths is performed by dry-state analytical characterization techniques, which mostly include SEM, mercury intrusion porosimetry, and gas adsorption analysis. A combination of USANS and SANS techniques could offer new opportunities for the characterization of pore dimensions in these multihierarchical materials on the length scale of 10–10,000 Å. In addition to the dry-state structural characterization, USANS and SANS allow for detecting sample areas that are not accessible to the solvent using contrast matched mixtures of deuterated and protonated forms of a mobile phase. Such studies can be performed *in situ* at the flow through conditions corresponding to those used in real devices. Benefits offered by USANS and SANS in this area have not yet been fully appreciated by the community. So far, we are aware of only one paper in which USANS and SANS were employed to study the structure of polymer monoliths in combination with TEM and gas adsorption methods [72].

7.6 Ceramics, Alloys, and Composite Materials

SAS has been widely used to explore the morphology of ceramic systems of technological importance in order to extract microstructure information that complements the information obtained from diffraction experiments, x-ray or neutron radiography, tomography, and electron microscopy. SAS provides important contributions to research supporting advanced technological developments in various areas that include structural and electronic ceramics and glasses, pervoskites, high T_C superconductors, fuel cells, sintering, cavitation and damage effects, alloy/polymer/ceramic composites, archeology, as well as cement and concrete science. Detailed analysis of SAS applications in these areas is outside the scope of this book, and the interested reader is referred to reviews that summarize the contributions of USAS, SAS, and some other x-ray and neutron techniques in materials science [73, 74].

7.7 Structure of Sedimentary Rocks

Sedimentary rocks (coal, shales, sandstones, mudrocks, clays, and others) are formed from a mixture of organic and inorganic debris deposited in anoxic conditions. After the initial water expulsion and mechanical compression in response to increased geostatic loading, the sediments are gradually compacted into grains. The following stages of the rock formation include an extensive diffusive transport of the formation brine within the rock matrix, which results in the rock-pore interface with the highly developed internal surface area. Consequently, the microstructure of sedimentary rock becomes much more complex than just a collection of deformed grains [75]. The microstructure and evolution of porosity in time and space plays a critical role in many geological processes, including the migration and retention of water, gas, and hydrocarbons, weathering and diagenesis, as well as in technological processes such as CO_2 sequestration, as well as gas and oil recovery [76, 77].

Because the scale of pore/grain structures in rocks often covers seven orders of magnitude or more, no single experimental technique can provide morphological information on the entire system. As discussed in [78, 79], traditional techniques for characterizing porosity are either limited to the pore spaces accessible to fluids. In most cases they are unable to monitor structural variations as a function of pore size. Standard petrographic techniques, such as backscattering SEM, and TEM cover only certain ranges of length scales (Fig. 7.7), and quantification of pore structures for thin-section-sized samples is difficult, especially at high image resolutions [80].

Application of nondestructive USAS and SAS to the study of the rock morphology has become of great importance as the scattering data can deliver structural information on pore/grain geometry and the specific surface area for a single sample at 10–10,000 Å length scale. It has been discovered that sedimentary rocks display the most extensive fractal behavior observed in nature, with self-similarity expanding over at least three orders of magnitude in the length scale [76]. Surface fractal dimension is a sample specific parameter. It can vary even for similar rocks, which suggests that structures of pore/grain interfaces depend on the detailed history of a given rock, as well as on the materials forming the interfaces [80]. Both surface and mass fractal structures have been documented, the latter describing the statistical distribution of pores within the sample.

The upper spatial resolution the USAS technique is limited to 10,000 Å. The most promising way of increasing the resolution of USAS and SAS is use of these techniques in combination with the autocorrelation function and digital image analysis of the electron microscopy images [77, 80 83]. Such approach allows for examining and quantifying the porosity over a seven decade range of length scales from ~ 10 Å to ~1 cm.

References

1. Sengers, J.V., Kayser, R.F., Peters, C.J., White, H.J. (eds.): Equations of State for Fluids and Fluid Mixtures. Elsevier, Amsterdam (2000)
2. Ramanathan, M., Kilbey, S.M., Ji, Q., Hill, J.P., Ariga, K.: Materials self-assembly and fabrication in confined spaces. J. Mater. Chem. **22**, 10389 (2012)
3. Broom, D.P., Thomas, K.M.: Gas adsorption by nanoporous materials: future applications and experimental challenges. MRS Bull. **38**, 412 (2013)
4. Lu, A.H., Dai, S.: Themed issue on porous carbon materials. J. Mater. Chem. A. **1**, 9326 (2013)
5. Lee, J., Kim, J., Hyeon, T.: Recent progress in the synthesis of porous carbon materials. Adv. Mater. **18**, 2073 (2006)
6. He, L., Melnichenko, Y.B., Gallego, N.C., Contescu, C.I., Guo, J., Bahadur, J.: Investigation of morphology and hydrogen capacity of disordered carbons. Carbon **80**, 82–90 (2014)
7. Ruppert, L.F., Sakurovs, R., Blach, T.P., He, L., Melnichenko, Y.B., Mildner, D.F.R., Alcantar-Lopez, L.: A USANS/SANS study of the accessibility of pores in the barnett shale to methane and water. Energy Fuel **27**, 772–779 (2013)
8. Hoinkis, E.: Small-angle scattering of neutrons and X-rays from carbons and graphites. In: Thrower, P.A. (ed.) Chemistry and Physics of Carbon. Marcel Dekker, New York (1997)
9. Halasz, I., Liang, J.J.: In: Fitzgerald G., Govind N. (eds.) Molecular Modeling Aspects of Exploring Silica Properties. Applications of Molecular Modeling to Challenges in Clean Energy. ACS Symposium Series, vol. 1133, pp. 113–134 (2013)
10. Melnichenko, Y.B., Mayama, H., Cheng, G., Blach, T.: Monitoring phase behavior of sub- and supercritical CO_2 confined in porous fractal silica with 85% porosity. Langmuir **26**, 6374 (2010)
11. Enke, D., Janowski, F., Schweiger, W.: Porous glasses in the 21st century—a short review. Micropor. Mesopor. Mater. **60**, 19 (2003)
12. Striemer, C.C., Gaborski, T.R., McGrath, J.L., Fauchet, P.M.: Charge- and size-based separation of macromolecules using ultrathin silicon membranes. Nature **445**, 749 (2007)
13. Giraldo, L.F., Lopez, B.L., Perez, L., Urrego, S., Sierra, L., Mesa, M.: Mesoporous silica applications. Macromol. Symp. **258**, 129 (2007)
14. Ferey, G., Serre, C., Devic, T., Maurin, G., Jobic, H., Llewellyn, P.L., De Weireld, G., Vimont, A., Daturi, M., Chang, J.S.: Why hybrid porous solids capture greenhouse gases? Chem. Soc. Rev. **40**, 550 (2011)
15. MacGillivray, L.R. (ed.): Metal-Organic Frameworks: Design and Application. Willey, New Jersey (2010)
16. Jiang, J.X., Cooper, A.I.: Microporous organic polymers: design, synthesis, and function. Top. Curr. Chem. **293**, 1 (2010)
17. Xie, Y., Wang, T.-T., Liu, X.-H., Zou, K., Deng, W.-Q.: Capture and conversion of CO_2 at ambient conditions by a conjugated microporous polymer. Nat. Commun. **4**, 2013 (1960). doi:10.1038/ncomms2960
18. Cao, G., Wang, Y.: Nanostructures and Nanomaterials. World Scientific, New Jersey (2011)
19. Lowell, S., Shields, J.E., Thomas, M.A., Thommes, M.: Characterization of Porous Solids and Powders: Surface Area, Pore Size and Density. Springer (2006)
20. Riikonen, J., Salonen, J., Lehto, V.P.: Utilizing thermoporometry to obtain insights into nanostructured materials. J. Therm. Anal. Calorim. **105**(811) (2011)
21. Wiltzius, P., Bates, F.S., Dierker, S.B.: Structure of porous Vycor glass. Phys. Rev. A **36**, 2991 (1987)
22. Fratzl, P., Vogl, G., Klaumuzer, S.: Small-angle scattering from porous amorphous substances. J. Appl. Crystallogr. **24**, 588 (1991)
23. Cahn, J.W.: On spinodal decomposition. Acta Metall. **9**, 795 (1961)
24. Levitz, P., Ehret, G., Sinha, S.K., Drake, J.M.: Porous Vycor glass—the microstructure as probed by electron microscopy, direct energy-transfer, small-angle scattering, and molecular simulation. J. Chem. Phys. **95**, 6151 (1991)

25. Agamalian, M., Drake, J.M., Sinha, S., Axe, J.D.: Neutron diffraction study of the pore surface layer of Vycor glass. Phys. Rev. E **55**, 3021 (1997)
26. Kim, M.H., Glinka, C.J.: Ultra small-angle neutron scattering study of the nanometer to micrometer structure of porous Vycor. Micropor. Mesopor. Mater. **91**, 305 (2006)
27. Schaefer, D.W., Keefer, K.D.: Structure of random porous materials: silica aerogel. Phys. Rev. Lett. **56**, 2199 (1986)
28. Rojanski, D., Huppert, D., Bale, H.D., Dacai, X., Schmidt, P.W., Farin, D., Seri-Levy, A., Avnir, D.: Integrated fractal analysis of silica—adsorption, electronic energy transfer, and small-angle X-ray scattering. Phys. Rev. Lett. **56**, 2505 (1986)
29. Vacher, R., Woignier, T., Pelous, J., Courtens, E.: Structure and self-similarity of silica aerogels. Phys. Rev. B **37**, 6500 (1988)
30. Wang, J., Shen, J., Zhou, B., Wu, X.: SAXS investigation of silica aerogels derived from TEOS. NanoStruct. Mater. **7**, 699 (1996)
31. Takeda, M.W., Kirihara, S., Miyamoto, Y., Sakoda, K., Honda, K.: Localization of electromagnetic waves in three-dimensional fractal cavities. Phys. Rev. Lett. **92**, 093902 (2004)
32. Onda, T., Shibuichi, S., Satoh, N., Tsujii, K.: Super-water-repellent fractal surfaces. Langmuir **12**, 2125 (1996)
33. Mayama, H., Tsujii, K.: Menger-sponge-like fractal body created by a novel template method. J. Chem. Phys. **125**, 124706 (2006)
34. Yamaguchi, D., Mayama, H., Koizumi, S., Tsujii, K., Hashimoto, T.: Investigation of self-assembled fractal porous silica over a wide range of length scales using a combined small-angle scattering method. Eur. Phys. J. B **63**, 153 (2008)
35. Hurd, A.J., Schaefer, D.W., Martin, J.E.: Surface and mass fractals in vapor-phase aggregates. Phys. Rev. A **35**, 2361 (1987)
36. Zhao, D., Feng, J., Huo, Q., Melosh, N., Fredrikson, G.H., Chmelka, B.F., Stucky, G.D.: Triblock copolymer syntheses of mesoporous silica with periodic 50 to 300 Å pores. Science **279**, 548 (1998)
37. Zhao, D., Sun, J., Li, Q., Stucky, G.D.: Morphological control of highly ordered mesoporous silica SBA-15. Chem. Mater. **12**, 275 (2000)
38. Zhang, F., Yan, Y., Yang, H., Meng, Y., Yu, C., Tu, B., Zhao, D.J.: Understanding effect of wall structure on the hydrothermal stability of mesostructured silica SBA-15. Phys. Chem. B **109**, 8723 (2005)
39. Guo, W.P., Li, X., Zhao, X.S.: Understanding the hydrothermal stability of large-pore periodic mesoporous organosilicas and pure silicas. Micropor. Mesopor. Mater. **93**, 285 (2006)
40. Pollock, R.A., Walsh, B.R., Fry, J., Ghampson, I.T., Melnichenko, Y.B., Kaizer, H., Pynn, R., DeSisto, W.J., Clayton, M.C., Frederick, B.G.: Size and spational distribution of micropores in SBA-15 using CM-SANS. Chem. Mater. **23**, 3828 (2011)
41. Pollock, R.A., Gor, G., Walsh, B.R., Fry, J., Ghampson, I.T., Melnichenko, Y.B., Kaiser, H., Pynn, R., DeSisto, W.J., Wheeler, M.C., Frederick, B.G.: The role of liquid vs vapor water in the hydrothermal degradation of SBA-15. J. Phys. Chem. C **116**, 22802 (2012)
42. Kresge, C.T., Leonowicz, M.E., Roth, W.J., Vartuli, J.C., Beck, J.S.: Ordered mesoporous molecular sievs synthesized by a liquid-crystal template mechanism. Nature **359**, 710 (1992)
43. Gouze, B., Cambedouzou, J., Parres-Maynadie, S., Rebiscoul, D.: How hexagonal mesoporous silica evolves in water on short and long term: role of pore size and silica wall porosity. Micropor. Mesopor. Mater. **183**, 168 (2014)
44. Beguin, F., Frackowiak, E.: Carbons for Electrochemical Energy Storage and Conversion Systems. CRC Press, Boca Raton (2009)
45. Gogotsi, Y., Nikitin, A., Ye, H., Zhou, W., Fischer, J.E., Yi, B., Foley, H.C., Barsoum, M.W.: Nanoporous carbide-derived carbon with tunable pore size. Nat. Mater. **2**, 591 (2003)
46. He, L., Chathoth, S.M., Melnichenko, Y.B., Presser, V., McDonough, J., Gogotsi, Y.: Small-angle neutron scattering characterization of the structure of nanoporous carbons for energy-related applications. Micropor. Mesopor. Mater. **149**(46) (2012)

47. Kalliat, M., Kwak, Y., Schmidt, C.W.: Small-angle X-Ray Investigation of the Porosity in Coals (volume 169 of ACS Symposium series New Approaches in Coal Chemistry). American Chemical Society, Pittsburgh (1981)
48. Gibaud, A., Xue, J.S., Dahn, J.: A small angle X-ray scattering study of carbons made from pyrolyzed sugar. Carbon **34**, 499 (1996)
49. Kirste, V.R., Pord, G.: Rontgenkleinwinkelstreuung an kolloiden systemen—asymptotisches verhalten ger streukurven. Kolloid-Zeitschrift Zeitschrift fur Polymere **184**, 1 (1962)
50. Auvray, L., Auroy, P.: Scattering at interfaces: variations on Porod's Law. In: Lindner, P., Zemb, T. (eds.) Neutron, X-Ray and Light Scattering: Introduction to an Investigative Tool for Colloidal and Polymeric Systems. North-Holland, Amsterdam (1991)
51. Laudisio, G., Dash, R.K., Singer, J.P., Yushin, G., Gogotsi, Y., Fischer, J.E.: Carbide-derived carbons: a comparative study of porosity based on small-angle scattering and adsorption isotherms. Langmuir **22**, 8945 (2006)
52. Dash, R., Chmiola, J., Yushin, G., Gogotsi, Y., Laudisio, G., Singer, J., Fischer, J., Kucheyev, S.: Titanium carbide derived nanoporous carbon for energy-related applications. Carbon **44**, 2489 (2006)
53. Dris, V.A., Tchoubar, C.: X-Ray Diffraction by Disordered Lamellar Structures. Springer, Berlin (1990)
54. Jenkis, G.M., Kawamura, K.: Polymeric Carbons—Carbon Fiber, Glass and Char. Cambridge University Press, Cambridge (1976)
55. Braun, A., Bartsch, M., Schnyder, B., Kotz, R., Haas, O., Haubold, H.-G., Goerigk, G.: X-ray scattering and adsorption studies of thermally oxidized glassy carbon. J. Non-Cryst. Solids **260**, 1 (1999)
56. Braun, A., Seifert, S., Ilavsky, J.: Highly porous activated glassy carbon film sandwitch structure for electrochemical energy storage in ultracapacitor applications: Study of the porous film structure and gradient. J. Mater. Res. **25**, 1532 (2010)
57. Ilavsky, J., Jemian, P.R.: Irena: tool suite for modeling and analysis of small-angle scattering. J. Appl. Crystallogr. **42**, 347 (2009)
58. Pekala, R.W.: Organic aerogels from the polycondensation of resorcinol with formaldehyde. J. Mater. Sci. **24**, 3221 (1989)
59. Chathoth, S.M., He, L., Mamontov, E., Melnichenko, Y.B.: Effect of carbon dioxide and nitrogen on the diffusivity of methane confined in nano-porous carbon aerogel. Micropor. Mesopor. Mater. **148**, 101 (2012)
60. Hinde, A.L.: PRINSAS—a Windows-based computer program for the processing and interpretation of small-angle scattering data tailored to the analysis of sedimentary rocks. J. Appl. Crystallogr. **37**, 1020 (2004)
61. Melnichenko, Y.B., Wignall, G.D., Cole, D.R., Frielinghaus, H.: Adsorption of supercritical CO_2 in aerogels as studied by small-angle neutron scattering and neutron transmission techniques. J. Chem. Phys. **124**, 204711 (2006)
62. Cohaut, N., Thery, A., Guet, J.M., Rouzand, J.N., Kocon, L.: The porous network in carbon aerogels investigated by small angle neutron scattering. Carbon **45**, 1185 (2007)
63. Sousa, C.T., Leitao, D.C., Proenca, M.P., Ventura, J., Pereira, A.M., Araujo, J.P.: Nanoporous alumina as templates for multifunctional applications. Appl. PHys. Rev. **1**, 031102 (2014)
64. Marchal, D., Bourdillon, C., Deme, B.: Small-angle neutron scattering by highly oriented hybrid bilayer membranes confined in anisotropic porous alumina. Langmuir **17**, 8313 (2001)
65. Deme, B., Marchal, D.: Polymer-cushioned lipid bilayers in porous alumina. Eur. Biophys. J. **34**, 170 (2005)
66. Benfield, R.E., Grandjean, D., Dore, J.C., Esfahanian, H., Wu, Z., Kroll, M., Geerkens, M., Schmidt, G.: Structure of assemblies of metal nanowires in mesoporous alumina membranes studied by EXAFS, XANES, X-ray difraction, and SAXS. Faraday Discuss. **125**, 327 (2004)
67. Turkevych, I., Ryukhtin, V., Garamus, V., Kato, S., Takamasu, T., Kido, G., Kondo, M.: Studies of self-organization process in nanoporous alumina membranes by small-angle neutron scattering. Nanotechnology **23**, 325606 (2012)

68. Grigor'ev, S.V., Grigor'eva, N.A., Syromyatnikov, A.V., Napol'skii, K.S., Eliseev, A.A., Lukashin, A.V., Yu, D., Tret'yakov, D., Eckerlebe, H.: Two-dimensional spatially ordered Al_2O_3 systems: small-angle neutron scattering investigation. JETP Lett. **85**, 449 (2007)
69. Kainourgikas, M., Steriotis, T., Charalambopoulou, G., Strobl, M., Stubos, A.: Determination of the spatial distribution of multiple fluid phases in porous media by ultra-small-angle neutron scattering. Appl. Surf. Sci. **256**, 5329 (2010)
70. Arrua, R.D., Strumia, M.C., Alvarez Igarzabal, C.I.: Macroporous monolithic polymers: preparation and applications. Materials **2**, 2429 (2009)
71. Nischang, I.: Porous polymer monolyths: morphology, porous properties, polymer nanoscale gel structure and their impact on chromatographic performance. J. Chromatogr. A **1287**, 39 (2013)
72. Ford, K.M., Konzman, B.G., Rubinson, J.F.: A more informative approach for characterization of polymer monolithic phases: small angle neutron scattering/ultrasmall angle neutron scattering. Anal. Chem. **83**, 9201 (2011)
73. Fratzl, P.: Small-angle scattering in materials science—a short review of applications in alloys, ceramics and composite materials. J. Appl. Crystallogr. **36**, 397 (2003)
74. Allen, A.J.: Characterization of ceramics by x-ray and neutron small-angle scattering. J. Am. Ceram. Soc. **88**, 1367 (2005)
75. Cohen, M.H.: In: Johnson, D.L. (ed.) The Morphology of Porous Sedimentary Rocks. Physics and Chemistry of Porous Media. AIP Conference Proceedings No. 154, pp. 3–16 (1984)
76. Radlinski, A.P.: Small-angle neutron scattering and the microstructure of rocks. Rev. Mineral. Geochem. **63**, 363 (2006)
77. Anowitz, L.M., Cole, D.R., Rother, G., Allard, L.F., Jackson, A.J., Littrell, K.C.: Diagenetic changes in macro- to nano-scale porosity in the St. Peter sandstone: an (ultra) small angle neutron scattering and backscattering electron imaging analysis. Geochim. Cosmochim. Acta **102**, 280 (2013)
78. Radlinski, A.P., Mastalerz, M., Hinde, A.L., Hainbuchner, M., Rauch, H., Baron, M., Lin, J.S., Fan, L., Thiagarajan, P.: Application of SAXS and SANS in evaluation of porosity, pore size distribution and surface area of coal. Int. J. Coal Geol. **59**, 245 (2004)
79. Anowitz, L.M., Lynn, G.W., Cole, D.R., Rother, G., Allard, L.F., Hamilton, W.A., Porcar, L., Kim, M.H.: A new approach to quantification of metamorphism using ultra-small and small angle neutron scattering. Geochim. Cosmochim. Acta **73**, 7303 (2009)
80. Wang, H.W., Anovitz, L.M., Burg, A., Cole, D.R., Allard, L.F., Jackson, A.J., Stack, A.G., Rother, G.: Multi-scale characterization of pore evolution in a combustion metamorphic complex, Hatrurim basin, Israel: combining (ultra) small-angle neutron scattering and image analysis. Geochim. Cosmochim. Acta **121**, 339 (2013)
81. Radlinski, A.P., Ioannidis, M.A., Hinde, A.L., Hainbucher, M., Baron, M., Rauch, H., Kline, S. R.: Angstrom-to-milimiter characterization of sedimentary rock microstructure. J. Colloid Interface Sci. **274**, 607–612 (2004)
82. Berryman, J.G.: Measurement of spatioal correlation functions using image processing techniques. J. Appl. Phys. **57**, 2374–2384 (1985)
83. Berryman, J.G., Blair, S.C.: Use of digital image analysis to estimate fluid permeability of porous materials: application of two-point correlation functions. J. Appl. Phys. **60**, 1930–1938 (1986)

Chapter 8
Neutron and X-Ray Porosimetry

This Chapter considers principles of the neutron- and x-ray scattering-based porosimetry, which exploits contrast variation and contrast matching method. The coherent scattering from porous solids arises largely from the contrast between the empty and filled pores. A porous medium is exposed to a solvent vapor having the same neutron scattering length density or electron density as the matrix skeleton. The smallest pores are filled with a condensed liquid already at low relative pressure, followed by filling of progressively larger pores. Thus, measured as a function of the relative pressure, SAS probes sub-populations of pores of increasing average size, which is calculated using the Kelvin equation. Illustrative examples are given of how SAS can be used for structural characterization of various types of porous materials including low-dielectric constant polymer films, porous silicas, carbonaceous materials, and Nafion membranes.

8.1 Principles of the Scattering-Based Porosimetry

The mechanism of vapor adsorption on flat surfaces begins with the formation of a monolayer followed by multilayer adsorption of vapor molecules. The specifics of vapor adsorption in porous materials depend on the pore size and geometry. Radius of curvature of meso- and macropores is large, and therefore the mechanism of fluid adsorption in such pores is similar to that on flat surfaces. Radius of surface curvature of micropores with sizes less than few nanometers is small. Such pores become completely filled with liquid-like adsorbate after the vapor injection even at pressure P lower than saturated vapor pressure P_0 ($P/P_0 < 1$) (micropore or cooperative filling adsorption mechanism) [1]. Many real materials contain pores with sizes ranging from micropores to macropores. After the fluid injection, such materials have always two coexisting populations of pores characterized by the surface and pore filling adsorption mechanisms at low relative pressures P/P_0.

Y.B. Melnichenko, *Small-Angle Scattering from Confined and Interfacial Fluids*, DOI 10.1007/978-3-319-01104-2_8

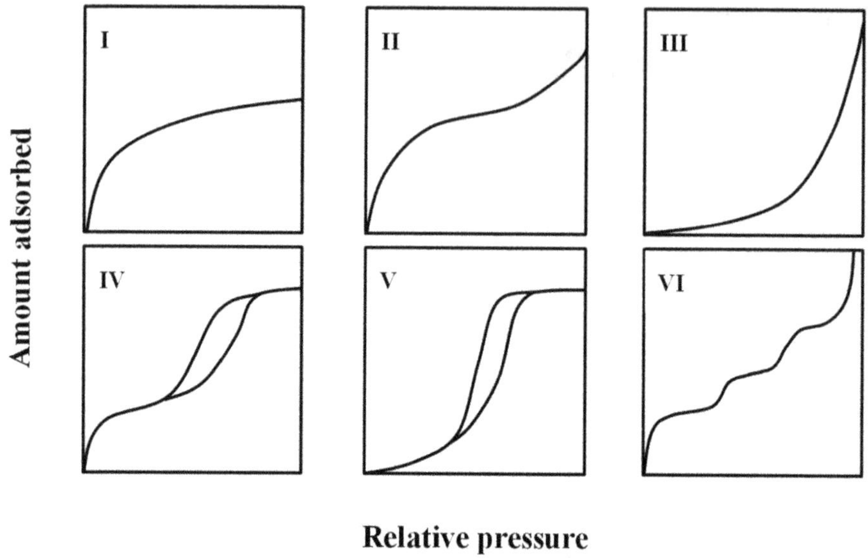

Fig. 8.1 The IUPAC classification of adsorption isotherms

Amount of the adsorbed fluid in a porous material increases with increasing relative pressure P/P_0 until the gas-liquid transition (capillary condensation) occurs in large pores at pressures below the bulk fluid pressure [1]. Capillary condensation is a well-documented and extensively studied phenomenon. It occurs at the *capillary condensation pressure* at temperatures below the liquid–gas critical temperature of the bulk fluid. Such transition appears in the adsorption isotherms (amount of adsorbed fluid vs. relative pressure) as a sudden and large jump in the amount of adsorbed fluid. In a porous material with monodisperse pore size distribution and identical shape of the pores, such a condensation represents a sharp transition. In real systems with pores of variable sizes, condensation is revealed as more continuous variation of the amount of adsorbed fluid as a function of pressure (Fig. 8.1).

In a traditional gas adsorption experiment, the adsorption of the fluid in the porous material is measured as a function of the vapor pressure. The mass uptake or volume fraction of the fluid due to adsorption is evaluated by a suitable analytical technique as the relative pressure P/P_0 is increased stepwise from 0 to 1. The pore size distribution is obtained indirectly by invoking a geometric model of adsorption to convert the adsorbate mass or volume to a size. In a desorption experiment, P/P_0 is decreased stepwise from 1 to 0. As discussed earlier, during the adsorption process the smallest pores fill first with the adsorbate at low values of P/P_0. As P/P_0 increases, progressively larger pores become completely filled with adsorbate, until all accessible pores are filled at $P/P_0 = 1$. Pore emptying during desorption generally occurs in reverse order from largest pores to smallest. In many cases the adsorption and desorption branches do not coincide (exhibit the hysteresis). The hysteresis is usually attributed to the metastability of the desorption branch but can also be caused by pore blocking or cavitation. The characteristics of the hysteresis loop depend on the molecule-surface interaction parameter, the distribution of pore sizes, as well as pore shape and connectivity of the pore network [2]. Quantitative information on the pore size distribution and other structural parameters may be inferred from the adsorption-desorption isotherms by application of computer simulations as well as the density functional theory (DFT) [3].

Capillary condensation in sufficiently large pores may be described using classical thermodynamic theory. For cylindrical pores, the Kelvin equation offers the simplest relation between the relative pressure and the pore radius R_P:

$$\ln\left(\frac{P}{P_0}\right) = -\frac{2}{R_P RT}, \tag{8.1}$$

where σ is the surface tension, R is the gas constant, and T is the absolute temperature. The derivation of Eq. (8.1) is based on few assumptions: near surface density oscillations may be neglected; pores are large enough so that the surface tension has physical meaning and can be taken to be equal to that of bulk fluid; the gas phase is described by the ideal gas law, and the liquid phase is incompressible. The Kelvin equation can be modified to account for most of these factors. More accurate interpretations, however, require computer modeling and or simulations of the isotherms such as Cage Cell Monte Carlo simulations or non-linear density functional theory methods. The smallest pore size below which the Kelvin equation becomes grossly inaccurate can be evaluated using both DFT and molecular simulation and is estimated to be $R_P \cong 75$ Å [2]. Given these constrains, the Kelvin equation in its simplest form can be used to provide a qualitative estimate of the pore size distribution for meso- and microporous materials.

There exist several non-destructive, scattering based porosimetry techniques that can be used for characterizing pore structures, including focused positron beam technique, ellipsometric porosimetry (EP), SAXS and SANS, specular x-ray and neutron reflectometry (SXR and NR, respectively), as well as grazing incidence SAXS (Gi SAXS) and grazing incidence SANS (Gi SANS). Each of these methods has its advantages and disadvantages as well as the inherent limitations (see e.g., [4] and references therein). SANS porosimetry exploits neutron contrast matching method and is based on the phenomenon of adsorption and condensation in pores. A porous film is exposed to a solvent vapor having the same neutron SLD as the porous matrix. At any given relative pressure, pores are filled with the condensed liquid solvent with the SLD matching that of the matrix. The coherent scattering therefore largely arises from the contrast between the empty and filled pores. Because the smallest pores fill first (at low relative pressure) followed by progressively larger pores, the SAS characterizes sub-populations of pores of increasing average size. When all pores are filled, the coherent part of the scattering intensity drops to 0 over all Q as the material becomes homogeneous in terms of the neutron SLD. SAS experiments at increasing pressure (sorption) or decreasing (desorption) pressure can be performed at different temperatures and the pore radius at which capillary condensation occurs can be calculated for each P/P_0 using the Kelvin equation. If the volume fraction of liquid solvent ϕ_{Solv} is independently measured as a function of P/P_0, the dependence of the average pore size on ϕ_{Solv} completely characterizes pore size distribution [5].

To more completely characterize the pore filling process, complementary measurements of solvent uptake vs. relative pressure can be performed using SXR. SXR porosimetry can be applied to measure film thickness, pore volume fraction, matrix mass density, coefficient of thermal expansion, and solvent/moisture uptake in thin porous films [6]. The SXR measurement yields an average electron density for the film, including any adsorbed solvent molecules. Fits to SXR data provide a critical angle θ_C, below which total external reflection occurs. The angle θ_C (in rad) is related to the bulk electron density in the films:

$$\rho_X^* = \frac{\pi\theta_C^2}{\lambda^2 r_{el}}, \tag{8.2}$$

where λ is the x-ray wavelength, and r_{el} is the radius of an electron, taken as 2.82×10^{-5} Å. To determine solvent uptake at a given value of P/P_0, the electron density of the dry film is measured under vacuum and after that as a function of P/P_0. Knowing the number of electrons per solvent molecule, the number density of solvent molecules in the film can be calculated. The increase in the film electron density upon filling the pores is proportional to the total number of solvent molecules in the film. Assuming that density of confined and bulk fluid are the same, the volume fraction of the solvent ϕ_{Solv} can be calculated:

$$\phi_{Solv} = \upsilon\left[\frac{\left(\rho_X^*\right)^{P/P_0} - \left(\rho_X^*\right)^{vac}}{n_{el}}\right], \tag{8.3}$$

where υ is a molecular volume defined in Eq. (1.17), $(\rho_X^*)^{vac}$ and $\left(\rho_X^*\right)^{P/P_0}$ is the electron density of the film before and after exposure to solvent at relative pressure P/P_0, respectively, and n_{el} is the number of electrons per solvent molecule. At any given P/P_0, the volume fraction of solvent in the film is determined using Eqs. (8.2) and (8.3). The overall pore volume fraction in the film ϕ_0 is calculated from Eq. (8.3) by measuring the film electron density at $P/P_0 = 1$ (liquid-filled pores) when $\phi_{Solv} = \phi_0$. At any P/P_0, ϕ_{Solv} is the volume fraction of solvent-filled pores, the quantity $\phi_0 - \phi_{Solv}$ is the volume fraction of empty pores, and $(1 - \phi_0)$ gives the volume fraction of the matrix [5].

8.2 Structure of Nanoporous Low-Dielectric-Constant Films

The demand for increased signal transmission speed and device density in the next generation of multilevel integrated circuits requires use of low dielectric constant (low-k) materials with permittivities of less than 2.0. The lowered dielectric constant has the effect of increasing signal propagation speed, decreasing the energy needed to propagate a signal, and decreasing the crosstalk between adjacent conductors. One avenue to low-*k* materials is the introduction of nanometer size pores into a solid film to lower it effective dielectric constant. For example, the dielectric constant of bulk silica, nominally about 4, reduces down to ~2 if the film porosity approaches three-quarters by volume. At the same time, introduction of voids may affect important material properties such as mechanical strength, moisture uptake, coefficient of thermal expansion, and adhesion to different substrates.

Due to obvious reasons, the accuracy of standard characterization techniques (e.g. gravimetric, volumetric, etc.) becomes very limited when dealing with extremely small samples. This makes SAS techniques increasingly popular for structure evaluation of a variety of thin films of practical importance, such as low-k polyphenylene, ordered mesoporous silica, and proton exchange membranes [7].

The first attempt to apply SANS for structural characterization of porous thin films with the thickness of the order of 1 μm was undertaken by Wu et al. [8] who evaluated the structural properties of porous silica (xerogel) thin films supported by silicon wafer substrates. To enhance the scattering signal, the samples were composed of a stack of six pieces of the wafer supported films. Since the silicon wafer substrate is virtually transparent to neutrons and does not cause significant scattering in the small angle region, the measured scattering intensities are exclusively from the porous films due to neutron contrast between the SLD of the pore walls and the SLD of pores. The neutron contrast was varied by preparing three different samples, one with pores saturated with air to evaluate the dry film structure, one with pores saturated with deuterated toluene to determine the pore connectivity, and one with pores saturated with deuterated water to measure moisture uptake.

A suitable model to describe the film structure is a random two-phase model described by Eq. (6.27) [9]. The value of $8\pi\phi_1\phi_2\left(\rho_1^* - \rho_2^*\right)^2$ and the Debye correlation length a can be determined from the slope and intercept of the SANS data plotted as $[d\Sigma(Q)/d\Omega]^{1/2}$ vs Q^2 in the intermediate Q-range $Q^2 > 0.0002$ Å^2 (Fig. 8.2). The slope and the $Q = 0$ intercept in the Fig. 8.2 are 163.2 ± 0.9 cm$^{1/2}$ Å^2 and 0.175 ± 0.002 cm$^{1/2}$, respectively. For the mass density of the sample 0.55 g/cm^3 and the density of the wall material 1.16 g/cm^3 this gives the porosity 53 ± 1 % and $a = 30.55$ Å, which allows calculating a pore chord length of 65 ± 1 Å from Eq. (6.43). The deviation (downturn) from the linear dependence $[d\Sigma(Q)/d\Omega]^{1/2}$ vs. Q^2 at small Q often referred to as "excess forward scattering" is observed in many porous materials. It is believed to occur due to a nonuniform distribution of pores (e.g. clustering of pores) or a presence of large scale chemical heterogeneities.

Homogeneous saturation of pores with a wetting liquid (e.g. D-toluene), should increase $\Delta\rho^*$ and result in a proportional increase of neutron scattering signal leaving the shape of the function $[d\Sigma(Q)/d\Omega]$ unchanged. However, the SANS from the films immersed in D-toluene has a qualitatively different shape (Fig. 8.2) and smaller than expected increase in scattering, which suggests that the porous matrix is only partially filled with the liquid. Calculations based on a two-layer model showed that the volume fraction of interconnected pores

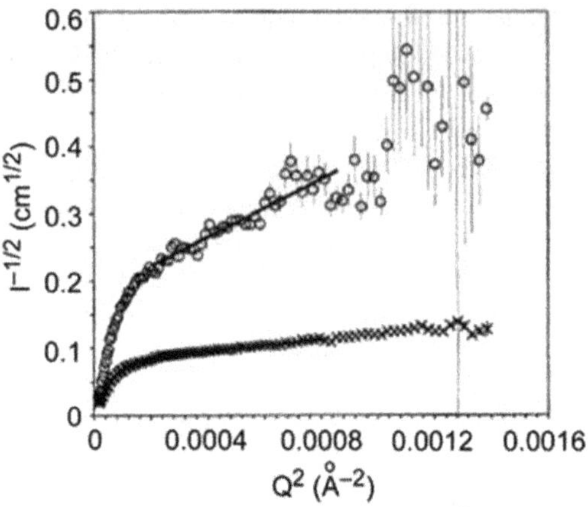

Fig. 8.2 Debye plot derived from a SANS experiment on a stack of Nanoglass™ thin films (with the silicon substrate attached). The *open circles* are for the as-received thin films and the crosses are for the films immersed in D-toluene. Reprinted with permission from [8]. Copyright [2000], AIP Publishing LLC

accessible to D-toluene is 22.1 ± 0.05 %. Application of the same two-layer model for SANS from films immersed in a non-wetting liquid (deuterated water) showed that water uptake is rather small (3.10 ± 0.05 %) at room temperature.

SANS and x-ray porosimetry were used in [5] to characterize properties of thin film low-*k* materials (methylsilsesquioxane, MSQ, and hydrogensilsesquioxane, HSQ). Utilizing the vapor adsorption is much more complicated than simple immersion of a porous sample in a liquid as it requires a setup capable of delivering vapor mixtures with a variable composition of hydrogenated and protonated molecules. Contrast variation was conducted by filling the pores with mixtures of deuterated and protonated toluene via vapor adsorption. Two streams containing saturated D- and H-toluene vapor were combined and passed through a SANS sample cell. The composition of the vapor was varied incrementally by controlling the relative flow rates of D- and H-toluene streams. For MSQ film, the scattered intensity passed through a minimum at 0.22 D-toluene/0.78 H-toluene. At this point, little coherent scattered intensity remained over all measured Q-range. The contrast match point of MSQ corresponded to SLD of $2.0 \pm 0.11 \times 10^{10}$ cm^{-2}, which translates into the matrix mass density 1.97 ± 0.11 g/cm^3.

In contrast, the scattering behavior of the HSQ dielectric film was found to be different in various Q-domains. The broad peak in the intermediate Q-region $0.03 - 0.1$ Å^{-1} disappeared near a solvent fraction 0.5 D-toluene/0.5 H-toluene, however the strong upturn at low $Q < 0.02$ Å^{-1} remained present at all solvent compositions. As was mentioned earlier, the upturn at small Q may be due to compositional inhomogeneities or closed pores. In the absence of detailed knowledge of the physical reasons for the upturn, the density of HSQ matrix may be estimated considering the data only at $Q > 0.4$ Å^{-1}, which gives a matrix SLD $= 2.9 \times 10^{10}$ cm^{-2} and a corresponding matrix mass density 2.12 g/cm^3. This value may be close to the actual average matrix mass density if the compositional inhomogeneities occupy only small volume fraction of the sample.

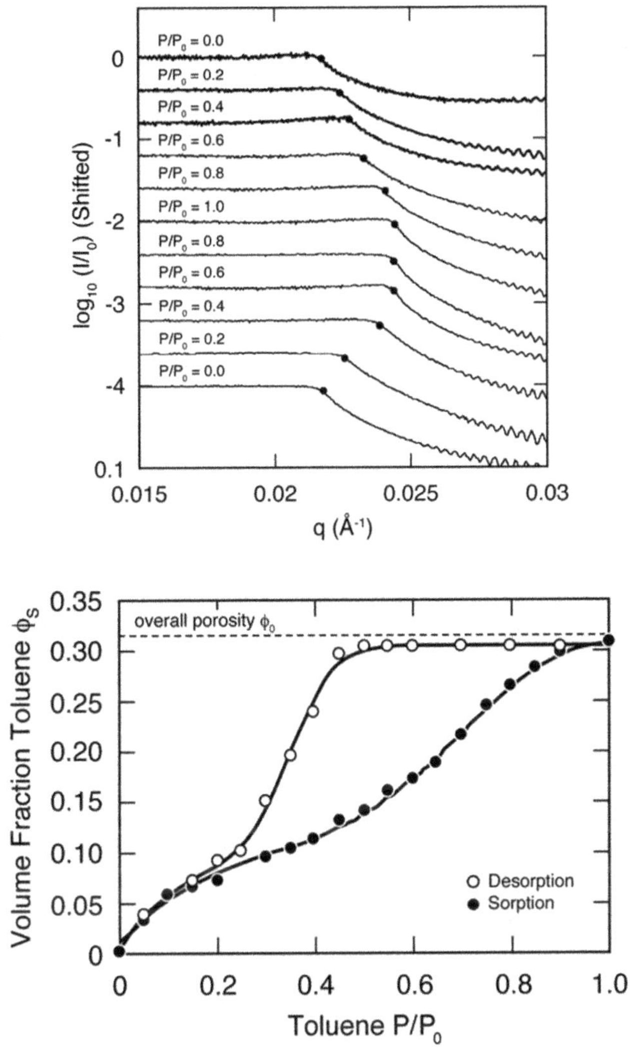

Fig. 8.3 *Top*: SXR porosimetry data for the MSQ dielectric. *Connecting lines* are shown instead of individual data points for clarity. *Black dots* represent best-fit values for Q_C. *Bottom*: Adsorption data for toluene-h_8 in the nanoporous MSQ thin film obtained by SXR. Reprinted from [5], not subject to US copyright

Figure 8.3 shows SXR data [log(reflected intensity)/incident intensity] vs. scattering vector Q for the toluene porosimetry experiment. The critical value of Q_C (the value of Q at which total external reflection of the x-rays occurs) is the point below which the total external reflection occurs and above which the x-rays penetrate the film. The measured values of Q_C give corresponding set of critical angles θ_C in Eq. (8.2) via $Q_C = (4\pi \sin\theta_C)/\lambda$ and the volume fraction of solvent at each P/P_0 is calculated from Eq. (8.3). As may be seen in Fig. 8.3, the data sets exhibit a significant hysteresis loop and are representative of "Type IV" isotherms, characteristic of most of mesoporous solids (Fig. 8.1).

Neutron porosimetry was used to study of the structure of low-*k* MSQ thin films [10]. A previously determined "contrast match" solvent mixture of 0.22 D-toluene/ 0.78 H-toluene was chosen as a "probe" solvent. The relative pressure of the solvent was varied by mixing the air and solvent vapor streams at different flow rates at constant temperature (PV porosimetry) or by changing the temperature of the saturated film (TV porosimetry). The measured cross section was fitted to Eq. (6.27). It was found that the model adequately describes the experimental data at all relative pressures and temperatures. A clear hysteresis in sorption–desorption behavior of the Debye correlation length as a function of relative pressure was observed using PV method. The behavior of correlation length measured using TV porosimetry also exhibited a hysteresis, however the absolute values of the Debye correlation length were found to be significantly different from those obtained via PV method, which indicated that the pore filling processes in PV and TV types of experiments must follow different pathways as a function of relative pressure.

The described experiments as well as the related studies [4, 11–15] demonstrate the power of neutron and x-ray porosimetry in characterizing the structure of thin films. Such information is needed by materials engineers to help understand correlations between processing conditions and the resulting physical properties and thus optimize and develop future low-*k* materials.

8.3 Vapor Adsorption in Porous Silica

8.3.1 Contrast Matching SANS

In early SANS studies, porous solids were exposed *ex situ* to a vapor at different relative pressures, sealed in quartz cells and transferred to the SANS instrument for the structural characterization. Such procedure is most adequate for systems with equilibration times long compared to the available neutron beam time. For many macro- and mesoporous materials equilibration time is short enough to allow in situ monitoring of the adsorption using scattering techniques. A series of measurements can be conducted using the same sample at both equilibrium and non-equilibrium conditions.

The controlled porous glass (CPG 10-75, Electro-Nucleonics, Fairfield, NJ) consists of porous SiO_2 grains with a size of 74–125 μm. According to the manufacturer specifications, the pore size is 77 ± 7 Å, the pore volume is 0.47 ml/g, and the surface area is 182 m^2/g. In situ SANS study of benzene adsorption in CPG 10-75 was conducted in [16]. The SLD of the solid phase is 3.59×10^{10} cm^{-2} was calculated assuming a density of silica 2.27 g/cm^3.

As adsorbate, a mixture of 59 vol.% C_6D_6 and 41 vol.% C_6H_6 was used in order to match the SLD of the SiO_2 matrix. However, complete contrast matching was not achieved, which was interpreted as due to incoherent scattering from impurities and short range density fluctuations in SiO_2 matrix.

The experiments were performed using the SANS instrument V4 at the BENSC using the apparatus SANSADSO [17] (Sect. 4.1). The neutron wavelength was $\lambda = 6$ Å and the range of scattering vectors covered in the experiments was $0.003 \leq Q \leq 0.18$ Å^{-1}. Initially, the scattering curve for the degassed and evacuated sample was measured. After that, the sample was exposed to benzene vapor and scattering was measured at constant temperature 36 ± 0.2 °C at different values of P/P_0. In addition to equilibrium measurements, few kinetic scans after step-like increase or decrease in P/P_0 were recorded.

The scattering curve of the dry sample shows two peaks: the main peak (1) observed to 0.0316 Å^{-1} and the secondary peak (2) observed at 0.008 Å^{-1}. Adsorption of one benzene monolayer at $P/P_0 = 0.26$ has almost no effect on peak (1), however it reduces the amplitude of peak (2). The condensation in the pores occurs when P/P_0 is increased from 0.74 to 0.84, which results in a factor of 15 decrease in the amplitude of peak (1). At the same time, the amplitude of peak (2) remains higher than that for the dry sample, which indicates that at $P/P_0 = 0.84$ significant volume of voids between the grains remains not completely filled with benzene.

Scattering from the dry sample in the limit of high Q is described by a surface fractal dimension $D_S = 2.3 \pm 0.01$, in a good agreement with the previously measured SAXS data ($D_S = 2.2 \pm 0.05$) [18]. Adsorption of one benzene monolayer at $P/P_0 = 0.26$ does not change the slope, which indicates that the molecular monolayer simply replicates the original surface. Significant surface smoothing is observed at $P/P_0 = 0.65$ at which $D_S = 2.08 \pm 0.02$ and finally at $P/P_0 = 0.74$ D_S reaches value of 2.02 ± 0.02, which indicates formation of the smooth liquid-like benzene film on the pore surface. The process is completely reversible, as demonstrated by desorption experiment during which removal of the adsorbate brings the fractal dimension back to the initial value $D_S = 2.28$.

Adsorption of n-pentane on MCM-41 silica was studied using SANS in [19]. The synthesized sample of MCM-41 had spacing between the pore centers of 44.2 Å estimated using X-ray and neutron diffraction. The BET measurements gave the value of specific sample area 1160 ± 19 m^2/g. A BJH pore size distribution analysis gave a narrow distribution with an average pore diameter of 24 Å. Combining these results, the total pore length was estimated to be of the order of 10^{11} m/g.

Neutron scattering experiments were carried out at the NRU center, Chalk River, using the C5 spectrometer and a home-built vacuum rack similar to that described in Sect. 4.1. The neutron wavelength was 2.4 Å and the Q range covered in the experiments was from 0.040 to 0.410 Å^{-1}. For the purpose of contrast matching, a mixture of n-pentane with isotopic ratio $C_5H_{12}/C_5D_{12} = 41.6/58.4$ was used. The contrast matched condition was calculated assuming SiO_2 density of 2.05 g/cm^3, which is by 12 % lower than the density of crystalline SiO_2 as the calcinated SiO_2

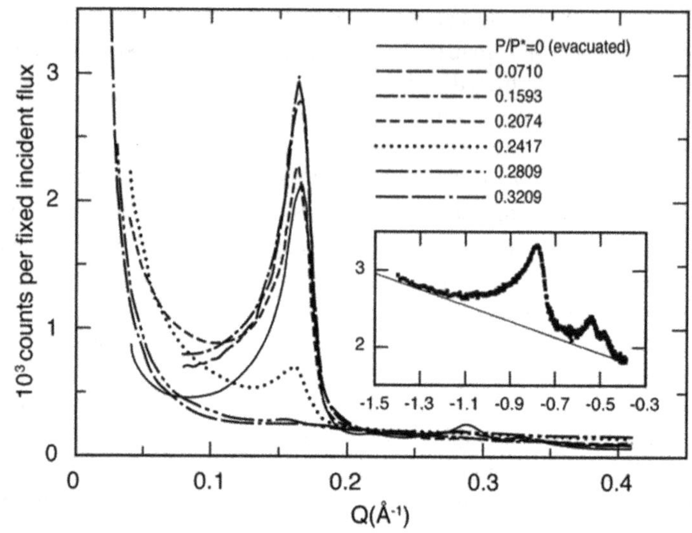

Fig. 8.4 Neutron scattering profiles obtained from MCM-41 as a function of the relative pressure. The *inset* shows the in-vacuo profile plotted on a log-log scale along with an *underlying straight line*, whose slope is -1 [19]. 2008 Canadian Science Publishing or its licensors. Reproduced with permission

skeleton of MCM-41 is expected to be an amorphous material with defects and micropores.

The intensity profiles measured as a function of relative pressure are shown in Fig. 8.4. Because of the hexagonal symmetry of the long cylindrical pores in MCM-41, the observed Bragg peaks can be indexed in terms of the 2D Miller indices (*hk*). The main peak at $Q = 0.164$ Å^{-1} is the first-order peak (10), while the broad and weaker peaks at ~0.28 and ~0.33 Å^{-1} are the higher-order peaks (11) and (20), respectively. With increasing amount of the adsorbed pentane, the amplitude of the (10) peak increases initially by ~40 % before decreasing almost to zero as the relative pressure reaches $P/P_0 = 0.28$. At the same time, the amplitude of the higher order peaks does not show an initial increase and monotonically decreases with P/P_0. The disappearance of (10) peak at $P/P_0 = 0.28$ indicates complete filling of the mesopores with the contrast matched mixture C_5H_{12}/C_5D_{12} as well as full accessibility of pores to n-pentane.

The non-monotonic behavior of the amplitude of the (10) peak was observed in a number of other SANS studies [20]. The increase of the (10) peak was documented in [21] in early stage of adsorption of contrast matched benzene in SBA-15 silica. The authors of [21] ascribed this effect to thickening of the effective pore walls. Another experimental observation of the initial increase of the (10) peak was reported in [22]. In this work, two mesoporous silicas designated as SE1010 and SE3030 were investigated by *in situ* adsorption of nitrogen with SANS. At high relative pressures, the amplitude of all peaks was practically annulled due to contrast matching between condensed nitrogen and solid SiO_2 (Sect. 8.5).

According to [22], the non-monotonic variation of the (10) peak amplitude was due to filling of micropores of the host material leading to the average contrast enhancement between the host and the remaining empty pores. Similar explanation of this phenomenon was also given in other papers describing SANS studies of the vapor adsorption in low-dielectric-constant and silica films [5, 15]. Such explanation, however, seems to have limited applicability to MCM-41, as a higher contrast would result in the amplitude enhancement of all peaks, whereas in [21] only the peak (10) showed the initial increase. The question if the non-monotonic behavior of the (10) peak is a universal phenomenon for a certain class of materials remains open for discussion. Evidently, finding a definitive answer will require systematic SANS studies of vapor adsorption in porous materials of different composition and pore geometry.

It was suggested recently that pores may experience a measurable deformation during the adsorption and condensation of fluids due to capillary pressure [23–26]. The hypothesis of sorption strain-induced long-range correlations during pore filling was tested using in situ SANS by studying capillary condensation and emptying of water and perfluoropentane (C_5F_{12}) in SBA-15 [27]. The BET specific surface area of the granulated SBA-15 sample was 826 m^2/g and the total specific pore volume 1.134 cm^3/g. In the liquid state, perfluoropentane and an H_2O/D_2O mixture with 62 % D_2O match the SLD of the silica matrix so that a two-phase model applies for the completely and partially filled pores upon condensation. Moreover, the two fluids exhibit a considerable difference in their surface tension and saturation vapor pressure, and therefore are characterized by different capillary pressures.

SANS experiments were conducted at the V4 instrument at the BER II reactor in Berlin. The wavelength of neutrons was $\lambda = 6$ Å with $\Delta\lambda/\lambda = 12$ % and the covered Q range was $0.003 < Q < 0.23$ Å^{-1}. The sample was sealed between the quartz windows and the *in situ* gas adsorption measurements were conducted in parallel with SANS at select equilibrium points along the adsorption/desorption isotherm. The total scattering intensity can be separated in three parts: grain scattering, which follows roughly the Q^{-4} behavior; Brag diffraction peaks at $Q = 0.063$, 0.109, and 0.166 Å^{-1}, corresponding to the (10), (11), and (21) reflections of the 2D hexagonal lattice, respectively, and diffuse small-angle scattering arising from the correlations between filled and unfilled pore domains.

The obtained SANS data were fitted to analytical model for spatially random (not correlated) pore filling described by equation that takes into account contribution from all three types of scattering [27]:

$$I(Q,\gamma) = K\gamma(1-\gamma)\frac{|F(Q)|^2}{Q}\left[1+\frac{(1-\gamma)}{\gamma}\frac{S(Q)}{Q}\right] \tag{8.4}$$

where K is a constant factor, γ is the pore filling fraction, $|F(Q)|^2/Q$ is the spherically averaged form factor of a single cylindrical pore, and $S(Q)/Q^2$ is the spherically averaged structure factor of the pore lattice. Experimental data taken at all relative pressures showed excellent agreement with the model. This suggests that pore-wall deformations due to capillary pressure do not play a significant role in the processes of pore condensation or evaporation in SBA-15. The adsorption branch of water is very similar to that of C_5F_{12}, whereas the desorption branches of both fluids are very different. The equilibrium film thickness for perfluoropentane is considerably thicker than for water, which is reflected by the much narrower hysteresis loop for C_5F_{12} adsorption isotherm. The observed different behavior of water and C_5F_{12} upon desorption can be tentatively associated with the differences in the molecule-molecule and molecule-surface interactions of both fluids.

In many neutron porosimetry studies, nitrogen was used as a contrast matched fluid for the structural characterization of silica. Use of nitrogen in conjunction with SANS is based on the fact that the SLD of liquid nitrogen at $T \cong 78$ K ($\rho^*_{LiqN2} = 3.22 \times 10^{10}$ cm^{-2}) is very close to that of amorphous silica ($\rho^*_{SiO2} = 3.47 \times 10^{10}$ cm^{-2}, Table 1.2). Hence, the scattering contrast between liquid nitrogen-filled pores and SiO_2 matrix is close to zero and this fluid can be used for contrast matching SANS similarly to the vapor or liquid isotopic mixtures. Use of nitrogen as a contrast matched fluid provides additional benefits due to much shorter equilibration times and lower incoherent background.

Complete filling with nitrogen at 77.7 K of pores in CPG 10-75 [28, 29], Gelsil 50 [30], Gelsil 75 [31], and SBA-15 [32] shows that the ratio of the intensity measured from the samples after complete pore filling ($I(Q)_{filled}$) and the intensity ($I(Q)_{VAC}$) from evacuated samples $I(Q)_{filled}/I(Q)_{VAC} = (1 - \rho^*_{LiqN2}/\rho^*_{SiO2})$ is close to ~0.02 for Gelsil 50 and ~0.01 for CPG 10-75, Gelsil 75, and SBA-15. The actual ratio $I(Q)_{filled}/I(Q)_{VAC}$ may be even lower if closed pores, inaccessible to liquid nitrogen are present in the studied samples. This shows that the contrast matching condition $(1 - \rho^*_{LiqN2}/\rho^*_{SiO2}) \ll 1$ is nearly met for capillary-condensed nitrogen in silica. Thus, assuming that the density of the adsorbed film formed on pore walls is similar to that of bulk liquid nitrogen, SANS from completely or partially filled pores can be approximated by the two-phase model: the solid silica matrix with liquid nitrogen (phase 1), and the liquid-free space (phase 2) [33].

Nitrogen adsorption in silica was studied using SANS in [22]. Predominantly mesoporous granulated silica SE3030 had average pore size of 95 Å, and relatively low pore size polydispersity, thus ensuring capillary condensation in mesopores as a single adsorption process. In addition, the sample had limited microporosity, which offered an opportunity to study the different subsequent stages of sorption (micropore filling, formation of nitrogen layers on mesopore walls, and capillary condensation in mesopores) in a single sample. Another sample studied in this work was a silica sample representing a random porous network synthesized using sol–gel method with polyethylene oxide (PEO) as a molecular template. This sample had a broad pore size distribution ranging between micro- and mesopores up to ~50 Å. It was chosen in order to explore the differences of the adsorption in samples with relatively monodisperse and polydisperse pore size distribution.

SANS experiments were conducted using the V4- SANS instrument at BENSC using the SANSADSO apparatus [17] (Sect. 4.1). SANS patterns were measured at T = 78 K in the range of scattering vectors $0.005 < Q < 0.27$ Å^{-1}. SANS data

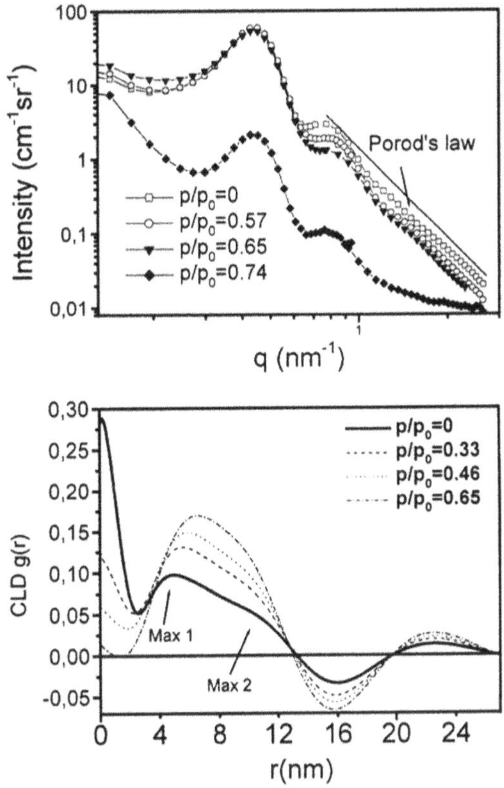

Fig. 8.5 *Top*: SANS curves of SE3030 silica recorded at different relative pressures P/P_0 during sorption of nitrogen. *Bottom*: chord length distribution *g(r)* obtained from SANS data of SE3030 silica for each relative pressure during nitrogen sorption. The changes of the slope of *g(r)* at small chords are caused by filling of micropores and smoothing of the surface roughness. The shift of the first maximum at $r \cong 50$ Å (Max 1) to larger r can be attributed to the formation of nitrogen layers on the mesopore walls. The broader shoulder at $r \cong 100$ Å (Max 2) corresponds to the pore size. Reprinted with permission from [22]. Copyright 2001 American Chemical Society

obtained from SE3030 are shown in Fig. 8.5. A pronounced interference peak is observed at $Q = 0.08$ Å^{-1}, and an extended range of Porod scattering is observed at $Q > 0.01$ Å^{-1}. After condensation in mesopores, the peak intensity drops by almost two orders of magnitude, confirming that the contrast matched condition was achieved. SANS curves obtained from the disordered porous PEO silica look qualitatively different (Fig. 8.6). There is no interference peak in *I(Q)*, which indicates randomness of the sample structure. Similarly to SE3030, *I(Q)* drops significantly after the capillary condensation point is reached at $P/P_0 \cong 0.45$.

A significant increase of *I(Q)* was observed for both SE3030 and porous PEO silica samples at small P/P_0 in the low Q region $Q < 0.13$ Å^{-1}. This effect was interpreted as due to filling of micropores and corresponding enhancement of the

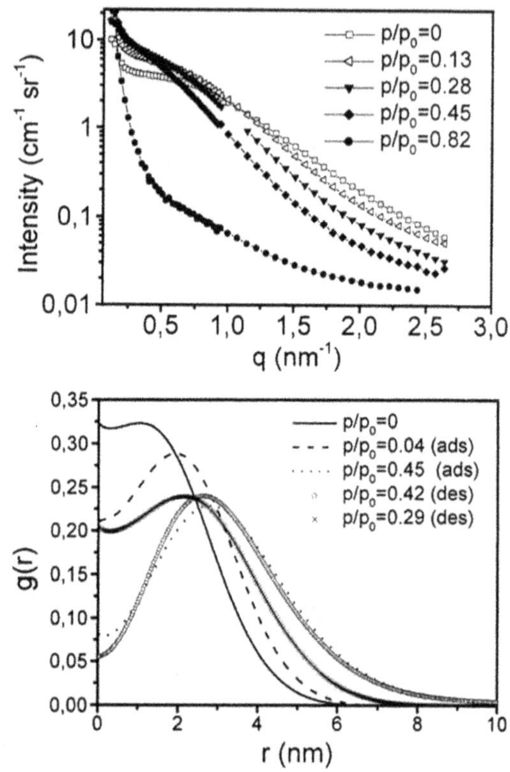

Fig. 8.6 *Top*: SANS curves of PEO silica measured at different relative pressures P/P_0 during sorption of nitrogen. *Bottom*: chord length distributions calculated from SANS data at different P/P_0 during adsorption and desorption for PEO silica. The graphs show the filling of micropores and smoothing out of angular structures during the sorption of nitrogen. The shift of the maximum indicates an increase of the average size of the unfilled pores. Reprinted with permission from [22]. Copyright 2001 American Chemical Society

neutron contrast between the SLD of still empty pores and the SLD of the wall material impregnated with contrast matched fluid. For the same reason, in the high Q domain, $I(Q)$ decreased simultaneously due to the reduction of the neutron contrast between micropores filled with liquid-like nitrogen and the SLD of the solid (non-porous) silica matrix.

The conducted analysis of the chord length distribution (CLD, Sect. 6.6) in SE3030 silica as a function of P/P_0 illustrates the different processes occurring during nitrogen sorption (Fig. 8.5). The behavior of the CLD $g(r)$ at small r is determined by the structure of the interface and micropores. Small structural elements disappear rapidly already at small P/P_0, which reflects smoothing of the wetted interface and the filling of micropores, with the estimated sizes $R < 15$ Å. The two maxima observed at $r \cong 50$ Å (Max 1) and 95 Å (Max 2) were related to the wall thickness and the mean pore size. The observed shift of the first maximum with P/P_0 corresponds to the increase of the average wall thickness, which is the sum of the thickness of the silica walls, and the thickness of the adsorbed nitrogen layer, which gradually increases from ~3 Å at $P/P_0 = 0.33$ to ~10 Å at $P/P_0 = 0.65$.

A similar analysis was performed for the nitrogen adsorption in PEO silica (Fig. 8.6) characterized by a disordered system of pores. In this case, the CLDs were also calculated for the desorption branch. A maximum, which corresponds to the superposition of the chord distributions of walls and pores, is observed in all calculated CLDs. In agreement with SE3030 silica, the CLDs show filling of micropores at low P/P_0. The shift of the maximum to higher r is related to the gradual filling of smaller pores and the corresponding increase of the average size of the remaining empty pores. As nitrogen is partially removed from the porous network during desorption, the mean pore size of empty pores decreases gradually. These observations are in a reasonable agreement with the Kelvin Eq. (6.1).

Neutron porosimetry using nitrogen adsorption provides the unique possibility to obtain structural characteristics, which are hardly accessible by other analytical techniques. In particular, it allows to study all stages of the fluid adsorption in porous media: micropore filling, formation of adsorbed layers, and capillary condensation in mesopores [34–38].

8.3.2 Synchrotron SAXS

Synchrotron SAXS is a powerful tool for studying vapor adsorption in porous systems. Structural information about the adsorbed film can be derived from the intensities of the diffraction peaks, which result from the regular packing of the pores. The accuracy of film thickness determination depends on the number of diffraction peaks that can be resolved. In this respect, use of x rays is beneficial due to a higher flux and energy/wavelength resolution compare to neutrons, allowing a better resolution of the individual peaks.

Adsorption and capillary condensation of perfluoropentane (C_5F_{12}) in periodic mesoporous silica SBA-15 was studied in [39] by in situ synchrotron diffraction. The specific surface area of the sample was 785 m^2/g and the total specific pore volume was 1.033 cm^3/g. The radius of the cylindrical pores $R = 41$ Å was determined from the desorption branch of the hysteresis loop by the nonlocal DFT. For in situ small-angle synchrotron x-ray diffraction in a tandem with adsorption, a special cell was designed. SAXS measurements were performed at the beamline A2 at Deutsches Elektronen-Synchrotron (DESY) facility in Hamburg, Germany. Synchrotron radiation was monochromatized at energy 8.27 keV ($\Delta E/E = 10^{-3}$) and the Q range covered in the experiments was $0.03 < Q < 0.24$ Å^{-1}.

A series of SAXS profiles of SBA-15 measured at various relative pressures of C_5F_{12} are shown in Fig. 8.7. Five diffraction peaks corresponding to the (10), (11), (20), (21), and (30) diffractions from the 2D hexagonal lattice could be resolved for the evacuated sample. After reaching the pore condensation point, the amplitude of the peaks was strongly reduced. In addition to the diffraction peaks, a diffuse small-angle scattering was observed, which also changed with the relative pressure. This scattering signal was interpreted as due to weakly ordered heterogeneities in the SBA-15 matrix, which are also affected by adsorption.

The integrated intensities of the high-resolution data obtained were analyzed using a structural model involving four levels of electron density (solid silica

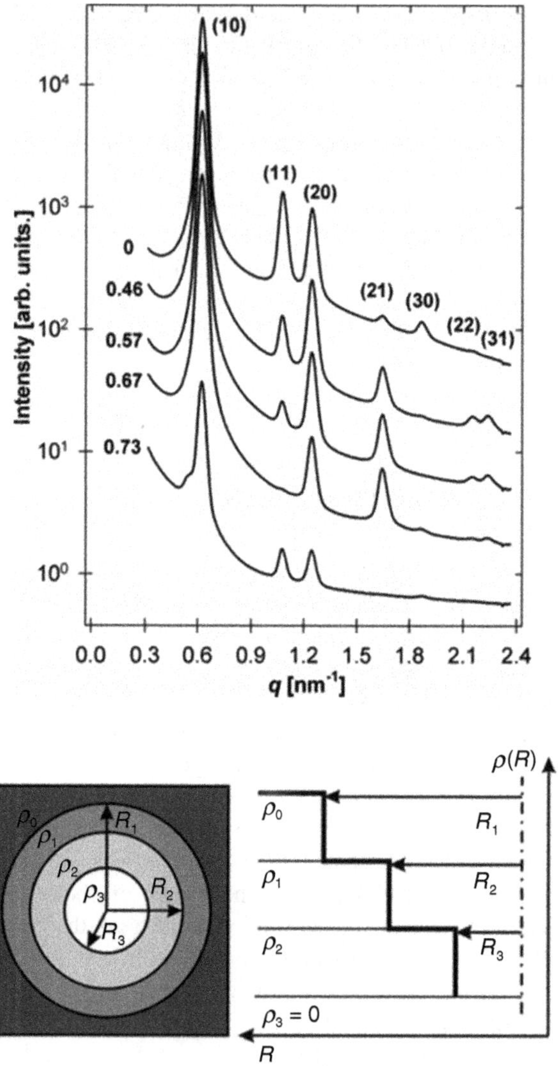

Fig. 8.7 *Top*: SAXS profiles during adsorption of C_5F_{12} in SBA-15 at different relative pressures shown in the *inset*. For the sake of clarity, the curves are shifted vertically by a factor of three with respect to each other. The four top profiles were measured at pressures below the pore condensation point, and the remaining profile was measured above the condensation point. *Bottom*: model of the pore structure of SBA-15 with four density levels: silica matrix (electron density ρ_0), microporous corona (electron density ρ_1), adsorbed liquid film (electron density ρ_2), and vapor in the core of the pore (electron density ρ_3) [39]

matrix, microporous corona around the mesopores, adsorbed film, and inner pore space (Fig. 8.7). The model provided quantitative information about the structure of the evacuated sample, the filling of the corona, and the growing thickness of the adsorbed liquid film with increasing relative pressure. Fits obtained from evacuated sample confirmed results of the previous x-ray diffraction studies [40] in that cylindrical mesopores of SBA-15 are surrounded by a corona with relatively uniform density (Sect. 7.2.4). The determined values of $R_1 = 51.5 \pm 0.15$ Å and $R_2 = 39.5 \pm 0.35$ Å (Fig. 8.7) yielded a total wall thickness of 37.4 Å and a corona thickness of ~12 Å, which implies that the micropore volume of the corona represents ~22 % of the total pore volume. The results showed that the processes of corona filling and the film growth on the pore walls occurred simultaneously in a wide range of pressures. Such behavior is expected for pore walls with highly corrugated surface due to the simultaneous smoothening of the surface and the growth of the thickness of the adsorbed film with pressure. The step-density model used in [40] for the data analysis can be extended to account for the possible finite density gradients in the adsorbed films [41–43].

8.4 Carbonaceous Materials

Access of liquids and gases to pores in porous carbons and the development of porosity during activation or gasification are critical in the selection and or production of these materials. In this context, the existing pores may be divided into two classes: interconnected pores that have access to the exterior of the particles, and pores that are disconnected from the surface and are effectively inaccessible to fluid transport from the exterior [44]. During the activation or gasification process, the volume of accessible pores may increase, previously inaccessible porosity may be opened, and new porosity may be developed. Reduction of the number of closed pores is generally desirable as it allows improving the material performance.

Contrast-matching SANS (CM-SANS) and SAXS is one of the few techniques, which can be used for non-invasive direct investigation and evaluation of the closed porosity in a variety of materials. The technique consists of performing SAS on dry samples, where the total scattering comes from all available pores, including both accessible (open) and inaccessible (closed) pores. The pores accessible from the exterior of the sample are filled with a contrast matched liquid solution and thus scattering from open porosity is eliminated. Consequently, scattering from contrast matched samples yields information about closed porosity in the materials. The difference between the scattering of dry and contrast matched sample gives information about the open porosity. Contrast matching pores with liquids was used for investigation of both classes of porosity in different materials including polymer resins [45], char [46], porous graphitic materials [47, 48], activated carbons [49, 50], and coal [51–56]. Contrast matching of open pores can be also conducted using pressurized gases and supercritical fluids [57]. Examples of such CM-SANS applications are given in Sect. 10.7.

Information on the filling degree of sub-nanometer carbon pores with water is of interest to various applications, such as supercapacitors and water desalinization. Filling degree and accessibility of pores to D_2O in carbide-derived carbons was investigated in [58]. TiC-CDC powder was produced by chlorination of TiC powder (Alfa Aesar, Ward Hill, USA) with an average particle size of about 2 μm. This precursor was placed in a furnace, purged in argon, and heated to the synthesis temperatures (300, 400, 600, and 800 °C) under flowing chlorine. The samples are referred to as sample 300, 400, 600, and 800, respectively, representing the synthesis temperature.

Pure D_2O was drawn into CDC pores by exposing CDC powders to D_2O vapor in a closed oven at 80 °C for 24 h. Hydration levels were calculated from the difference in weight of dry and wet CDC powders. It shows that the intake of D_2O in 1 g of CDC-300, CDC-400, CDC-600 and CDC-800 powders was 0.32, 0.34, 0.53 and 0.61 g, respectively. This corresponds to the larger pore volume of CDC obtained at higher chlorination temperatures. The wet CDC powders were sealed into quartz cells with internal thickness 1 mm.

SANS experiments were conducted at ORNL on the General Purpose SANS instrument with a neutron wavelength of $\lambda = 4.8$ Å ($\Delta\lambda/\lambda \sim 0.13$). The sample-to-detector distance was chosen to cover an overall range of scattering vectors $0.01 < Q < 0.8$ Å^{-1}. Average acquisition time for each scattering curve was ~20 min. Figure 8.8 shows the background corrected scattering data from dry and

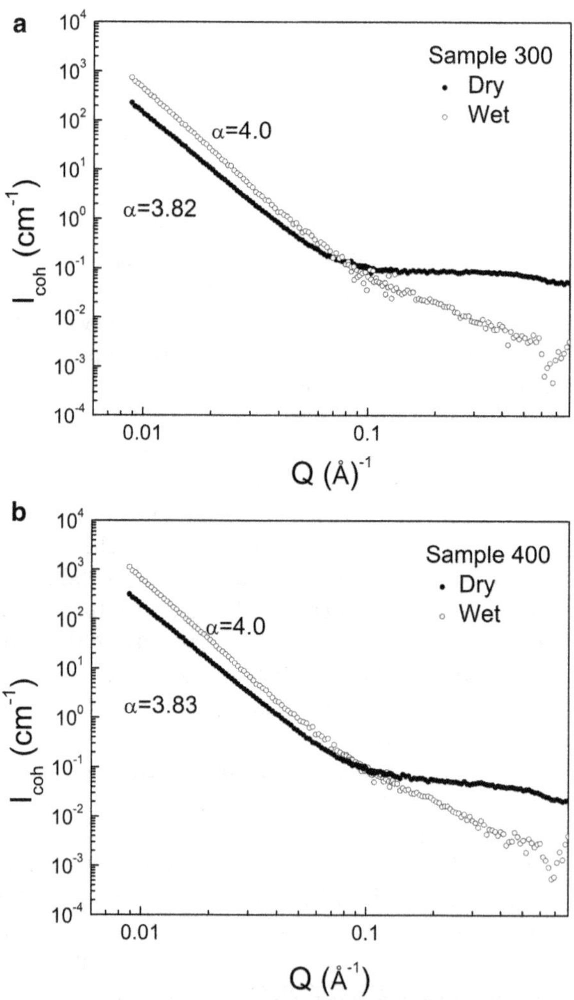

Fig. 8.8 Background subtracted SANS cross section *I(Q)* of the dry (*filled circles*) and D_2O saturated carbons (*open circles*). Exponents of the power-law scattering are shown in the *insets*. Reprinted from [58]. Copyright (2012), with permission from Elsevier

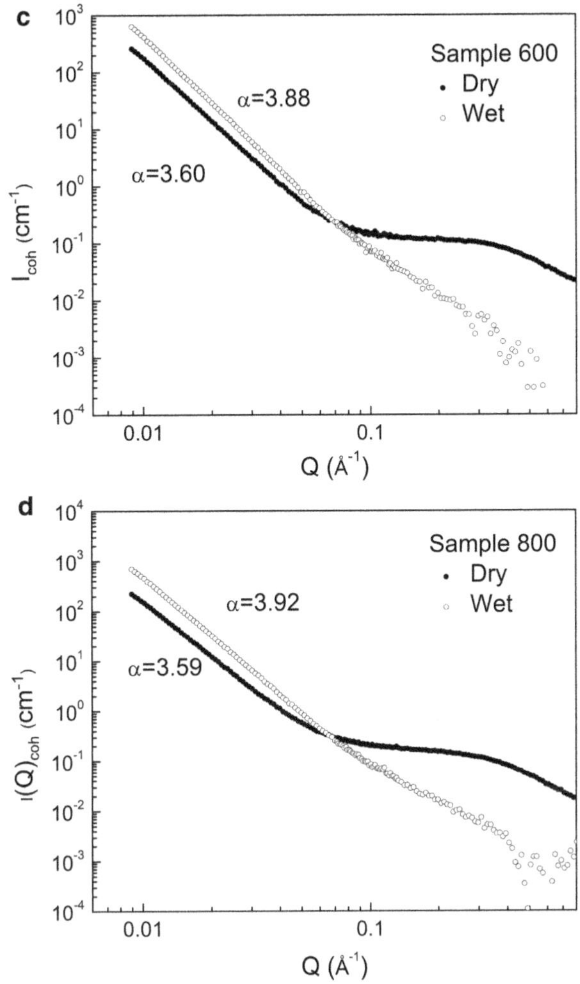

Fig. 8.8 (continued)

D_2O saturated (wet) carbons. As may be seen in Fig. 8.8, saturation of carbons with D_2O leads to two major effects for all samples. On the one hand, the intensity of scattering slightly *increases* in the low Q domain $Q < 0.07$ Å^{-1}, and the exponent values in $I(Q) \sim Q^{-\alpha}$ increase from $3.59 \leq \alpha \leq 3.86$ to $3.88 \leq \alpha \leq 4.0$. On the other hand, the intensity of scattering from pores for $Q > 0.07$ Å^{-1} *decreases* by a factor of ~25–30 for samples 300 and 400, respectively and by a factor of ~50 for samples 600 and 800.

The power law scattering from porous carbons at small values of Q may be due to scattering from either internal surface of macropores or from the external surface of porous carbon particles. Wetting of carbon samples with D_2O delineates the source of this scattering. The SLD of D_2O is 6.4×10^{10} cm^{-2}, which is close to that of carbon (6×10^{10} cm^{-2}, Table 1.2). Thus, saturation of pores with D_2O should reduce significantly neutron contrast between carbon matrix and pores and *decrease* the intensity of scattering from both macro- and micropores by a factor of $(6)^2/(6-6.4)^2 = 225$. Hence, the observed *increase* in *I(Q)* in the low-*Q* domain (<0.07 Å^{-1}) indicates that this scattering signal is coming from the outer and relatively rough surface of carbon particles. The increased values of α after wetting (Fig. 8.8) can be interpreted as due the formation of the D_2O film, which smoothens the surface roughness leading to the values of α, which approached the Porod law exponent $\alpha = 4$. Saturation of meso- and micropores in carbon matrix with a fluid that has much higher SLD compare to that of air leads to the increase of the average SLD of the porous matrix and corresponding enhancement of the total scattering signal at $Q < 0.07$ Å^{-1}.

Compared to the low-*Q* range, at values of *Q* larger than ~0.07 Å^{-1} the intensity of D_2O saturated samples is lower than *I(Q)* from dry CDC. The drop of *I(Q)* from micropores is less than expected from completely water-filled pores based on the neutron contrast calculations. The observed relatively moderate drop in the intensity may be attributed to the partial filling of small pores due to their limited accessibility and or due to their relative hydrophobicity. We note that bigger pores in carbons prepared at higher chlorination temperatures (samples 600 and 800) show about a factor of two higher accessibility to D_2O molecules than samples 300 and 400, which is related to the larger pore diameter in samples 600 and 800 (Fig. 7.14, [58]).

Qualitatively similar results were obtained for *liquid* water saturated microporous activated carbon fibers (ACF), which represents a binder-free conductive porous carbon used as electrodes in double layer capacitors [59]. Commercially available activated carbon fibers (ACC-507-20) were purchased from Kynol (USA). SANS experiments were conducted at ORNL General Purpose SANS instrument with a neutron wavelength of $\lambda = 4.75$ Å with a wavelength spread, $\Delta\lambda/\lambda$ of 0.13. Two sample-to-detector distances, 10 and 0.3 m, were used to cover a *Q* range between $Q = 0.008$ Å^{-1}, and $Q = 0.8$ Å^{-1}. Average acquisition time for each scattering curve was approximately 60 min. The measurements were conducted at room temperature. SANS patterns were recorded from dry ACC electrodes, and ACF immersed and kept in ultra-pure H_2O or D_2O for more than 24 h.

SANS patterns measured from dry ACF and ACF infiltrated with H_2O and D_2O are shown in Fig. 8.9a. A negative SLD of H_2O (-0.56×10^{10} cm^{-2}) increases the neutron contrast between carbon and H_2O saturated pores, leading to stronger scattering from H_2O-saturated ACF. At the same time, a positive SLD of D_2O (6.4×10^{10} cm^{-2}) being close to that of carbon (4×10^{10} cm^{-2}) leads to a weaker scattering from D_2O-saturated ACF (Fig. 8.9a). By deconvolution of the total scattering patterns into the scattering from the fiber outer surfaces, scattering from pores and incoherent background it was possible to identify the degree of pore filling over a broad pore size range (Fig. 8.9b).

The saturation of the ACF pores with H_2O leads to a parallel shift of the ACF scattering curve at all *Q* values with a factor of about 1.3, which is a ratio of the

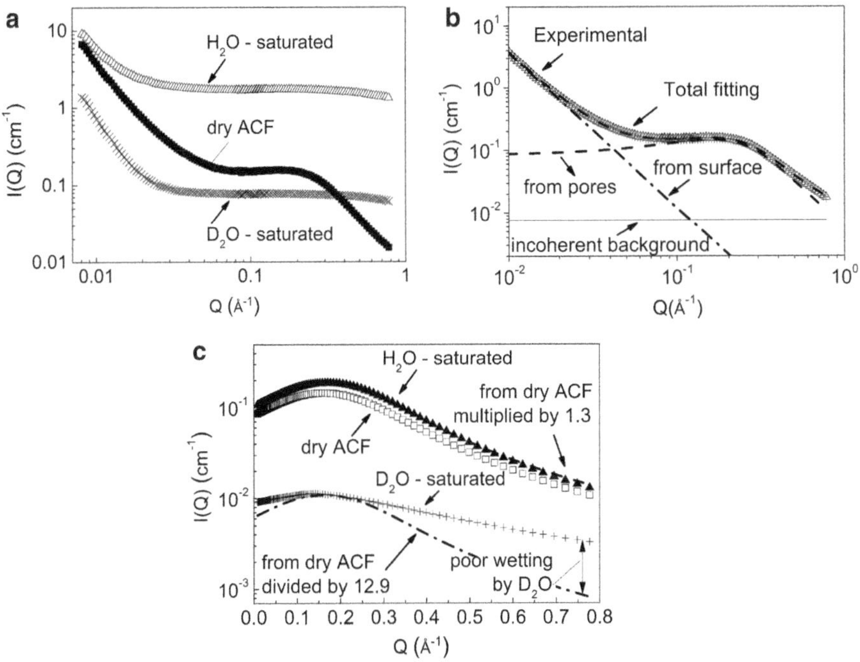

Fig. 8.9 Neutron scattering experiments on dry ACF electrodes as well as on ACF immersed into H_2O and D_2O: (**a**) SANS patterns; (**b**) deconvolution of the total scattering from dry ACF into scattering from the incoherent background, fiber surface, and pores; (**c**) background subtracted neutron scattering from meso- and micropores in the studied samples. *I(Q)* for H_2O saturated sample is shifted up parallel to *I(Q)* for dry ACF due to an increased contrast between H_2O-filled pores and carbon matrix. Parallel shift implies that all pores are accessible to H_2O molecules. *Dash-dotted line* would correspond to the experimentally measured *I(Q)* for D_2O saturated sample if all pores were completely accessible to D_2O [59]. Angewandte Chemie by Gesellschaft Deutscher Chemiker Reproduced with permission of WILEY–VCH Verlag GMBH & CO. KGAA in the format Republish in a book via Copyright Clearance Center

scattering intensities from H_2O-saturated and dry ACF at the value of the scattering vector corresponding to the maximum scattering intensity (Fig. 8.9c). This indicates that all the pores are evenly filled with H_2O. In contrast, saturation with D_2O leads to a Q-dependent decrease in the scattering intensity, indicating that a significant portion of the smallest pores ($Q > 0.3$ Å^{-1}) is not completely filled with D_2O. Vapor adsorption isotherms collected at 293 K in a separate experiments further confirmed the SANS results. H_2O adsorption was observed at slightly lower relative pressures, indicating stronger interactions with carbon pore walls. More impressively, nearly 250 cm^3 g^{-1} less D_2O was adsorbed on the ACF surface at the relative pressure of 0.8, indicating incomplete filling of pores by D_2O, as suggested by Fig. 8.9c. The observed differences in the filling degree helped to understand the dramatic discrepancy between the electroadsorption of D_2O- and H_2O-based electrolyte ions in the smallest carbon nanopores revealed in [59].

Recent SAXS studies of the adsorption of water vapor on highly activated microporous carbons revealed physical mechanisms responsible for the degree of pore filling in these materials [60, 61]. Scattering experiments demonstrated that the adsorption process as well as the pore filling degree depends strongly on the surface pre-treatment. Lightly oxidized carbon samples display formation of a significant amount of water clusters in the macroporous region and a small degree of filling ~10 %. At the same time, in the highly oxidized samples cluster formation is limited and the pore filling degree can reach ~70 %. The observed differences were attributed to the blockage of micropores due to the formation of water clusters around functional groups at the pore entrances. These experiments underline the importance of the surface chemistry that can change dramatically the affinity between the carbon surface and the fluid phase due to introduction of heteroatoms, such as oxygen, which has trend to bond along the edges of the graphene layers.

8.5 Kinetics of Sorption and Desorption

Information about the rate at which porous media absorb or release fluids is of great importance to a variety of applications including heterogeneous catalysis, gas storage, CO_2 capture and sequestration, as well as the enhanced oil and coalbed methane recovery. In some cases, the equilibration time of the adsorption kinetics is long (of the order of weeks [61]), and the kinetics can be studied by the conventional macroscopic (gravimetric and volumetric) methods. In many instances, however, the equilibration time of the pore saturation or release processes is fast enough to prevent obtaining accurate experimental information using the conventional techniques. On the other hand, microscopic methods such as NMR and Quasi-elastic neutron scattering (QENS) can provide information about adsorbate diffusivity on a very short length scale, from few Angstrom to few nanometers. As soon as the diffusion coefficient is the observation-time-dependent parameter, the rate of kinetic processes occurring on the length scales of millimeters and centimeters cannot be extrapolated from the diffusivities measured on the length scales accessible to NMR and QENS.

Advent of the high-flux nuclear reactors and especially synchrotron sources has offered new opportunities in applying SAS for studying the kinetics of fluid sorption, condensation and drainage in porous media as the acquisition time for SANS (of the order of minuets [62]) and for synchrotron SAXS (of the order of seconds [63]) is short enough to enable measurements of the evolution of scattering patterns as a function of time for relatively fast processes [16, 64]. In addition, due to the capability of SAS techniques to simultaneously register scattering from pores of different sizes, they allow to study separately the kinetics of different sorption processes such as micropore filling, surface adsorption, and capillary condensation.

8.5.1 Dynamic Micromapping of CO_2 Sorption in Coal

Synchrotron SAXS in a tandem with x-ray transmission radiography was used in [65] to study the kinetics of CO_2 sorption into coal. Coal sample was from the Bowen Basin in Australia, retrieved from 1035 m depth (Baralaba). SAXS experiments were performed on the ChemMatCARS beam line, Advanced Photon Source, Argonne National Laboratory, USA. The x-ray photon energy was tuned to 12.4 keV ($\lambda = 1.00$ Å) and the beam footprint on the sample was 0.2×0.2 mm^2. The Baralaba coal sample (0.7 mm thick, 13 mm diameter) was mounted in a SAXS pressure cell (Sect. 4.3, Fig. 4.6), and an x-ray radiography transmission map and SAXS patterns were obtained for the central region of the sample. The cell was then pressurized with CO_2 up to 50 bar and depressurized back down to 2 bar at 35 °C over a period of about 18 h. The CO_2 pressure was changed stepwise and allowed to equilibrate for 40–80 min after each step, during which time SAXS data were continuously recorded in the Q-range corresponding to the micropore and lower mesopore region (pore diameters 5–176 Å). The SAXS acquisition time was 5 s at each of five locations, completing the cycle of five sample movements and five SAXS data acquisitions in about 60 s. The 2D SAXS spectra were azimuthally averaged to produce SAXS profiles (absolute intensity versus Q).

Figure 8.10 presents the x-ray radiography (relative transmission) map for the Baralaba sample. The variation of transmission across the sample is caused by spatial variability of the chemical composition, attributed to varying mineral matter (MM) content. In this sample, which was cut with its plane perpendicular to the bedding plane, horizontal depositional layers are clearly discernible. To account for this inhomogeneity, the SAXS measurements were performed at several fixed

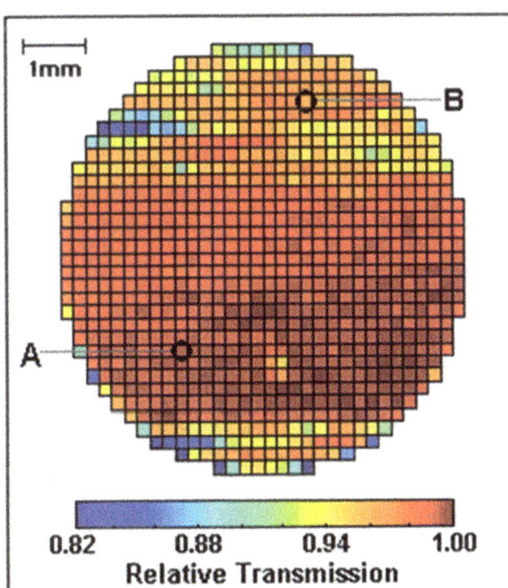

Fig. 8.10 12.4-keV radiography image for the Baralaba coal. X-ray transmission decreases in regions of high mineral matter content owing to the generally heavier elements introduced Reprinted with permission from [65]. Copyright 2009 American Chemical Society

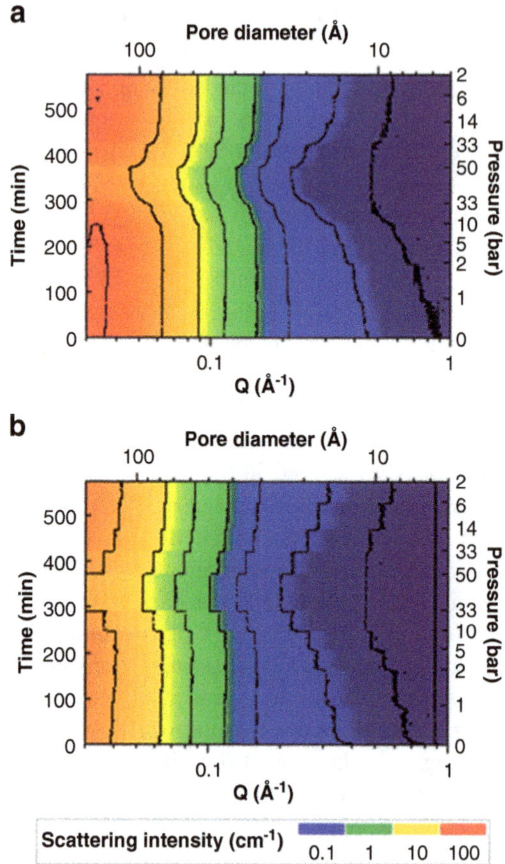

Fig. 8.11 Evolution of SAXS intensity profiles versus Q with increasing CO_2 pressure acquired at location A, with low mineral matter content (**a**), and location B, with higher mineral matter content (**b**). Reprinted with permission from [65]. Copyright 2009 American Chemical Society

locations. Location A had low MM content relative to the average for the sample (1.1 vol.%) indicated by the relatively high transmission, whereas location B had much higher MM content.

Figure 8.11 shows the time series of 575 SAXS intensity profiles, acquired in the Q range from 0.03 to 1.0 Å^{-1} at locations A and B for the Baralaba coal during stepwise pressurization from vacuum to 50 bar CO_2 pressure followed by depressurization. Such a graphical representation can be considered a SAXS signature of the sorption process in the micro- and meso-pore regions for the illuminated $0.2 \times 0.2 \times 0.7$ mm^3 volume element. In particular, a vertical section (constant Q value) provides information about the kinetics of CO_2 sorption into pores of a corresponding size. More specifically, the temporal evolution of the pore size specific CO_2 fluid density in confinement can be calculated using Eq. (1.16).

Figure 8.12 shows the CO_2 diffusion half-times determined for location A for various pore sizes at several CO_2 pressures during adsorption. The kinetics results are consistent with a coal sorption model [66], according to which the diffusion process inside the macropores has a volume and a surface component, the latter

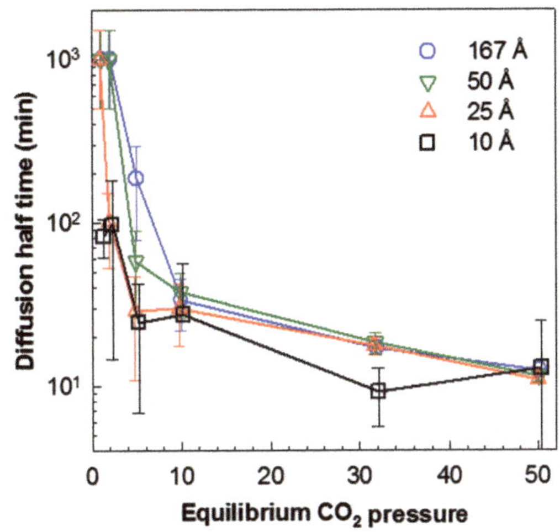

Fig. 8.12 Diffusion half times of CO_2 (CO_2 density) versus final CO_2 pressure load for Baralaba coal at location A (low mineral matter content region) for pores of specified diameter. Reprinted with permission from [65]. Copyright 2009 American Chemical Society

being pressure-dependent owing to the increasing molecular CO_2 coverage of the internal coal-pore interface with increasing pressure of CO_2. The apparent diffusion coefficient in such a two-tiered system is $D_{\text{app}} = C \times R^2/t_{½}$, where C is a constant of the order of 1–10, R is half of the smallest dimension of the adsorbing sample volume and $t_{½}$ is the experimental diffusion half-time.

Three important inferences can be drawn from the SAXS kinetics data. Firstly, the fact that the sorption process is much faster at location B than at location A suggests that mineral matter dispersed in the coal matrix provides conduits for the movement of fluid towards the adsorption sites located inside the coal matrix. Secondly, for both MM-rich and MM-poor regions, the sorption kinetics accelerates as the CO_2 pressure increases, as can be seen both indirectly, from the evolution of the iso-intensity profiles in Fig. 8.11 in response to a step change in the CO_2 pressure, and directly, from the variation of diffusion half times with pressure (Fig. 8.12). Thirdly, adsorption of CO_2 by coal is pore-size-specific: this remarkable phenomenon is evident in Fig. 8.12: on initial exposure to carbon dioxide (corresponding to the smallest pressure of CO_2) the smallest pores are flooded preferentially, followed by successively larger pores, which is similar to capillary condensation.

Based on these results, the Baralaba coal sample can be pictured as an interconnected fractal porous system with a distributed molecular pumps preferentially loading the smallest micropores. Invading CO_2 molecules initially diffuse along the high-permeability network of macropores with relatively small adsorbate capacity and are sorbed into the smallest micropores first. Only then are larger pores invaded to their sorption capacity.

8.5.2 Vapor Adsorption in MCM-41

CM-SANS was used in [67] to study the kinetics of vapor adsorption in cylindrical mesopores of MCM-41 with pore diameter 23 Å. The adsorbates were contrast matched mixtures of n-hexane (50 % C_6D_{14}) and cyclohexane (50 % C_6D_{12}) having different kinetic diameter of the molecules (4 and 6.4 Å for n-hexane and cyclohexane, respectively). SANS measurements were made at ILL, Grenoble, and LLB, Saclay, using D22 and PAXY spectrometers, respectively. Acquisition times were not reported, however examination of the experimental data suggests that it did not exceed several minutes. Controlled doses of liquid adsorbate were injected with a calibrated microsyringe into quartz cells with the path length of 2 mm, containing accurately weighed amounts of outgassed samples at 298 K. The volume of injected adsorbate was just sufficient to saturate the mesopores of MCM-41 leaving the much larger inter-granular voids unfilled.

The evolution of the SANS with time after exposure MCM-41 to the saturated vapor pressure of contrast matched cyclohexane is shown in Fig. 8.13a. The amplitude of the (10) diffraction peak progressively decreases with time after induction period of about 9 min and becomes time-independent after ~25 min.

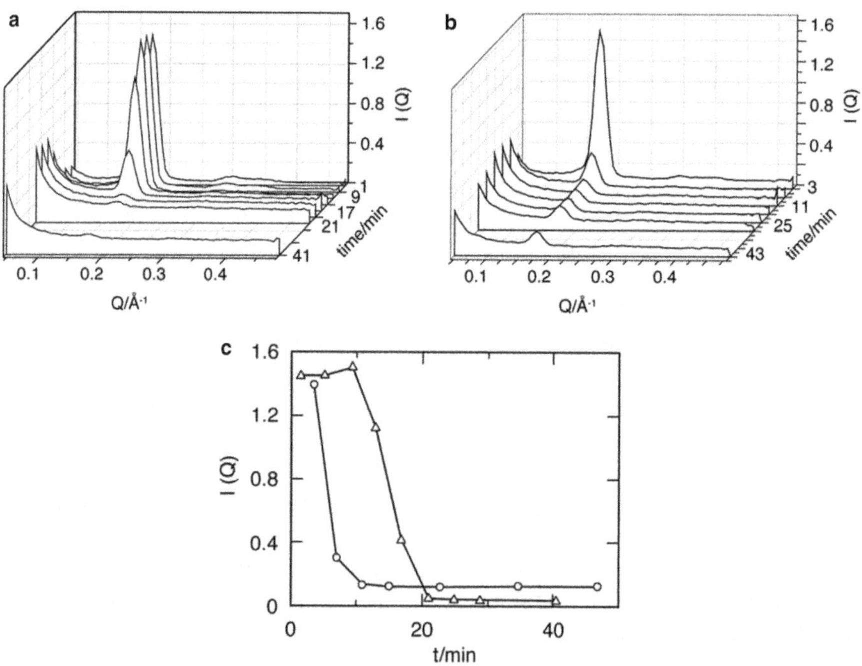

Fig. 8.13 Kinetic evolution of SANS at high Q of silica after exposure to matched cyclohexane (50 % C_6D_{12}) (**a**) and to matched n-hexane (50 % C_6D_{14}) (**b**) at 298 K. (**c**) Kinetic changes in the intensity of the (10) diffraction peak of silica after exposure to cyclohexane (*open triangles*) and n-hexane (O) [67]. With kind permission from Springer Science and Business Media

The rate of adsorption of contrast matched n-hexane, as determined from the suppression of the diffraction peak is markedly faster (Fig. 8.13b). No induction period was observed for this fluid, and the uptake was completed after the period of 7 and 11 min.

A comparison of the adsorption rates of both fluids in MCM-41 is shown in Fig. 8.13c. The smaller kinetic diameter of n-hexane correlates with the faster rate of the pore filling in MCM-41. There are, however, some other effects that may contribute in the observed difference of the vapor adsorption, such as differences in the surface-molecule interactions. Future systematic studies of the adsorption rate of various fluids in pores of variable sizes may provide important information about features of the gas separation processes using membranes and molecular sieves.

8.5.3 Vapor and Water Uptake in Nafion Membranes

Nafion is a perfluorinated sulfonic acid (PFSA) ionomer used in fuel cells as proton conducting membranes. Water molecules are produced at the cathode and transported to the anode, generating water concentration profiles across the membrane. The cells can operate properly only when the membranes are highly conductive (fully hydrated) as the ion conductivity drops by few orders of magnitude if a part of the membrane dries out. Understanding the water adsorption kinetics and mechanism in Nafion membranes is crucial in the design of more efficient energy conversion devices and new proton-conductive materials [68].

Water concentration profiles were investigated using SANS in [69] in order to get a better insight of the sorption mechanism and especially separate the surface and bulk contributions. Nafion 112 membranes, 50 μm thick were obtained from DuPont. They were treated in boiling nitric acid and rinsed in deionized water. For sorption experiments, Nafion membranes were equilibrated to room humidity (relative humidity, RH, ~10 %) and loaded in a custom designed chamber connected to a water filled tank to control the relative humidity.

SANS experiments were performed at the LLB (Saclay, France) on the PAXY spectrometer. The wavelength of neutrons was 5 Å and the range of the scattering vectors covered was $0.005 < Q < 0.4$ Å^{-1}. SANS data were recorded as a function of RH and time and the acquisition time required to obtain reasonable statistics was 15 min. The SANS data recorded during the sorption process are shown in Fig. 8.14a. SANS spectra exhibit the ionomer peak, which position and intensity are directly related to water content. The scattering intensity decreases with water content and the ionomer peak shifts toward smaller values of Q due to the swelling of the ionic domains [70]. The sorption process was very slow and the swelling equilibrium was not achieved within 38 h. The peak position varied almost linearly as a function of time. This indicated that water sorption proceeded as a single process and could not be split into a short time thermodynamically driven sorption followed by a longer time polymer matrix reorganization.

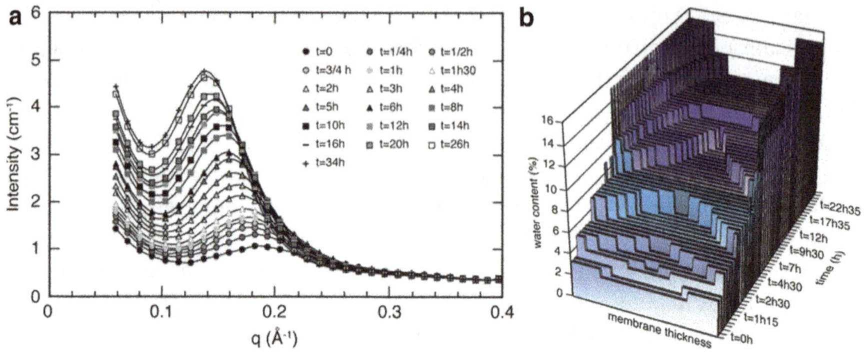

Fig. 8.14 (**a**) Evolution of the SANS spectra of a Nafion 112 membrane during the sorption of water vapor. (**b**) Evolution of the water concentration profiles during the sorption process [69].

Solid lines in Fig. 8.14a represent fits to a model that uses a linear combination of the reference spectra and the coefficients representing the relative thickness of selected layers with different water content [71]. From these fits, the evolution of the water concentration profiles across the membrane during the sorption process may be calculated (Fig. 8.14b). The results show that water is heterogeneously distributed across the membrane volume, especially during the first 8 h of the sorption process. The integration of the water concentration profiles agrees with classical sorption kinetics determined by gravimetric experiments. This suggests that the water sorption kinetics in Nafion may be viewed as a Fickian diffusion process.

Initial fast stages of water sorption in Nafion membranes were studied in [63] using synchrotron SAXS with sub-second time resolution. The experiments were carried out at the Advance Light Source, Lawrence Berkeley National Laboratory, beam line 7.3.3. The wavelength of the incident x-ray beam was 1.24 Å, and the scattering vector was in the range of $0.008 < Q < 0.4$ Å^{-1}. Nafion 117 membranes were cleaned and dried before placing them into custom-made sample holder containing a small opening for the injection of water. Images were taken with time resolution as low as 0.2 s during the membrane water sorption.

Figure 8.15a shows the time evolution of SAXS profiles during *water* uptake of dry Nafion membrane brought in a contact with liquid water. The ionomer peak Q_{peak} shifts to lower values of Q. In less than a minute, d-spacing between water domains, $d = 2\pi/Q_{peak}$, increases from 34.5 to 52 Å, and does not show any change after the first 3 min of hydration, suggesting that the sorption process had reached equilibrium. Similar SAXS experiments conducted with *vapor* sorption in Nafion membrane revealed much slower kinetic rate and the equilibrium sorption was not achieved after more than 1 week (Fig. 8.15b). These results provide evidence that the kinetics of hydration depends crucially on the phase state of water. In liquid water, the membrane surface is covered with hydrophilic groups, and water uptake occurs within seconds. In water vapor, the presence of hydrophilic groups on the

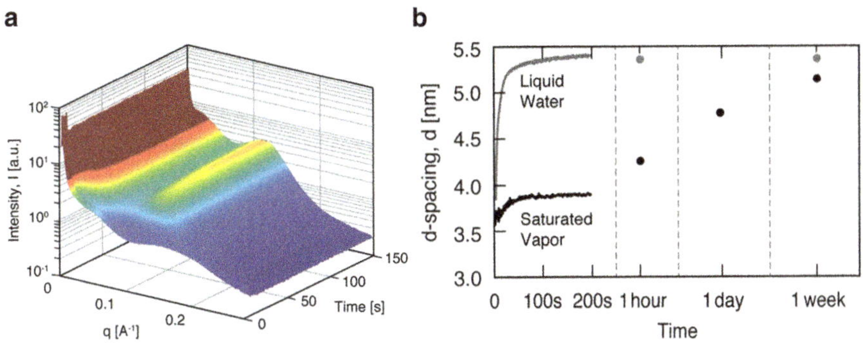

Fig. 8.15 (**a**) Radial average scattering profiles of the Nafion membrane during sorption of liquid water. (**b**) Increase in domain spacing with time for a membrane equilibrating in saturated vapor and liquid water. Reprinted with permission from [63]. Copyright 2012 American Chemical Society

surface increases the interfacial resistance to molecular transport, resulting in a much slower sorption. The results of the SAS experiments [70, 71] provide the insights on the fast and slow structural changes in Nafion that occur during water sorption, which can help to optimize device operations and develop new materials for energy conversion devices.

References

1. Horikawa, T., Do, D.D., Nicholson, D.: Capillary condensation of adsorbates in porous materials. Adv. Colloid Interface Sci. **169**, 40 (2011)
2. Gelb, L.D., Gubbins, K.E., Radhakrishnan, R., Sliwinska-Bartkowiak, M.: Phase separation in confined systems. Rep. Prog. Phys. **62**, 1573 (1999)
3. Rouquerol, F., Rouquerol, J., Sing, K.: Adsorption by Powders & Porous Solids. Academic, London (1999)
4. Silverstein, M.S., Bauer, B.J., Hedden, R.C., Lee, H.-J., Landes, B.G.: SANS and XRR porosimetry of a polyphenylene low-*k* dielectric. Macromolecules **39**, 2998 (2006)
5. Hedden, R.C., Lee, H.-J., Soles, C.L., Bauer, B.J.: Characterization of pore structure in a nanoporous low-dielectric-constant thin film by neutron porosimetry and X-ray porosimetry. Langmuir **20**, 6658 (2004)
6. Lee, H.-J., Soles, C.L., Liu, D.-W., Bauer, B.J., Wu, W.-L.: Pore size distributions in low-k dielectric thin films from x-ray porosimetry. J. Polym. Sci. B. **40**, 2170 (2002)
7. Melnichenko, Y.B., Wignall, G.D.: Small-angle neutron scattering in materials science: recent practical applications. J. Appl. Phys. **102**, 021101 (2007)
8. Wu, W.L., Wallace, W.E., Lin, E.K., Lynn, G.W., Glinka, C.J., Ryan, E.T., Ho, H.M.: Properties of nanoporous silica thin films determined by high-resolution x-ray reflectivity and small-angle neutron scattering. J. Appl. Phys. **87**, 1193 (2000)
9. Debye, P., Anderson, H.R., Brumberger, H.: Scattering by an inhomogeneous solid. 2. The correlation function and its application. J. Appl. Phys. **28**, 679 (1957)
10. Hedden, R.C., Lee, H.J., Bauer, B.J.: Characterization of nanoporous low-k thin films by small-angle neutron scattering contrast variation. Langmuir **20**, 416 (2004)

11. Lee, H.-J., Soles, C.L., Liu, D.-W., Bauer, B.J., Lin, E.K., Wu, W.-L., Grill, A.: Structural characterization of porous low-*k* thin films prepared by different techniques using x-ray porosimetry. J. Appl. Phys. **95**, 2355 (2004)
12. Silverstein, M.S., Shach-Caplan, M., Bauer, B.J., Hedden, R.C., Lee, H.-J., Landes, B.G.: Nanopore formation in a polyphenylene low-k dielectric. Macromolecules **38**, 4301 (2005)
13. Klotz, M., Roussac, V., Rebiscouul, D., Ayral, A., van der Lee, A.: Adsorption-desorption isotherms of nanoporous thin films measured by X-ray reflectometry. Thin Solid Films **495**, 214 (2006)
14. Goh, T.K., Wong, T.K.S.: Investigation of structure, thermal and oxygen plasma stability of mesoporous methylsilsesquioxane films by X-ray reflectivity and small angle scattering. Microelectron. Eng. **75**, 330 (2004)
15. Vogt, B.D., Pai, R.A., Lee, H.-J., Hedden, R.C., Soles, C.L., Wu, W.-L., Lin, E., Bauer, B.J., Watkins, J.J.: Characterization of ordered mesoporous silica films using small-angle neutron scattering and X-ray porosimentry. Chem. Mater. **17**, 1398 (2005)
16. Hoinkis, E.: In situ small angle neutron scattering study of benzene adsorption in the porous SiO_2 glass CPG-10-75. Adv .Colloid Interface Sci. **76–77**, 39 (1998)
17. Hoinkis, E.: Small angle neutron scattering study of C6D6 condensation in a mesoporous glass. Langmuir **12**, 4299 (1996)
18. Hohr, A., Neumann, H.-B., Schmidt, P.W., Pfeifer, P., Avnir, D.: Fractal surface and cluster structure of controlled-pore glasses and Vycor porous glass as revealed by small-angle X-ray and neutron scattering. Phys. Rev. B **38**, 1462 (1988)
19. Brown, R.J.C., Evans, M.J.B., Tun, Z.: Neutron scattering study of adsorption in porous MCM-41 silica. Can. J. Phys. **88**, 707 (2010)
20. Hoinkis, E.: Small-angle scattering studies of adsorption and of capillary condensation in porous solids. Part. Part. Syst. Charact. **21**, 80 (2004)
21. Ramsay, J.D.F., Kallus, S., Hoinkis, E.: SANS characterisation of mesoporous silicas having model structures. In: Unger, K.K., Kreysa, G., Baselt, J.P. (eds.) Characterisation of Porous Solids V. Studies in Surface Science and Catalysis, vol. 128, p. 439. (2000)
22. Smarsly, B., Goltner, C., Antonietti, M., Ruland, W., Hoinkis, E.: SANS investigation of nitrogen sorption in porous silica. J. Phys. Chem. B **105**, 831 (2001)
23. Guenther, G., Prass, J., Paris, O., Schoen, M.: Novel insights into nanopore deformation caused by capillary condensation. Phys. Rev. Lett. **101**, 086104 (2008)
24. Grosman, A., Ortega, C.: Influence of elastic deformation of porous materials in adsorption-desorption process: a thermodynamic approach. Phys. Rev. B **78**, 085433 (2008)
25. Kim, H.Y., Gatica, S.M., Stan, G., Cole, M.W.: Effects of substrate relaxation on adsorption in pores. J. Low Temp. Phys. **156**, 1–8 (2009)
26. Prass, J., Mueter, D., Fratzl, P., Paris, O.: Capillarity-driven deformation of ordered nanoporous silica. Appl. Phys. Lett. **95**, 083121 (2009)
27. Erko, M., Wallacher, D., Brandt, A., Paris, O.: In-situ small-angle neutron scattering study of pore filling and pore emptying in ordered mesoporous silica. J. Appl. Crystallogr. **43**, 1 (2010)
28. Hoinkis, E., Rohl-Kuhn, B.: In situ small-angle neutron scattering study of nitrogen adsorption and condensation in mesoporous silica glass CPG-10-75. J. Colloid Interface Sci. **296**, 256 (2006)
29. Hoinkis, E.: In: Berichte des Hahn-Mainer-Institut Berlin. BENSC Experimental Report, pp. 212–213 (1998)
30. Hoinkis, E.: In: Berichte des Hahn-Mainer-Institut Berlin. BENSC Experimental Report, pp. 295 (2000)
31. Hoinkis, E., Rohl-Kuhn, B.: In edited by K. Kaneko and Y. Hanzawa. Proceedings of the Conference on the Fundamentals of adsorption, pp. 601–607. International Adsorption Society; IK International: Chiba-City, Japan (2002)
32. Schreiber, A., Ketelsen, I., Findenegg, G.H., Hoinkis, E.: In: Studies in Surface and Surface Science and Catalysis. Proceedings of the 7th International Symposium on the Characterization of Porous Solids (COPS-VII), pp. 17–24. Aix-en-Provence, France (2005)

33. Eschricht, E., Hoinkis, E., Madler, F.: Nitrogen distribution at 77.7 K in mesoporous Gelsil generated via evolutionary minimization with statistical descriptors derived from adsorption and in situ SANS. Langmuir **23**, 2145 (2007)
34. Smarsly, B., Thommes, M., Ravikovitch, P.I., Neimark, A.V.: Characterization of worm-like micro- and mesoporous silicas by small-angle scattering and high-resolution porosimetry. Adsorption **11**, 653 (2005)
35. Mascotto, S., Wallacher, D., Brandt, A., Hauss, T., Thommes, M., Zickler, G.A., Funari, S.S., Timmann, A., Smarsly, B.: Analysis of microporosity in ordered mesoporous hierarchically structures silica by combining physisorption with in situ small-angle scattering (SAXS and SANS). Langmuir **25**, 12670 (2009)
36. Mascotto, S., Wallacher, D., Kuschel, A., Polarz, S., Zickler, G.A., Timmann, A., Smarsley, B.: Adsorption in periodically ordered mesoporous organosilica materials studied by in situ small-angle X-ray scattering and small-angle neutron scattering. Langmuir **26**, 6383 (2010)
37. Sharifi, M., Marschall, R., Wihelm, M., Wallacher, D., Wark, M.: Detection of homogeneous distribution of functional groups in mesoporous silica by small angle neutron scattering and in situ adsorption of nitrogen or water. Langmuir **27**, 5516 (2011)
38. Sharifi, M., Wallacher, D., Wark, M.: Distribution of functional groups in periodic mesoporous organosilica materials studied by small-angle neutron scattering with in situ adsorption of nitrogen. Beilstein J. Nanotechnol. **3**, 428 (2012)
39. Zickler, G.A., Jahnert, S., Wagermaier, W., Funari, S.S., Findenegg, G.H., Paris, O.: Physisorbed films in periodic silica studied by in situ synchrotron small-angle diffraction. Phys. Rev. B **73**, 184109 (2006)
40. Imperor-Clerc, M., Davidson, P., Davidson, A.: Existence of a microporous corona around the mesopores of silica-based SBA-15 materials templated by triblock copolymers. J. Am. Chem. Soc. **122**, 11925 (2000)
41. Jahnert, S., Muter, D., Prass, J., Zickler, G.A., Paris, O., Findenegg, G.H.: Pore structure and fluid sorption in ordered mesoporous silica. I. Experimental study by in situ small-angle x-ray scattering. J. Phys. Chem. C **113**, 15201 (2009)
42. Muter, D., Jahnert, S., Dunlop, J.W., Findenegg, G.H., Paris, O.: Pore structure and fluid sorption in ordered mesoporous silica. II. Modeling. J. Phys. Chem. C **113**, 15211 (2009)
43. Findenegg, G.H., Jahnert, S., Muter, D., Paris, O.: Analysis of pore structure and gas adsorption in periodic mesoporous solids by in situ small-angle X-ray scattering. Colloids. Surf. A. **357**, 3 (2010)
44. Antxustegi, M.M., Hall, P.J., Calo, J.M.: The use of contrast-matching small-angle neutron scattering techniques to monitor closed porosity in carbons. J. Colloids Interface Sci. **202**(490) (1998)
45. Hall, P.J., Ruiz Machado, W., Galan, D.G., Barrientos Barria, E.L., Sherrington, D.C.: Use of contrast matching small-angle neutron scattering to monitor the presence of closed porosity in controlled porosity styrene–divinylbenzene resins. Faraday Trans. **92**, 2607 (1996)
46. Snook, I., Yavorsky, I., Hanley, H.J.M., Lin, M., Mainwaring, D., Rogers, H., Zulli, P.: Characterization of metallurgical chars by small angle neutron scattering. Energy Fuel **16**, 1009 (2002)
47. Hoinkis, E., Allen, A.J.: A small angle neutron scattering study of porous graphitic materials before and after adsorption and condensation of C_6D_6 within the accessible pores. J. Colloids Interface Sci. **145**, 540 (1991)
48. Mergia, K., Stefanopoulos, K.L., Ordas, N., Garcia-Rosales, C.: A comparative study of the porosity of doped graphites by small-angle neutron scattering, nitrogen adsorption and helium pycnometry. Micropor. Mesopor. Mater. **134**, 141 (2010)
49. Hoinkis, E., Ziehl, M.: A small-angle neutron scattering study of activated carbon fibers. Carbon **41**, 2047 (2003)
50. Mileeva, Z., Ross, D.K., Wilkinson, D., King, S.M., Ryan, T.A., Sharrock, H.: The use of small angle neutron scattering with contrast matching and variable adsorbate partial pressures in the study of porosity in activated carbons. Carbon **50**, 5062 (2012)

51. Gethner, S.: The determination of the void structure of macroporous coals by small-angle neutron scattering: void geometry and structure in Illinois No-6 bituminous coal. J. Appl. Phys. **59**, 1068 (1986)
52. Hall, P.J., Antxustegi, M., Ruiz, W.: Contrast-matching small-angle neutron scattering evidence for the absence of a connected pore system in Pittsburgh No 8 coal. Fuel **77**, 1663 (1998)
53. Antxustegi, M., Hall, P.J., Calo, J.: Development of porosity in Pittsburgh No. 8 coal char as investigated by contrast-matching small-angle neutron scattering and gas adsorption techniques. Energy Fuel **12**, 542 (1998)
54. Hall, P.J., Brown, S.D., Calo, J.M.: The pore structure of the argonne coals as interpreted from contrast matching small angle neutron scattering. Fuel **79**, 1327 (2000)
55. Mitropoulos, A.C., Stefanopoulos, K.L., Kanellopoulos, N.K.: Coal studies by small-angle X-ray scattering. Micropor. Mesopor. Mater. **24**, 29 (1998)
56. Favvas, E.P., Sapalidis, A.A., Stefanopoulos, K.L., Romanos, G.E., Kannellopoulos, N.K., Kargiotis, E.K., Mitropoulos, A.C.: Characterization of carbonate rocks by combination of scattering, porosimetry and permeability techniques. Micropor. Mesopor. Mater. **120**, 109 (2009)
57. Melnichenko, Y.B., He, L., Sakurovs, R., Kholodenko, A.L., Blach, T., Mastalerz, M., Radlinski, A.P., Cheng, G., Mildner, D.F.R.: Accessibility of pores in coal to methane and carbon dioxide. Fuel **91**, 200 (2012)
58. He, L., Chathoth, S.M., Melnichenko, Y.B., Presser, V., McDonough, J., Gogotsi, Y.: Small-angle neutron scattering characterization of the structure of annoporous carbons for energy-related applications. Micropor. Mesopor. Mater. **149**, 46 (2012)
59. Boukhalfa, S., He, L., Melnichenko, Y.B., Yushin, G.: Small-angle neutron scattering for in situ probing of ion adsorption inside micropores. Angew. Chem. Int. Ed. **52**, 4618 (2013)
60. Laszlo, K., Rochas, C., Geissler, E.: Water vapor adsorption and contrast-modified SAXS in microporous polymer-based carbons of different surface chemistry. Adsorption **14**, 447 (2008)
61. Laszlo, K., Czakkel, O., Dobos, G., Lodewyckx, P., Rochas, C., Geissler, E.: Water capor adsorption in highly porous carbons as seen by small and wide angle X-ray scattering. Carbon **48**, 1038 (2010)
62. Kim, M.-H., Glinka, C., Carter, R.N.: In situ vapor sorption apparatus for small-angle neutron scattering and its application. Rev. Sci. Instrum. **76**, 113904 (2005)
63. Kusoglu, A., Modestino, M.A., Hexemer, A., Segalman, R.A., Weber, A.Z.: Subsecond morphological changes in Nafion during water uptake detected by small-angle X-ray scattering. ACS Macro Lett. **1**, 33 (2012)
64. Ramsay, J.D.F.: Surface and pore structure characterization by neutron scattering techniques. Adv. Colloid Interface Sci. **76–77**, 13 (1998)
65. Radlinski, A.P., Busbridge, T.L., MacA Gray, E., Blach, T., Cheng, G., Melnichenko, Y.B., Cookson, D.J., Mastalerz, M., Esterle, J.: Dynamic micromapping of CO_2 sorption in coal. Langmuir **25**(2385) (2009)
66. Gray, P.G., Do, D.D.: A graphical method for determining pore and surface diffusivities in adsorption systems. Ind. Eng. Chem. Res. **31**, 1176 (1992)
67. Kallus, S., Hahn, A., Ramsay, J.D.: Gas adsorption in MCM-41 porous silicas dynamic measurements using SANS. Eur. Phys. J. **12**, s01 (2003)
68. Vielstich, W., Lamm, A., Gasteiger, H.A.: Handbook of Fuel Cells: Fundamentals, Technology, Applications. Wiley, New York, NY (2004)
69. Gebel, G., Lyonnard, S., Mendil-Jakani, H., Morin, A.: The kinetics of water sorption in Nafion membranes: a small-angle neutron scattering study. J. Phys. Condens. Matter **23**, 234107 (2011)
70. Gebel, G., Diat, O.: Neutron and X-ray scattering: suitable tools for studying ionomer membranes. Fuel Cells **5**, 261 (2005)
71. Xu, F., Diat, O., Gebel, G., Morin, A.: Determination of transverse water concentration profile through MEA in a fuel cell using neutron scattering. J. Electrochem. Soc. **154**, 1389 (2007)

Chapter 9
Individual Liquids and Liquid Solutions Under Confinement

This Chapter deals with applications of SAS to study the influence of confinement on phase transitions of confined liquids and liquid solutions. It describes results of neutron and x-ray scattering studies of various aspects of the confined liquid behavior including adsorption of electrolyte ions in porous carbons, detection of the oil generation in pores of hydrocarbon rocks, and formation of nanobubbles on nanostructured surfaces of variable hydrophobicity. Two competing theoretical approaches that predict how pore induced random disorder may affect fluid behavior are Random Field Ising Model and a single pore model. SAS studies of confined solutions near their critical demixing points have been used to verify theoretical predictions, and explore the specifics of phase transitions in confined liquid crystals and supercooled water.

9.1 Confined Electrolytes

Energy storage systems, such as batteries and electrochemical double layer capacitors (EDCLs), or supercapacitors, are used in numerous applications as well as in everyday life. The proper combination of the electrode materials and an electrolyte can significantly increase both the energy stored by the device and its power. The available porous electrode materials include activated, carbide-derived, as well as template carbons with high surface areas and pores with sizes ranging from subnanometer to few nanometers [1]. Exohedral nanoparticles, such as carbon nanotubes or onion-like carbons, can provide even higher power due to a faster speed of ion sorption/desorption processes on their surfaces. The commonly used types of electrolytes include aqueous and organic electrolytes, as well as ionic liquids. The energy density of EDCL can be significantly improved and made

Y.B. Melnichenko, *Small-Angle Scattering from Confined and Interfacial Fluids*, DOI 10.1007/978-3-319-01104-2_9

similar to that of batteries by increasing the capacitance. This can be achieved in particular by designing the structure of carbon pores and carbon–electrolyte interfaces that are optimized for the ion adsorption from the electrolyte to maximize the stored charge.

9.1.1 Ion Adsorption in Electrolyte Saturated Porous Carbons

Transport and adsorption of electrolyte ions in porous carbons under applied potential controls the performance of EDCLs for rapidly emerging high power energy storage applications [2], capacitive deionization devices for water purification and desalination [3], rotational motors for artificial muscles, as well as microfluidic devices and nanorobotics [4]. In situ experimental information on the specifics of ion adsorption in pores is essential for improving the efficiency of porous carbons and optimizing operation conditions of the relevant devices.

In this respect, electrochemical analysis techniques combined with structural and chemical characterization of porous carbon materials offer limited prognostic abilities. Several studies revealed significant impact of structural defects in carbon materials on their electrosorption properties, as well as the enhanced ion adsorption in sub-nanometer pores [5] in some carbon materials. The observed phenomena, however, were not always confirmed in other porous carbons or electrolytes, and the universality of such observations remains a topic of debates in the scientific community.

The capability of SANS to provide *in situ* information on the pore size specific adsorption of confined fluids was utilized in [6] to directly visualize changes in the ion concentration in pores of different sizes as a function of the potential applied to nanoporous carbon electrode in *aqueous electrolyte*. The majority of electrolytes contain hydrogen atoms in either their solvent molecules or ions or both. By monitoring changes in the distribution of hydrogen in the nano-confined electrolyte upon the application of external potential to porous carbon electrodes one can reveal where the ion adsorption takes place. A simplified schematic of the experimental setup is shown in Fig. 9.1. Two parallel porous carbon electrodes were made of microporous activated carbon fabric, ACF. They were immersed into an aqueous electrolytes based on 1 M H_2SO_4 solutions in H_2O or D_2O and placed in a hermetically sealed and transparent to neutrons quartz electrochemical cell. A neutron beam was directed only into a working electrode, the potential on which was controlled by a potentiostat. The experiments were conducted using ORNL General Purpose SANS instrument with a neutron wavelength $\lambda = 4.75$ Å^{-1} and the Q-range between 0.005 and 0.9 Å^{-1}.

The recorded SANS patterns are shown in Fig. 9.2. Under the application of a negative potential to the ACF working electrode in a H_2SO_4 solution in H_2O, H-containing cations (such as H_3O^+) replace anions (such as HSO_4^- and SO_4^{2-}) as

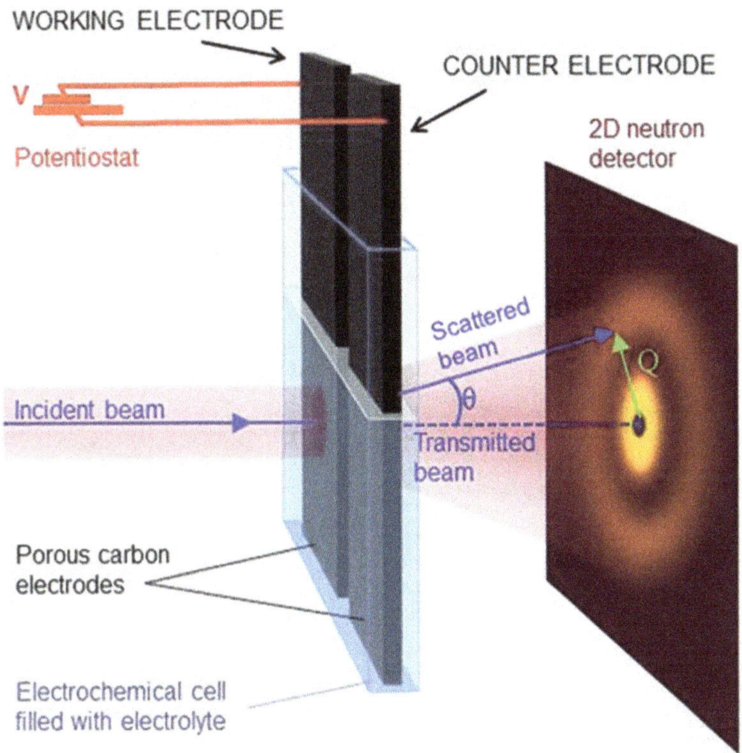

Fig. 9.1 Schematic of the experimental setup for *in-situ* studies of ion adsorption on the surface of microporous carbon electrodes [6]. Angewandte Chemie by Gesellschaft Deutscher Chemiker Reproduced with permission of Wiley-VCH Verlag GMBH & CO. KGAA in the format Republish in a book via Copyright Clearance Center

well as H_2O molecules, thereby increasing the concentration of H in the vicinity of the ACF surface. This leads to increasing intensity of scattering and shifting the intensity curve up to higher values (Fig. 9.2a). When a positive potential is applied to the working electrode, electrolyte anions replace both H-rich cations and H_2O molecules, thus reducing the concentration of H atoms in the vicinity of the ACF surface, and decreasing the scattering intensity (Fig. 9.2a).

The application of a negative potential to an ACF working electrode in a H_2SO_4 solution in D_2O increases the intensity of the scattered neutron beam even stronger because of a dramatically weaker scattering of neutrons from deuterium in D_2O (Fig. 9.2b). Under the application of a positive potential, the change in the H concentration becomes governed by the replacement of both cations as well as D_2O solvent molecules by electrolyte anions (such as HSO_4^- and SO_4^{2-}). The replacement of D_2O by the electro-adsorbed H-containing anions contributes to the increase in the H concentration in ACF pores. At the same time, the replacement of cations contributes to the decrease of the H concentration, therefore the eventual change in the scattering intensity is determined by the balance of both processes.

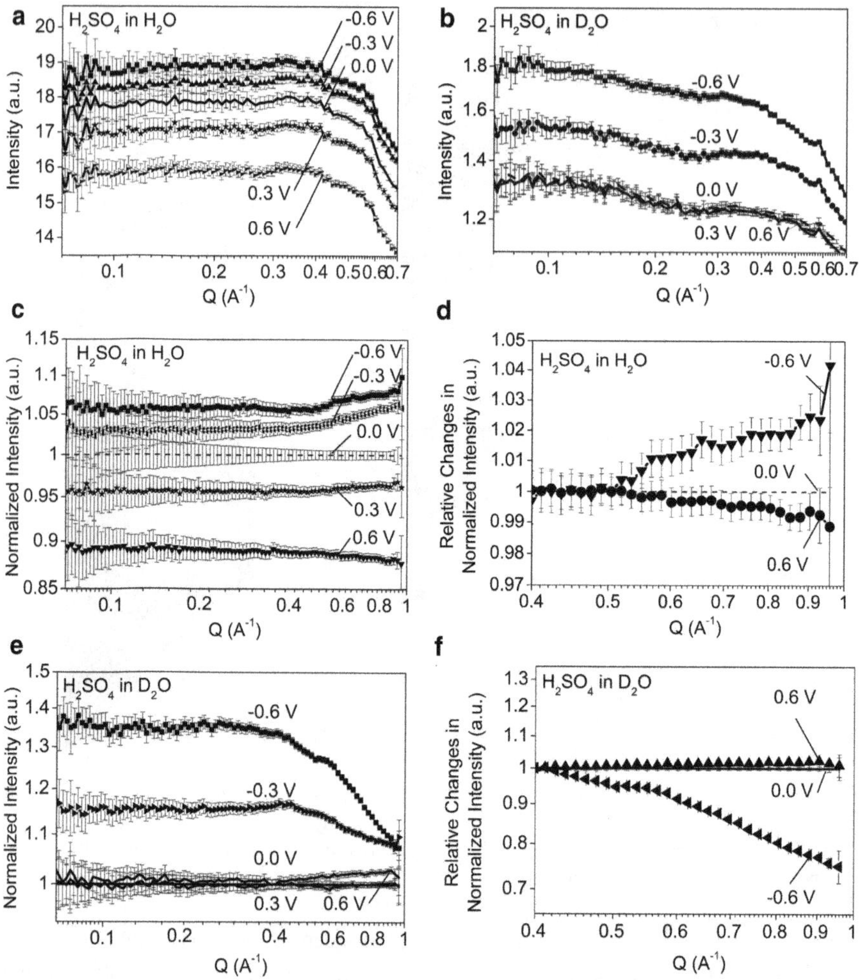

Fig. 9.2 In-situ neutron scattering experiments on ACF electrodes immersed into H_2O- (**a**, **c**, **d**) or D_2O-based (**b**, **e**, **f**) electrolytes under application of a potential between the working and counter electrodes: (**a**, **b**) SANS patterns, (**c**, **e**) SANS profiles normalized by the scattering curve measured at 0 V, (**d**, **f**) relative changes in the intensity of the normalized SANS profiles [6] Angewandte Chemie by Gesellschaft Deutscher Chemiker Reproduced with permission of Wiley-VCH Verlag GMBH & CO. KGAA in the format Republish in a book via Copyright Clearance Center

In the studied system, minimal changes in the scattering intensity were observed when the potential of the working electrode was increased from 0 to 0.6 V (Fig. 9.2b).

To facilitate the analysis, the SANS intensities were normalized to a zero-voltage SANS profile (Fig. 9.2c, e). In the case of the H_2O solvent, under the application of −0.6 V the scattering intensity increases evenly by about 6 % in the range of scattering vectors $Q < 0.5$ Å^{-1} (Fig. 9.3c). Such increase is caused by a

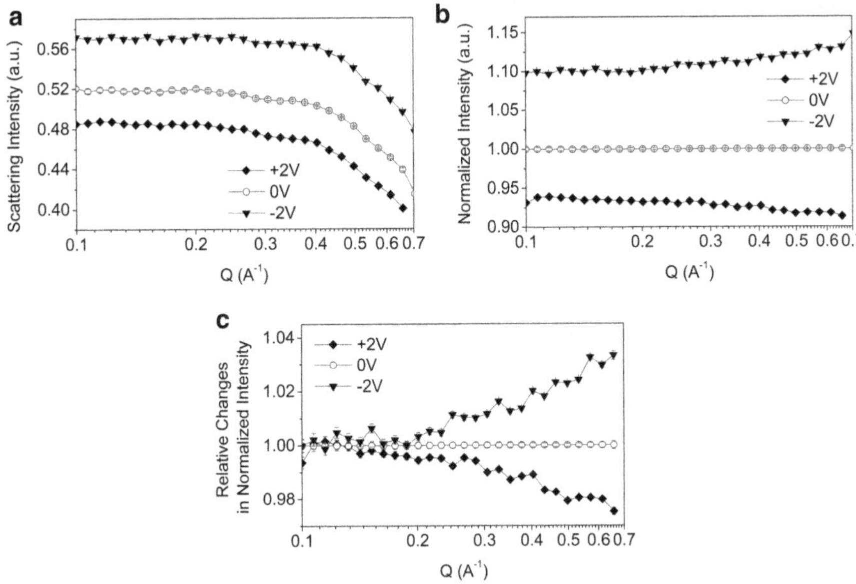

Fig. 9.3 *In-situ* neutron scattering experiments on ACF electrodes immersed into 1 M TEATFB/ d-AN electrolyte under application of a potential between the working and counter electrodes: (**a**) SANS patterns, (**b**) SANS profiles normalized by the scattering measured at 0 V, (**c**) relative changes in the intensity of the normalized SANS profiles. Reprinted with permission from [7]. Copyright 2014 American Chemical Society

6 % H enrichment of the relatively large pores. The smaller pores ($Q > 0.5$ Å^{-1}) of the ACF exhibit even higher, and Q-dependent ion adsorption capacity. Assuming that to the first approximation the pore radius R is related to Q as $R \cong 2.5/Q$ (Sect. 6.10, Eq. (6.96)), one can estimate the crossover value of the pore radius, below which the enhanced and pore size dependent H enrichment is observed as $R \cong 2.5/(0.5$ (Å^{-1})$) \cong 5$ Å.

The effect of the pore size on the ion electroadsorption can be seen more clearly in Fig. 9.2d, where the relative changes in the normalized scattering intensities are shown for both the negative and positive potentials applied to the ACF. Under the highest negative potential of −0.6 V, the H-enrichment remains constant for pores corresponding to $Q < 0.5$ Å^{-1} but increases in pores with smaller radii R corresponding to $Q > 0.5$ Å^{-1} ($R \leq 5$ Å). Similarly, under the highest positive potential of 0.6 V, the H density decreases more profoundly in the smallest pores (Fig. 9.2d). Such observations present unambiguous confirmation of a different degree of ion adsorption in pores of different sizes and support the hypotheses of the enhanced ion adsorption in the smallest pores [5]. In case of a D_2O solvent (Fig. 9.2b, e, f), the negative polarization increases the neutron scattering intensity more dramatically because the scattering contrast between carbon and D_2O is much smaller than between carbon and H_2O. Under the application of −0.6 V, the concentration of H increases by about 35 % in large pores (Fig. 9.2e). However, in a sharp contrast to H_2O (Fig. 9.2c, d), cation adsorption is significantly reduced in the smallest pores ($Q > 0.4$ Å^{-1}, $R \leq 6.3$ Å^{-1}) (Fig. 9.2e, f). The latter effect may be explained by a much higher filling degree of micropores with H_2O compare to D_2O (Sect. 8.4, Fig. 8.9).

The developed methodology was applied to study the electroadsorption of *organic electrolyte* ions in carbon pores of different sizes [7]. A 1 M solution of tetraethylammonium tetrafluoroborate (TEATFB) salt in deuterated acetonitrile (d-AN) was used in an activated carbon with the pore size distribution similar to that used in commercial EDCLs. The SANS experimental setup and parameters were similar to those used in the previous study of the ion adsorption in aqueous electrolytes [6]. Since d-AN was used in the study, the only source of hydrogen present in the electrochemical system was $N(C_2H_5)_4^+$ cations. The application of a positive potential and the application of a negative potential results in an overall decrease and increase of the scattering intensity, respectively (Fig. 9.3a). Such trends are anticipated since the H-containing TEATFB cations must be electrosorbed into the pores under application of a negative potential, and expelled (replaced) with the non-hydrogen containing anions and solvent molecules at a positive potential.

To analyze the adsorption of ions as a function of pores sizes, the SANS intensities were normalized to the SANS profile measured at zero voltage (Fig. 9.3b). Under the application of −2 V the scattering intensity increased by ~10 %, which could be approximated as the H enrichment of 7.5 % within the relatively large pores. Under the application of +2 V the scattering intensity decreased by ~7 %, which manifested the H depletion of 5 %. A slightly larger increase in the scattering intensity under a negative potential than the corresponding decrease under a positive potential could be related to slightly higher initial concentration of BF_4^- anions in carbon pores. The smallest pores ($Q > 0.2$ Å^{-1}, $R < 12.5$ Å) of the ACF exhibited even higher ion adsorption capacity, as manifested by higher scattering intensities from these pores, indicative of the higher H concentration at negative potentials, and lower scattering intensities from these pores, indicative of the lower H concentration at positive potentials. Figure 9.3c shows the relative changes in the normalized scattering intensities for both the negative and positive potentials applied to the ACF working electrode as a function of the pore size. The data obtained support the previously reported results that indicated the enhancement of organic electrolyte ion adsorption in the small sub-nanometer pores in porous carbon materials [5, 8, 9].

9.1.2 *Ionic Liquids Under Confinement*

Ionic liquids (ILs) represent an important class of fluids, which possess unique properties such as good electrolytic and solvation properties, low vapor pressure, thermal stability, nonflammability, and chemical stability. As such, the ILs can offer new opportunities when used in energy storage devices, heterogeneous catalysis, and gas separation technologies. The known drawbacks of the ILs (high viscosity, slow gas diffusivity, and high cost) may be offset by deposition of the ILs on the supporting surface of the large surface area porous materials (creation of supported ionic liquid phase (SILP) systems [10]). The development of well-stabilized thin IL films may help to reduce the amount of IL required for large scale applications, as well as to make use of the most of the deposited IL.

The interaction between ILs and solid surfaces as well as the influence of confinement on their structure and morphology of the resulting SILPs requires application of scattering techniques. SANS and contrast-matching SANS have been utilized in [11, 12] to investigate the confined IL [bmim][PF_6] in ordered mesoporous silicas MCM-41 and SBA-15. Materials MCM-41 (the specific surface area 1070 m^2/g, pore radius $R_p = 15$ Å) and SBA-15 (the specific surface area 700 m^2/g, $R_p = 28$ Å) were s by Sigma-Aldrich and Claytec Inc., respectively. The [bmim][PF_6^-] IL was prepared following routine outlined in [13]. The pore filling of the samples with IL (named MCM_IL and SBA_IL) was achieved by physical imbibitions. The SANS measurements were carried out at the PACE spectrometer (LLB, Saclay, France) in a Q range varying from 0.0035 to 0.33 $Å^{-1}$ corresponding to two sample–detector distances (4.56 and 1.06 m). All measurements were performed under atmospheric pressure and room temperature.

The powder samples under investigation were mounted in a 1-mm-thick sealed quartz containers. For the contrast-matched SANS measurements H_2O/D_2O mixtures of variable composition were used. The "wet" samples, MCM_w, MCM_IL_w, SBA_w, and SBA_IL_w, were allowed to come to equilibrium for 4 h before measurements. Absolute values of the intensity, $I(Q)$, were obtained by correcting the raw data for sample transmission, scattering of the empty cell, and instrumental background, and placed on absolute scale using a water standard.

Figure 9.4a shows the SANS curves from the dry MCM and MCM_IL samples, respectively. The hexagonal network of both samples produces one intense peak (10) and one weak peak (11). The high-order peaks were not observed because their position was outside of the experimental Q window. In the case of the dry SBA sample, in addition to the strong (10) reflection, the overlapping (11) and (20), as well as the weak (21) Bragg peaks were detected (Fig. 9.4b). The intensity of all Bragg peaks for the samples treated with IL was decreased. The calculated values of the lattice parameter, $d = 44$ Å and d = 103 Å for MCM and SBA, respectively) remained unchanged after the IL loading, which implied that the sample structure was not affected by IL.

The SANS data obtained were used to evaluate the filling degree of pores with the IL and water. By comparing the ratio of the integrated (10) peak intensities before and after the IL loading it was concluded that SBA samples were completely filled with the IL, whereas pores of the MCM samples were only partially filled with the liquid. The modulation of the of the MCM peak intensities was further investigated by calculating the form factor of pores with the IL using density models, which take into account coexisting dense silica matrix, a confined IL layer, and an

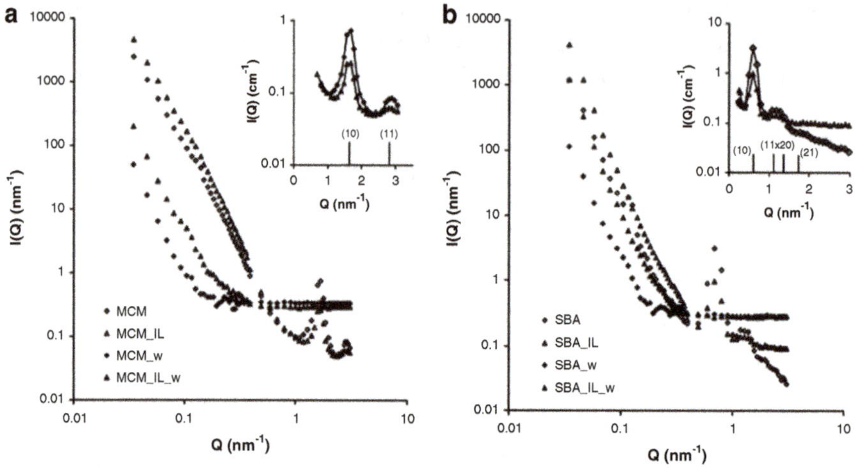

Fig. 9.4 SANS and silica contrast-matched SANS curves from pristine and IL-loaded (**a**) MCM and (**b**) SBA samples. (*Insets*) High-*Q* region spectral details of the pristine and IL-loaded (**a**) MCM and (**b**) SBA samples. Peak positions are indicated by *vertical lines*. Reprinted with permission from [11]. Copyright 2011 American Chemical Society

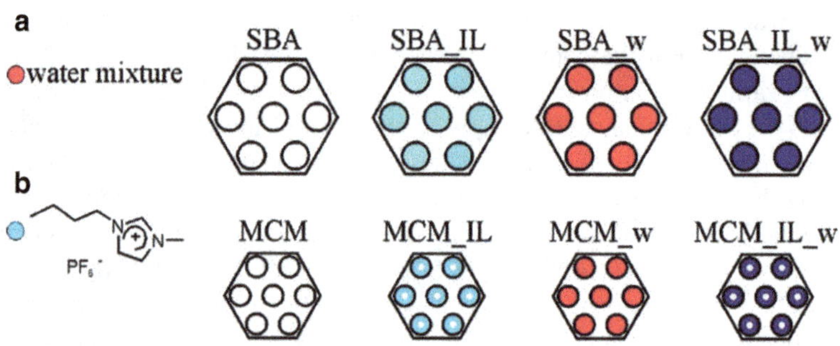

Fig. 9.5 Schematic drawing of the pore lattice for dry and "wet" pristine and IL-loaded (**a**) SBA and (**b**) MCM samples. Reprinted with permission from [11]. Copyright 2011 American Chemical Society

empty core inside the pores [14, 15]. Model fitting allowed estimating the thickness of the IL layer on cylindrical pores walls of MCM to be 14 Å, which implied that the radius of unfilled pore core was ~2 Å.

Immersion in the contrast matched mixture of H_2O/D_2O demonstrated that dry SBA and MCM samples were completely filled with water. At the same time, pores of the IL loaded MCM retained unfilled core, whereas pores of the IL loaded SBA were homogeneously filled with the IL-water mixture, which confirmed the mutual solubility of the studied IL and water (Fig. 9.5).

Although the diameter of the IL loaded MCM empty core (4 Å) is large enough to allow the diffusion of water molecules, the contrast-matched SANS measurements showed that actually water did not penetrate the core. This observation could be explained by the orientation of the IL molecules in the adsorbed layer with the negative ions located near the silica wall and the positive (hydrophobic) ions forming the walls of the pore core.

Information about the orientation of IL ions in pores depending on the sign of the surface charges is very important for their application in energy storage devices. Experiments [11, 12] demonstrate that contrast-matched SANS can be used to explore adsorption and orientation of the IL molecules in mesoporous matrices of different nature. SAS experiments may help to evaluate properties of ultrathin SILPs and their impact on the efficiency of gas separation and solid catalysis devices that utilize ionic liquid layers [16].

9.2 Detection of the Oil Generation in Hydrocarbon Source Rocks

Hydrocarbons are generated in organic-rich rocks due to chemical decomposition of the organic matter at elevated temperatures (in the range of 90–150 °C) over geological time of millions of years. The subsequent gradual increase in the molecular volume leads to the developing of the internal pressure, which is released through formation of microcracks and induces transport of generated hydrocarbons through the network of open pores out of the source rock. The length scale of the morphological changes related to the thermal maturation varies from nanometers to tens of micrometers. Later stages of the morphological changes can be studied using direct imaging techniques (SEM, STM, AFM, and optical microscopy). The initial stages of the primary hydrocarbon migration and a consequent microstructural evolution of the inorganic rock matrix requires application of techniques with much higher resolution. They can be explored by SANS/USANS due to the difference in contrast between the generated hydrocarbons and the pore space (Sect. 5.11, Fig. 5.17) [17].

In petroleum geology, the maturity of a rock is a measure of its state in terms of hydrocarbon generation. Source rocks are broadly categorized as *immature* (no hydrocarbon generation), *sub-mature* (limited hydrocarbon generation), *mature* (extensive hydrocarbon generation) and *overmature* (most hydrocarbons have been generated).

SANS study of the *naturally matured* source rocks was conducted in [18]. The source rock samples were obtained from several depths within the Velkerri Formation, Australia covering a wide range of maturity. The SANS data were collected on the W.C. Koehler 30 m SANS facility at the Oak Ridge National Laboratory. The neutron wavelength was 4.75 Å ($\Delta\lambda/\lambda \cong 5$ %) and a range of sample-detector distances (1.3–19.1 m) was used to cover a Q-range of $0.003 < Q < 0.5$ Å^{-1}.

Examples of SANS log-log plots for the immature and mature rocks are shown in Fig. 9.6. These spectra can be described by a power law for the low maturity

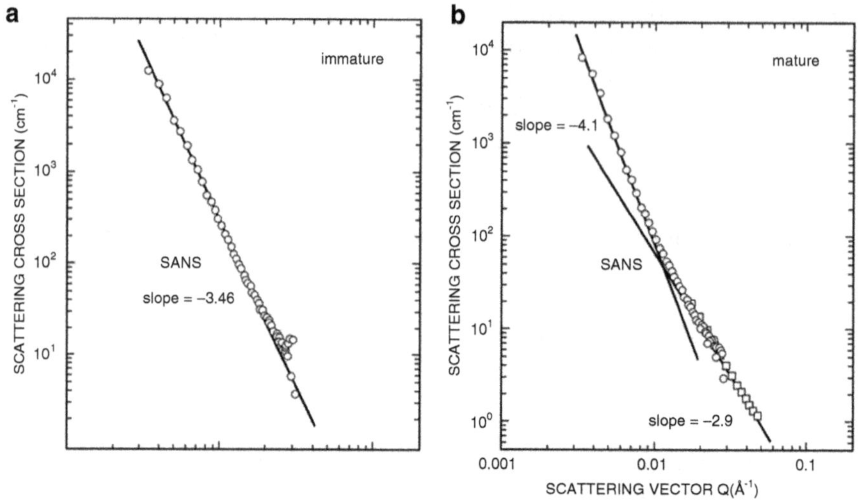

Fig. 9.6 SANS data for the immature (**a**) and the mature (**b**) source rock. Reprinted figure with permission from [18]. Copyright (1996) by the American Physical Society

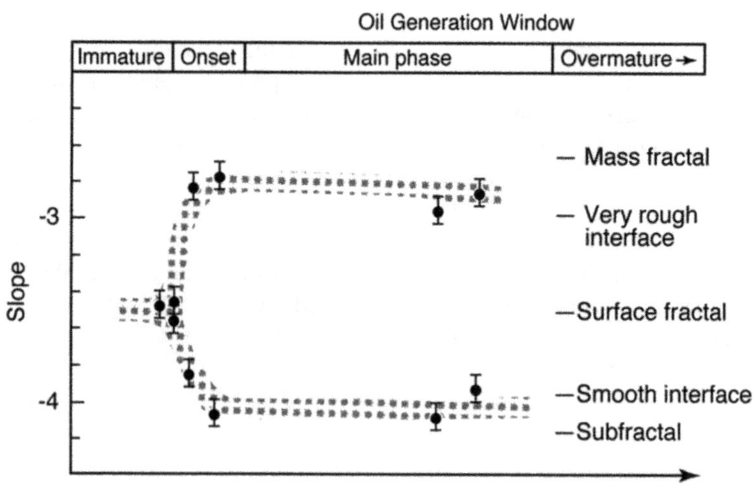

Fig. 9.7 Slope of SANS curves vs. maturity in the Q range as in Fig. 9.6. Reprinted figure with permission from [18]. Copyright (1996) by the American Physical Society

rocks and two different power laws for more mature samples. Similar variation of $I(Q)$ was observed for all samples studied. The plot of the slope versus maturity for the whole data set is shown in Fig. 9.7. The oil-generation window subdivision, as used in petroleum geology, is displayed above the plot. For SANS curves having two different slopes in the different Q ranges, both values are plotted for the corresponding value of maturity. The change of the slope for the SANS curves occurs for the more mature samples systematically near the value of $Q = 0.01$ Å^{-1}, which corresponds to the real space distance of about 600 Å.

As may be seen in Figs. 9.6 and 9.7, for the low maturity samples a single slope of -3.5 ± 0.1 is observed. It indicates surface fractal geometry of the solid–pore interface, which is practically not affected by the presence or absence of the organic matter at this early maturity stage of the rock evolution. With increased maturity of the source rocks, just at the point where the oil-generation window begins, there is a very significant branching of the slope values that become different on the length scale 50–600 and 600–2000 Å. For the smaller scale, the slope increases to -2.8 ± 0.1 from the original value of -3.5 for the immature source rock, thus crossing over the border value of -3 into the region of pore fractals (Sect. 6.8.1), which correspond to a system of interconnected microcracks created as a result of petroleum generation. On the scale 600–2000 Å, the slope approaches -4, which is the Porod limit indicating the smoothing of the solid-pore interface. Such smoothing may be explained either by wetting the inorganic rock matrix by the melting organic components (macerals) or redistribution of newly formed macerals at the boundaries of larger size pores [18].

SANS experiments were also performed on a series of *artificially matured* organic-rich sedimentary rocks of various maturity [19]. The rocks were subjected to hydrous pyrolysis under laboratory conditions equivalent to natural hydrocarbon generation. The rocks originated from the middle Proterozoic (1400 million years) sediments of the Velkerri Formation, McArthur Basin, Northern Territory, Australia. A 10 cm core from 110.0 m depth was sectioned into 1 cm thick pieces to be used in the closed system pyrolysis, and for subsequent SANS measurements. The closed system pyrolysis was performed in a mini-reactor under nitrogen atmosphere at pyrolysis temperatures (310–370 °C) for 48 h. The chosen pyrolysis environment and time helped to attain the equivalent natural hydrocarbon generation stage on the laboratory scale.

The SANS data were collected on the 80 m instrument D11 at the Institut Laue-Langevin (ILL), Grenoble, France. The neutron wavelength used was 14 Å. Three sample- detector distances (2.5, 12 and 35.7 m) were used in order to collect data in the Q-range $0.0004 < Q < 0.07$ Å^{-1}. In order to minimize the multiple scattering effects, samples were thinned down to 0.7 mm.

Figure 9.8 shows SANS data taken for the entire series of artificially pyrolysed rocks. The scattering cross section decreases by six orders of magnitude in the Q-range 0.001–0.1 Å^{-1}. In the intermediate Q-range the scattering intensity decreases according to a power law with the exponent -3.5 ± 0.05, which indicates surface fractality of the solid/pore interface.

The scattering intensity decreases by about a factor of two between the pyrolysis temperatures 320 and 330 °C. The observed decrease is related to the reduction of the pore-rock neutron contrast factor caused by the saturation of pores with generated hydrocarbons. The observed 56 % decrease in the scattering intensity for the artificially pyrolysed samples translates into the neutron contrast dropping to 0.66 of its original value for pores filled by water or formation gases. Thus, the

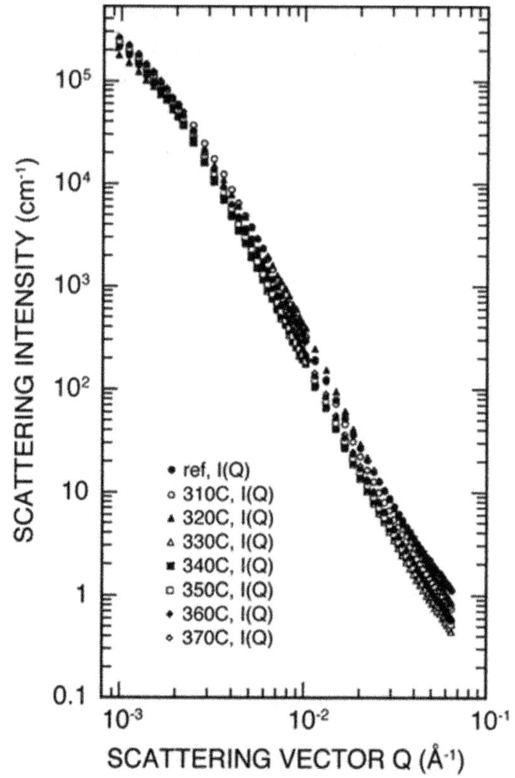

Fig. 9.8 SANS data for the artificially pyrolysed source rocks. Reprinted from [19]. Copyright (2000), with permission from Elsevier

estimated SLD for the mobile hydrocarbon phase at the beginning of the migration is ~1×10^{10} cm^{-2}, which is in a reasonable agreement with the SLD of aromatic hydrocarbons (Fig. 5.17).

The conducted SANS studies [18, 19] suggest an oil-generation scenario according to which hydrocarbons are produced from macerals finely dispersed inside the bulk of the source rock and also concentrated on the grain boundaries for grain sizes larger than 600 Å. After reaching the thermal conditions necessary for oil generation, the small grains crack and release oil. The intergranular maceral softens and its interface with the pore space becomes smooth. Simultaneously, an oil-wet network of conduits forms inside the rock.

The developed methodology was used to perform a comprehensive analysis of SANS and USANS data obtained from 165 potential source rocks recovered from nine exploration wells drilled in the Browse Basin, offshore Western Australia [20]. This study demonstrated that conclusions drawn from SANS and USANS experiments were consistent with the results obtained using traditional geochemical methods. They could be used for independent calibration and refinement of the hydrocarbon generation and expulsion scenarios derived from geochemical modeling [17].

9.3 Cavitation on Hydrophobic Nanostructured Surfaces

Cavitation, or formation of vapor cavities (or nanobubbles) on solid surfaces of "smart" materials, may affect significantly their self-cleaning [21, 22], anti-icing [23], enhanced heat transfer [24], or reduced fluid drag [25] properties. Better understanding of various scenarios at the initial stage of gas phase nucleation on solid surfaces is also important for predicting and controlling various boiling regimes, and maximizing cooling efficiency in nuclear reactors or rocket engines, where an exceedingly high amount of heat is generated in comparatively small volumes.

Although numerous models have been developed to predict the nucleation sites, detailed experimental information about how, where and how many nano- and mesoscopic bubbles form on various surface morphologies under different thermodynamic conditions is still lacking. Whereas there is a general agreement that the key processes governing gas phase nucleation occur on nano- and mesoscopic scales, the majority of experimental studies have been limited to macroscopic observations due to numerous challenges associated with direct experimental probing of the initial stages of cavitation at the nanoscale [26].

On the fundamental level, the most intriguing and poorly understood is the observed long-term stability of the vapor nanobubbles with radii of curvature less than 1 μm [27]. Based on standard diffusion model, such bubbles should be dissolved in ~1 μs in degassed water. However, surface bubbles have been found to persist for days, some 11 orders of magnitude longer than predicted by classical theory.

Synchrotron SAXS has been applied to investigate the in situ wetting of different rough superhydrophobic surfaces with extreme water contact angle ($\theta_C = 169°$) [28]. Figure 9.9a, b show two coatings derived from silica nanoparticles with a primary size of 12 nm mixed in 1:10 weight ratio with dimethylsiloxane polymer and sonicated for 15 min. A third ingredient, ethoxylated polysiloxane, was added to PS-B coating as pore size control filler. Glass capillaries were internally coated by drawing the liquid polymer/particle mixture inside, and then expelling the mixture. After allowing the solvent to evaporate, the type A coating (PS-A) was cured at 150 °C for 30 min in air, while the type B coating (PS-B) was cured at 450 °C for 30 min, giving a clearly different morphology. After curing, the thickness of the resulting coatings was between 2 and 4 μm (Fig. 9.9c). When recording SAXS patterns, the capillary could be moved a few millimeters vertically to allow either coated or uncoated surface to be irradiated. This provided scattering from blank capillaries, which could be subtracted as a background scattering.

The surface tension of the wetting liquid (initially pure water) was controlled by adding the surfactant sodium dodecylsulfate to 10 and 100 % of its critical micelle concentration (CMC) (1.013×10^{-2} M).

All SAXS measurements were conducted at the ChemMatCARS facility at the Advanced Photon Source, Chicago. Figure 9.9a compares the scattering of the superhydrophobic-type A and B coatings (PS-A and PS-B). The third coating

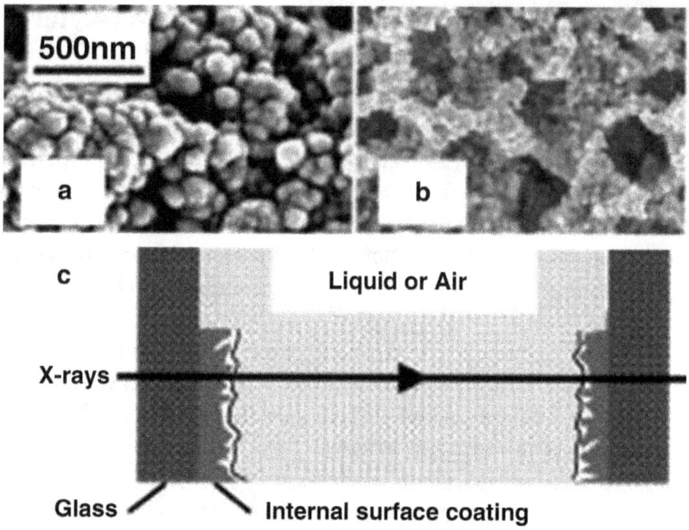

Fig. 9.9 SEM images of (**a**) PS-A and (**b**) PS-B silica/polysiloxane composite coatings. The capillaries (**c**) used for x-ray scattering had uncoated regions for blank scattering measurements. Reprinted with permission from [28]. Copyright [2007], AIP Publishing LLC

(polyacrylic) (PA) shown in this figure uses polyacrylic coat with an otherwise identical preparation method and morphology of the PS-A coating, but exhibits much less hydrophobicity overall. When measured as a smooth untextured surface, the polyacrylic polymer has a significantly lower inherent contact angle ($\theta_C = 20°$) than polysiloxane ($\theta_C = 75°$). This difference is amplified by roughness, making the PA surface superhydrophilic (i.e., $\theta_C = 0°$) and both PS surfaces superhydrophobic when wetted with pure water.

SAXS data shown in Fig. 9.10a where analyzed in terms of the SAXS Porod invariant, which is proportional to the mean square fluctuations of electron density in the irradiated volume (Sect. 6.2, Eq. (6.25)). The percentile changes in the invariant for each sample wetted with aqueous liquids are shown in Fig. 9.10b. Comparing ratios of the invariants from wet and dry samples helped to eliminate to the first approximation the effect of the coating thickness variability.

The invariant of superhydrophilic PA sample drops significantly after immersion in pure water and then becomes independent of the liquid surface tension within the experimental error. This indicates that wetting of the PA coating upon contact with pure water is complete and almost immediate. Drop in the invariant for coatings PS-A and PS-B looks similar, although the extent of the final decrease of the PS-B invariant is significantly greater, which indicates a much lower filling degree of the PS-B coated capillaries with water.

Further insight as to the structural changes in the wetting interface may be gained from the nonuniformity of intensity suppression over the measured Q range (Fig. 9.10a). Unlike the PA coating, which shows most of its wetting effect at $Q < 0.014$ Å^{-1}, the PS coatings show a relatively uniform wetting effect over the

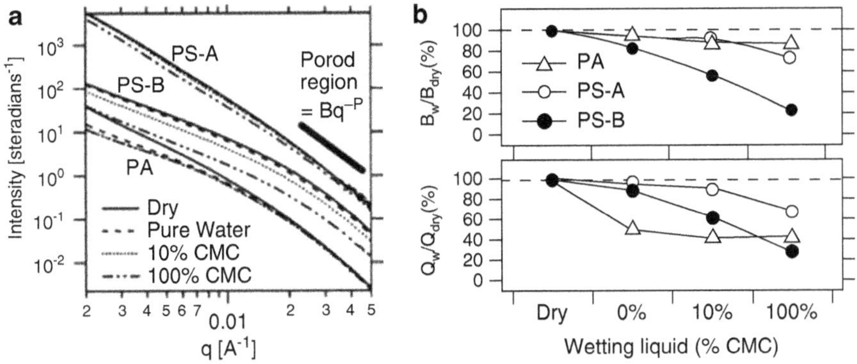

Fig. 9.10 (**a**) SAXS profiles of capillaries with internal surface coated with silica/polysiloxane composites of type A (PS-A) and type B (PS-B), as well as a PA composite. The PS-A curves shifted up by one decade for clarity. Each set of curves shows the scatter from dry, partially wetted, and fully wetted surfaces. (**b**) Percentile changes in the SAXS ratio of the invariants for the wet and dry samples (Q_W/Q_{dry}) for composite coatings of PA (*triangles*),polysiloxane PS-A (*open circles*), and PS-B (*filled circles*) composites. Reprinted with permission from [28]. Copyright [2007], AIP Publishing LLC

entire Q range. Using an estimate of length scale given by $2\pi/Q$, this suggests that structural moieties with length scales smaller than about 40 nm play a proportionately greater role in the wetting behavior of the superhydrophobic coatings.

The morphology of air bubbles trapped inside nanometer-scale cavities as a function of cavity depth was studied using synchrotron SAXS in [29]. For this purpose, large-area silicon surfaces (cm^2) were prepared containing highly uniform, periodic, densely packed 20 nm wide hydrophobic cavities. The ability to experimentally vary the cavity depth allowed control of the surface topography at sub-100-nm length scales. The high degree of nanostructured surface periodicity and uniformity facilitates the use of SAXS to provide a detailed view of the composite interface.

Nanostructured silicon surfaces were prepared from porous self-assembled templates using plasma etch, followed by an O_2 plasma etch to remove any remaining polymer from the silicon surface. These anisotropic plasma etch processes preferentially remove silicon from the pore bottoms, such that the resulting structured silicon surfaces contain hexagonally packed nanocavities as deep as 165 nm with average diameters only slightly wider than the initial polymer template ($w = 28 \pm 2$ nm). The nanocavity cross sections are approximately parabolic, independent of their depth. A significant advantage of this fabrication technique is that the nanocavity depth can be controlled by simply varying the etching time.

Nanostructured silicon surfaces were rendered hydrophobic by passivation with a 2.5 nm thick octadecyltrichlorosilane (OTS) monolayer after first cleaning in a mixture of sulfuric acid and hydrogen peroxide in a 3:1 volume ratio ("piranha" solution). Passivated silicon surfaces were finally rinsed by sonication in chloroform to remove unbound OTS molecules.

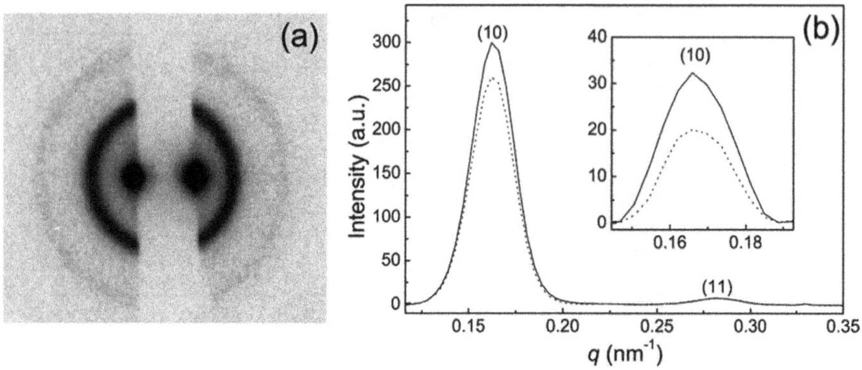

Fig. 9.11 (**a**) SAXS pattern from a silanized silicon surface patterned with 140 nm deep cavities. (**b**) Azimuthally integrated scattered intensity as a function of the scattering wave vector q from a surface with 140 nm deep cavities in contact with air (*solid line*) and water (*dotted line*). The *inset* shows similar data for a surface with 20 nm deep cavities. Reprinted with permission from [29]. Copyright 2010 American Chemical Society

SAXS measurements were carried out at beamline X6B of the National Synchrotron Light Source, Brookhaven National Laboratory. In these experiments, a collimated, high-energy (16 keV) x-ray beam incident perpendicular to the nanopatterned surface was scattered by the structure, and the scattered intensity was collected using a 2D CCD detector positioned approximately 1.2 m behind the sample. Compared to previous SAXS studies of ultrarough surfaces immersed in water [28], the spatial periodicity and uniformity of the nanotextures enhanced the scattered signal and simplified the analysis for obtaining detailed structural information from x-ray data. A representative SAXS pattern from a dry surface nanopatterned with 140 nm deep cavities (Fig. 9.11a) shows two powder diffraction rings characteristic of the random distribution of the 2D hexagonal lattice of cavities. The brightest ring represents the (10) peak, and the secondary ring at larger scattering angle is the (11) peak. The azimuthally integrated intensity is plotted in Fig. 9.11b as a function of the scattering vector Q. From the (10) peak position, $Q_{(10)} = 0.167 \pm 0.002\ \mathrm{nm}^{-1}$, the average nearest-neighbor distance of the cavities $d = 4\pi/3^{1/2}Q_{(10)} = 43.4 \pm 0.5$ nm was obtained, consistent with the scanning electron microscopy measurements.

Bringing a nanotextured surface into contact with water should result in a sensible reduction of electron density contrast (and therefore of the scattered x-ray intensity) if the liquid penetrates significantly into the cavities, due to the larger electron density of water compared to air. More specifically, the volume fraction of the cavity that is filled with water, ϕ_V, can be approximated by:

$$\phi_V = \frac{(\rho_X^*)_{Si}}{(\rho_X^*)_W}\left(1 - \sqrt{\delta I}\right), \tag{9.1}$$

where δI is the ratio between the scattered (10) peak intensity of the wet and dry surface, and $(\rho_X^*)_{Si}/(\rho_X^*)_W \cong 2.15$ is the ratio of the electron density of silicon and water. Complete cavity filling with water results in a 3.5 times decrease in (10) peak intensity, which corresponds to $\delta I = 29$ %.

To measure the SAXS scattering intensity from the nanotextured surface in contact with water, a drop of ultrapure water was placed between the patterned surface and a thin Mylar film supported by a rigid frame. The gap between the Mylar film and the patterned surface was reduced until the thickness of the sandwiched water layer reached ~100 μm. The thin water film was laterally unconfined and therefore subject only to atmospheric pressure. The x-ray absorption by the thin water layer and the Mylar film was negligible (<1 %), and therefore a correction for this effect was not required. The thickness of the water film varied slightly (in the range of a few hundred micrometers) during the course of the measurements, but this did not affect the results.

The analysis showed that the (10) integrated scattered x-ray intensity from all hydrophobic nanostructured surfaces decreased after the surfaces were put in contact with water. Figure 9.11b (dotted line) shows data from a surface patterned with 140 nm deep cavities, while the inset shows data from 20 nm deep cavities. The fractional intensity decrease, δI, for different cavity depth values h is reported in Table 9.1. Whereas for complete filling the calculated $\delta I = 29$ %, all measured δI are larger than ~60 %. This indicates that water only partially penetrates into the cavities and therefore air bubbles must be formed in the remaining volume (shown schematically in Fig. 9.12). The x-ray scattering intensity from surfaces in contact with water was stable over time (1 h) suggesting that the formation of nanobubbles is an equilibrium phenomenon.

This behavior was explained as due to a combination of capillary effects and the cavity geometry. For hydrophobic parabolic cavities, the liquid can reach a thermodynamically stable configuration by partially penetrating into the pores. Indeed, since the contact angle at the cavity wall, θ_w, is high and similar to the slope of the wall, the curvature of the liquid/air interface must be small when the water partially penetrates into the cavity (Fig. 9.12). This configuration minimizes the capillary pressure inside the bubble, stabilizing it against dissolution. Moreover, in the case of a flat water/air interface, the liquid penetrates uniformly into the cavities and Eq. (9.1) provides a reasonably accurate estimate of the volume fraction of water in the cavity from the measured changes in scattering intensity δI. This analysis

Table 9.1 The ratio of the scattered (10) peak intensity from the wet and dry nanotextured surface, δI, the volume fraction of the cavity that is filled with water, ϕ_V, the water penetration depth, δh, the contact angle on the cavity wall, θ_w, and the area fraction of the water/air interface, ϕ_A, as a function of the cavity depth h [29]

h (nm)	δI	ϕ_V	δh (nm)	θ_w (deg)	ϕ_A
20	0.65 ± 00.03	0. 42 ± 00.04	5 ± 1	108	0.19
34	0.83 ± 00.02	0.19 ± 00.02	4 ± 2	100	0.22
100	0.89 ± 00.06	0.12 ± 00.06	6 ± 3	93	0.23
140	0.89 ± 00.02	0.12 ± 00.03	9 ± 2	92	0.23
160	0.90 ± 00.02	0.11 ± 00.02	9 ± 2	92	0.23

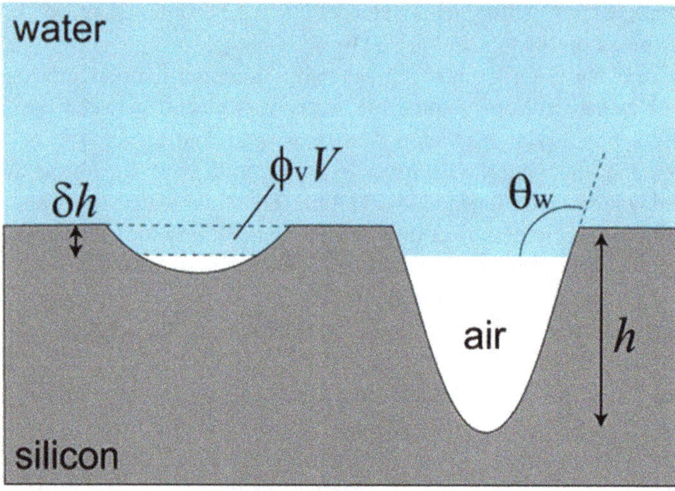

Fig. 9.12 Schematic representation of a partially filled shallow (*left*) and deep (*right*) cavity. ϕ_V is the fraction of the cavity volume V filled with water, δh is the penetration depth, and θ_w is the water contact angle on the cavity wall. Reprinted with permission from [29]. Copyright 2010 American Chemical Society

suggests that 42 % of the volume of the 20 nm deep cavities and 11 % of the volume of the 160 nm deep cavities is filled with water (Table 9.1).

Structure of vapor nanobubbles on deterministically nanopatterned surfaces created using a combination of electron beam lithography and reactive ion etching was evaluated using SANS as well as contrast matching SANS in [30]. The resulting patterned silicon wafers have a monolayer of highly uniform, periodic, densely packed cylindrical pores with axes oriented perpendicular to the wafer surface (Fig. 9.13). The pores are 100 nm deep and 100 nm in diameter forming a 2D hexagonal grid.

The key variable parameter in the study was surface wettability, which was tuned by modifying the patterned silicon surface with an ultrathin fluoropolymer layer. For the SANS experiments two samples were prepared. Sample 1 was a nanostructured Si wafer covered with an approximately 2 nm layer of naturally occurring silicon oxide (native oxide). Sample 2 had an additional 5 nm layer of a strongly hydrophobic plasma-deposited fluoropolymer. The native oxide and fluoropolymer thicknesses were measured ellipsometrically using a satellite silicon wafer with a smooth surface. Average contact angles measured for sample 1 and sample 2 after completing the SANS experiments were found to be 35° and 65°, respectively.

Custom built liquid tight copper sample holders were used in these experiments. Front windows of the holders were made of 1 mm thick unpatterned Si wafers. Wafers with the nanopatterned surfaces looking inside the cell served as exit windows. The internal cell volume may be filled with liquid with the total liquid thickness in the holder ~4 mm.

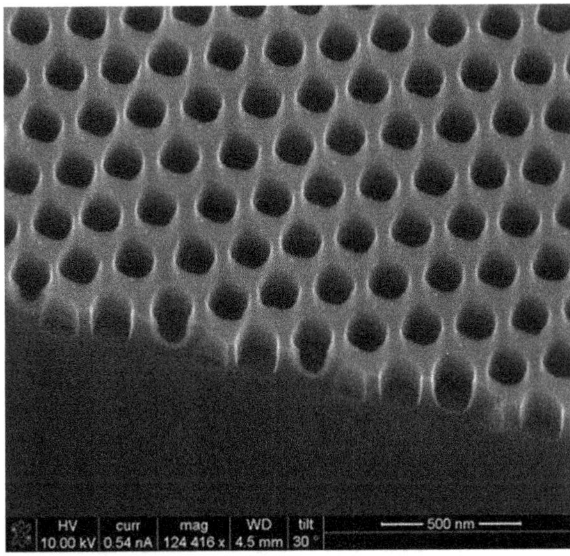

Fig. 9.13 SEM image of the cleaved nanostructured surface. Reprinted with permission from [30]. Copyright 2014 American Chemical Society [30]

Samples were aligned perpendicular to the beam (cavities parallel to the beam) to better than ±0.1°. SANS measurements with samples in air were conducted after drying them in a vacuum oven at 60 °C for 2 h. After completing these measurements, sample cells were filled with either degassed D_2O or a degassed contrast matched ($D_2O + H_2O$) mixture. The neutron beam size used was about 4.5 cm^2 in area and acquisition times were of the order of 3 h. The total data acquisition time of 3 h was split into six half hour runs, which were analyzed individually to check for possible changes of the bubble sizes with time. As long as the scattering data had no time dependence, the data were summed and averaged to increase the data quality.

Initial evaluation of the pore structure was conducted using the focusing mirror Very Small Angle Scattering (VSANS) diffractometer (KWS-3) (neutron wavelength $\lambda = 12.8$ Å, $\Delta\lambda/\lambda \sim 0.20$) available at Heinz Maier-Leibnitz FRM-II reactor. The experiments with wet samples were continued at Heinz Maier-Leibnitz FRM-II reactor on KWS-2 instrument (experiment 1) and at ORNL using the General Purpose SANS instrument (experiment 2) using $\lambda = 9.8$ Å ($\Delta\lambda/\lambda \sim 0.20$) and $\lambda = 12$ Å ($\Delta\lambda/\lambda \sim 0.13$), respectively. The sample-detector distance was chosen to cover an overall range of scattering vectors Q $0.0015 < Q < 0.015$ Å^{-1}. In experiment 1, SANS was measured from samples in air and samples immersed in D_2O. These measurements were repeated in experiment 2 in order to double check the reproducibility of the previous results. In addition, SANS was measured from both sample 1 and 2 in contact with contrast matched solution (CMS) of (40.85 wt% D_2O + 59.15 wt% H_2O) with an average scattering length density (SLD) matching that of Si wafer matrix.

Two-dimensional SANS patterns from both samples in air, in contact with D_2O, and immersed in CMS are shown in Fig. 9.14. Such patterns are typical for

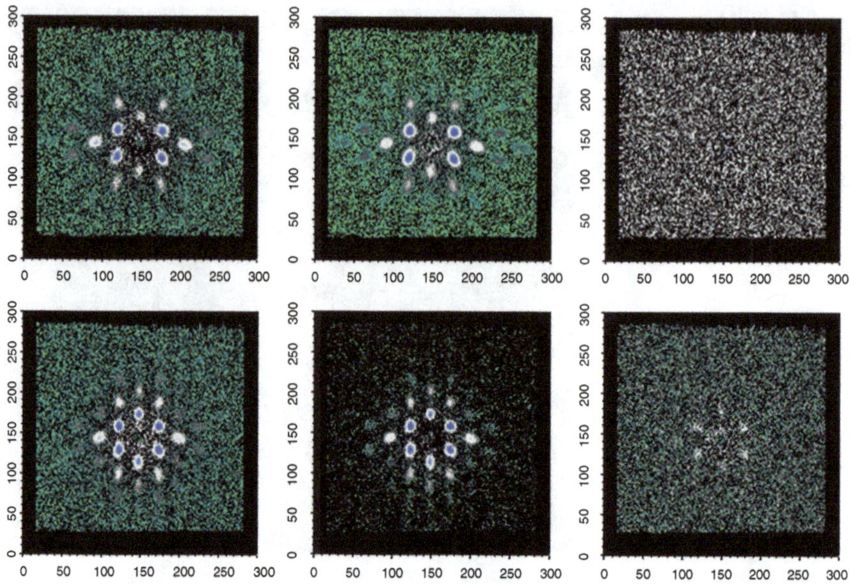

Fig. 9.14 Two-dimensional SANS patterns from sample 1 (**a**) and sample 2 (**b**) nanostructured wafers in contact with air, D_2O and contrast matched solution of $D_2O + H_2O$. Reprinted with permission from [30]. Copyright 2014 American Chemical Society

diffraction data obtained from a two dimensional array of hexagonally packed cylinders with the axes parallel to the beam. Diffraction peaks of up to three orders may be resolved for samples in contact with D_2O due to highly uniform patterning. Typical radially averaged scattering curves for sample 1 and 2 are shown in Fig. 9.15a, b, respectively. The position of the first order peaks for samples in air and in contact with D_2O is observed at $Q_{(10)} = 0.0037$ Å^{-1}, corresponding to the (10) reflection of the two-dimensional hexagonal lattice. This permits determination of the center-to-center distance between the cavities from the position of the main peak in Q-space by using the equation $d = 4\pi/(Q_{10}\sqrt{3}) = 1960$ Å, which is in a good agreement with the distance between cavities of 2000 Å estimated from the SEM image shown in Fig. 9.13. The secondary peaks are weaker but still detectable at positions $Q = 0.0061$ and 0.0092 Å^{-1}, corresponding to the (11) and (21) reflections, respectively.

As is seen in Table 1.2, neutron contrast between Si and D_2O $(2.08 \times 10^{10}\ \mathrm{cm}^{-2} - 6.39 \times 10^{10}\ \mathrm{cm}^{-2})^2 \approx 18.6 \times 10^{20}\ \mathrm{cm}^{-4}$ is much higher than that between Si and air filled cavities $(2.08 \times 10^{10}\ \mathrm{cm}^{-2})^2 \approx 4.3 \times 10^{20}\ \mathrm{cm}^{-4}$, which implies that water penetration into the cavities should lead to an increase of the amplitude of the observed peaks. The intensity of all peaks actually increases after immersing samples in D_2O, however the degree of the increase is smaller for the more hydrophobic sample 2 (see Fig. 9.15a, b). After immersing in contrast matching

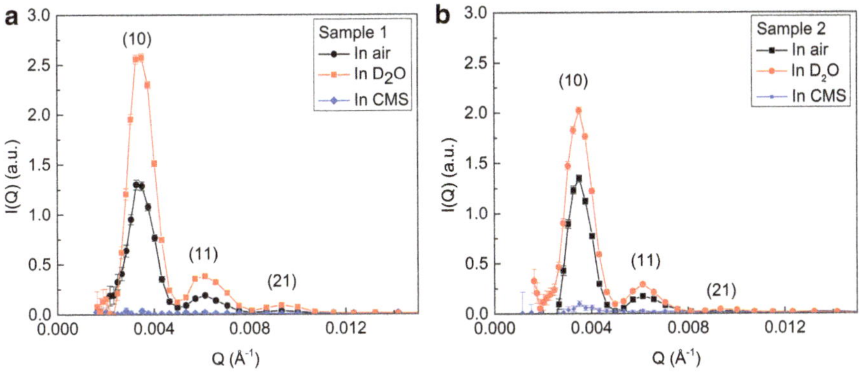

Fig. 9.15 Bragg peaks from (**a**) sample 1 and (**b**) sample 2 in air, in contact with D_2O, and in contact with contrast matched solution of $D_2O + H_2O$. Reprinted with permission from [30]. Copyright 2014 American Chemical Society

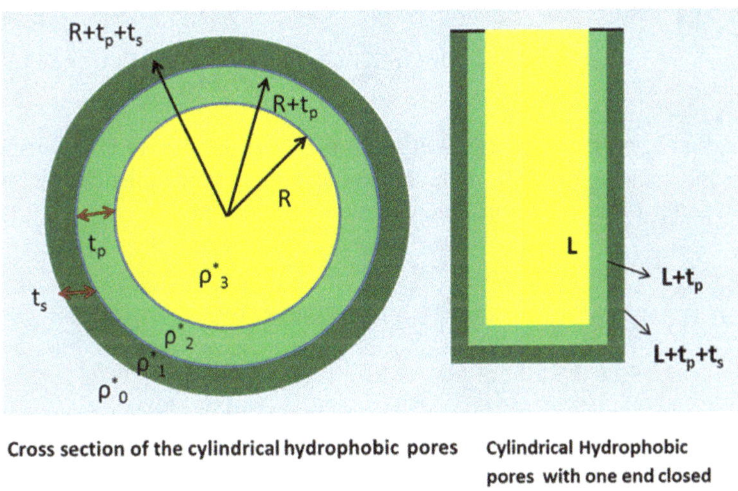

Fig. 9.16 Geometric model of the hydrophobically modified cylindrical pores used for calculation of the scattering amplitude and form factor of the hydrophobically modified cavities. *Blue*-Si matrix, *dark green*—SiO_2 layer, *light green*—fluoropolymer layer. Reprinted with permission from [30]. Copyright 2014 American Chemical Society

solution, the scattering from sample 1 becomes unmeasurable within the experimental error. At the same time, the SANS from sample 2 in contact with the CMS shows small but measurable residual scattering (Figs. 9.14 and 9.15).

Geometric model used for calculation of the scattering amplitude and form factor of the core-shell cylindrical cavities is shown in Fig. 9.16. For cylindrical cavities with a SiO_2 shell (sample 1) and same cavities having an additional shell of the fluoropolymer ($C_{10}H_4Cl_3F_{17}Si$) (sample 2), the scattering amplitude *F(Q)* is described by equation similar to *F(Q)* of spherical shells [31, 32]:

$$F(Q) = (\rho_3^* - \rho_2^*)\,\pi R^2 L f(Q,R,L) + (\rho_2^* - \rho_1^*)\,\pi (R+t_p)^2 (L+t_p) f(Q,R+t_p,L+t_p) + (\rho_1^* - \rho_0^*)\,\pi (R+t_p+t_s)^2 (L+t_p+t_s) f(Q,R+t_p+t_s,L+t_p+t_s) \quad . \tag{9.2}$$

In Eq. (9.2), $\rho_0^* = 2.08 \times 10^{10}\ \text{cm}^{-2}$ is the SLD of silicon, $\rho_1^* = 3.47 \times 10^{10}\ \text{cm}^{-2}$ is SLD of SiO_2, $\rho_2^* = 3.18 \times 10^{10}\ \text{cm}^{-2}$ is SLD of the fluoropolymer and ρ_3^* is SLD of the internal cylinder volume filled with air ($\rho_3^* = 0$), D_2O ($\rho_3^* = 6.39 \times 10^{10}\ \text{cm}^{-2}$) or contrast matching solution ($\rho_3^* = \rho_0^* = 2.08 \times 10^{10}\ \text{cm}^{-2}$) (Table 1.2). *R* and *L* is the radius and length of the internal volume of the cylinder, respectively. t_p and t_s is the thickness of the polymer and silica shell, respectively. For sample 1, it is assumed that $\rho_2^* = 0$, $t_p = 0$, $t_S = 2$ nm and the length of internal length $L = 100$ nm. Correspondingly, for sample 2 $t_p = 5$ nm and $L = 100 - 5 = 95$ nm. Function *f(Q,R, L)* is defined as:

$$f(Q,R,L) = 2J_0\left(\frac{QL\cos\alpha}{2}\right)\frac{J_1(QR\sin\alpha)}{QR\sin\alpha}, \tag{9.3}$$

where J_0 and J_1 is the first type Bessel function of zero and first order, respectively, and α is the angle between the cylinder axes and the normal to the incident beam.

Form factor of randomly oriented cylinders is given by:

$$P(Q) = \int_0^{\pi/2} F^2(Q,\alpha)\sin\alpha\, d\alpha. \tag{9.4}$$

In the case of cylinders oriented parallel to the beam (wafer placed perpendicular to the beam), $\alpha = \pi/2$ and Eq. (9.3) reduces to:

$$f(Q,R) = 2\,\frac{J_1(QR)}{QR}. \tag{9.5}$$

Hence, Eq. (9.2) can be rewritten as:

$$F(Q) = (\rho_3^* - \rho_2^*)\,\pi R^2 L f(Q,R) + (\rho_2^* - \rho_1^*)\,\pi (R+t_p)^2 (L+t_p) f(Q,R+t_p) + (\rho_1^* - \rho_0^*)\,\pi (R+t_p+t_s)^2 (L+t_p+t_s) f(Q,R+t_p+t_s). \tag{9.6}$$

The main Bragg peak $I(Q_{(10)} = 0.0037\ \text{Å}^{-1})$ is due to contrast between the scattering length density of the solid matrix ρ_2^* (Si with one (SiO_2) or two

(SiO_2 + fluoropolymer) shells and the medium filling the cavities ρ_3^* (air, D_2O or CMS). If the medium fills the cavity to a uniform depth, the intensity can be well approximated as $I \sim \left(\rho_0^* - \rho_3^*\right)^2$. For cavities filled with D_2O the peak intensity is expected to increase due to much higher contrast between Si and D_2O compare to that between Si and air. At the same time, for cavities filled with CMS the peak intensity should decrease due to decreasing contrast between the CMS and Si matrix. In the latter case, the volume fraction of cavities that is actually filled with CMS (ϕ_{CMS}) may be approximated by:

$$\phi_{CMS} = \frac{1 - \sqrt{(Measured\ pekratio}}{1 - \sqrt{\left(Peak\ ratio\ for\ completely\ filled\ cavities\right)}}. \tag{9.7}$$

Similar formula was used in [28] for calculation of the filling degree of Si cavities filled with H_2O measured using SAXS. In the two-phase approximation (Si pore walls filled with H_2O) (Eq. 9.7) reduces to:

$$\phi_{CMS} = \frac{\rho_{Si}^*}{\rho_{H_2O}^*}\left[1 - \sqrt{(Measured\ peak\ ratio)}\right.. \tag{9.8}$$

For D_2O-filled cylindrical cavities with composite matrix (additional shell of SiO_2 or two shells of SiO_2 + fluoropolymer) the filling degree is evaluated by calculating core-shell cylinder form factor (for the known sizes of the shells and corresponding SLDs) and comparing the ratios of the measured and calculated for the case of complete filling intensities at $I(Q)_{(10)}$:

$$\phi_{D2O} = \frac{\sqrt{(Measured\ peak\ ratio)} - 1}{\sqrt{\left(Peak\ ratio\ for\ completely\ filled\ cavities - 1\right)}}. \tag{9.9}$$

Equations (9.7) and (9.9) were used to calculate the experimentally measured filling degree of cavities with CMS and D_2O, respectively. Equation (9.6) was used to calculate the expected increase in the (10) peak amplitude for cavities completely filled with D_2O, which should be a factor of 4.83 and 6.68 for sample 1 and sample 2, respectively, compared to dry samples. The measured increase in scattering from cavities in contact with D_2O is actually much smaller than the predicted based on Eq. (9.6) for completely saturated cavities due to partial filling. The volume fraction of cavities that are completely filled with D_2O (ϕ_{D2O}) may be approximated by Eq. (9.10) in which *(Peak ratio for completely filled cavities)* is 4.83 or 6.68 for studied samples. Using values of the measured and calculated peak ratios it is possible to evaluate the pore filling degree ϕ_{D2O} for samples immersed in D_2O. The filling degree of cavities in sample 1 (ϕ_{D2O} = 0.31 and 0.36) measured in both experiments agree with each other. These numbers imply that about 1/3 of the cavity volume is filled with D_2O. Filling degree of pores with hydrophobic walls ($\phi_{D2O}=0.13$, experiment 1 and $\phi_{D2O}=0.14$, experiment 2) also agree with each other, however these numbers are about factor of three lower than ϕ_{D2O} of less hydrophobic cavities. Thus, in this sample 86–87 % of the cavity volume remains filled with vapor after immersing in the CMS.

The results of SANS experiments [30] demonstrate the high degree of sensitivity of SANS measurements for detecting and characterizing monolayers of nano- and mesoscale bubbles formed on the surface of different materials with the volume fraction as low as ~10^{-6} (volume of bubbles per cubic centimeter). The filling fraction of nanoscale pores (or size of trapped nanobubbles) depends strongly on the surface wettability. The volume of trapped nanobubbles may occupy as much as ~87 % of the cavity volume for strongly hydrophobic surfaces and ~2/3 of the pore volume in case of moderately hydrophobic surfaces. Filling fraction of the studied 100 nm cavities covered with fluoropolymer (sample 2) ($\phi_{D2O}=0.13$–0.14) is similar to that of the 100 nm deep cylindrical pores with much smaller diameter ~28 nm modified with an hydrophobic octadecyltrichlorosilane monolayer ($\phi_{D2O}=0.12$ [29]). This suggests that the surface wettability has a greater influence on the formation of nanobubbles in cavities of these length scales than the cavity geometry. At the same time, the SANS data demonstrate that relatively moderate hydrophobicity of nanostructured surfaces characterized by contact angles below 90° is sufficient for high density of interfacial nanobubbles to be formed. The detected nanobubbles are stable with time for at least several hours, which confirms the conclusion that dewetted and partially dewetted hydrophobic pores may be thermodynamically stable in contact with water under ambient pressure and temperature.

SAS experiments conducted in [28–30] demonstrated partial filling of small hydrophobic cavities with water at atmospheric pressure. Hydrophobic coatings repel liquids by trapping air inside microscopic surface textures. However, the trapped nanobubbles are expected to collapse and thus superhydrophobic state to vanish under external pressure, for which water infiltrates the surface texture completely [33, 34]. In particular, pressure-induced infiltration occurs because the liquid-air interface above the rough solid surface can sustain only a finite pressure approximately equal to the capillary pressure proportional to the water-air surface tension.

In situ synchrotron SAXS study of *forced* water infiltration into various hydrophobic nanotextures was conducted in [35] in order to understand the role of surface geometry and air-liquid exchange for the collapse and reversibility of superhydrophobicity at the nanoscale. Block-copolymer self-assembly and plasma etching were used to create large-area silicon surfaces containing features with uniform size and spacing on a length scale ~10 nm. Anisotropic silicon etching produced cylindrical pillars and lamellae with nearly vertical sidewalls (Fig. 9.17a, b). Tapered conical structures ≅10 nm wide at their tips (Fig. 9.17c, d) were obtained by introducing a slow lateral etch rate. Nanostructured silicon surfaces were rendered hydrophobic by silane coating. All nanostructures had a feature spacing

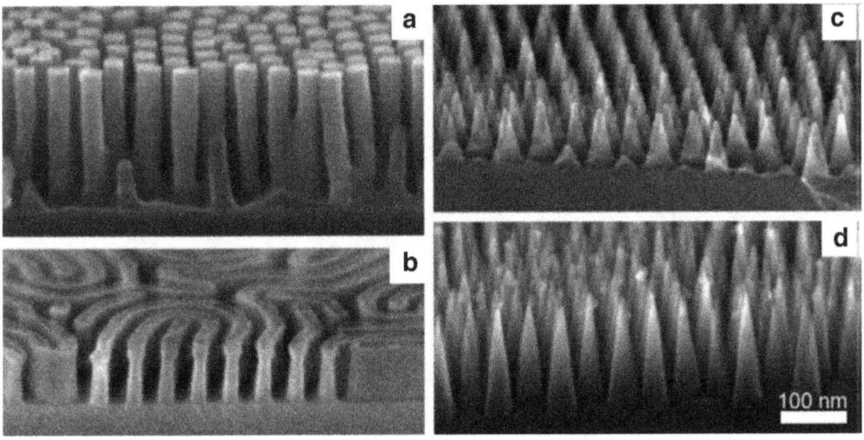

Fig. 9.17 Oblique side-view scanning electron microscope pictures of silicon surfaces patterned with (**a**) a hexagonal array of nanopillars, (**b**) nanolamellae, and (**c**, **d**) hexagonal arrays of nanocones with different aspect ratio (the scale bar is the same for all pictures). Reprinted figure with permission from [35]. Copyright (2014) by the American Physical Society

of $d \cong 50$ nm, whereas the feature size w at the top surface, the feature height h, and the sidewall angle α varied within the range 15 nm $< w <$ 35 nm, 75 nm $< h <$ 180 nm, and $1° < \alpha < 10°$, respectively.

SAXS experiments were carried out at beam line X9 of the National Synchrotron Light Source, Brookhaven National Laboratory. A collimated, high-energy (14 keV) x-ray beam was directed through the x-ray compatible pressure cell consisted of aluminum body with two windows made of a ~0.5 mm thick nanopatterned silicon wafer and a thin (~0.1 mm) foil of Kapton (Fig. 9.18a). The scattering patterns were collected using a 2D CCD detector positioned approximately 5 m behind the sample. The windows were sealed using Viton gaskets and the cell was filled with ultrapure water. The hydrostatic pressure was varied using a glass syringe and measured using a silicon pressure transducer.

A representative SAXS pattern generated by the array of nanopillars at atmospheric pressure consists of two powder diffraction rings—a brighter inner (10) ring and a fainter (11) outer ring—characteristic of the random distribution of the domains of the 2D hexagonal lattice (Fig. 9.18b). The azimuthally integrated, background-subtracted (10) peak intensity was found to have a nearly constant amplitude for hydrostatic pressure in excess of atmospheric pressure (ΔP) up to $\Delta P \approx 15$ atm and to decrease monotonously beyond that point up to 45 atm (Fig. 9.18c). A similar qualitative behavior was observed for all nanotextured surfaces.

The observed reduction of diffracted intensity reflects the loss of electron density contrast caused by water infiltration into the texture. If the liquid penetrates the texture to a laterally uniform depth, the volume fraction $\phi_W(\Delta P)$ of the texture that is filled with water is given in good approximation by [35]:

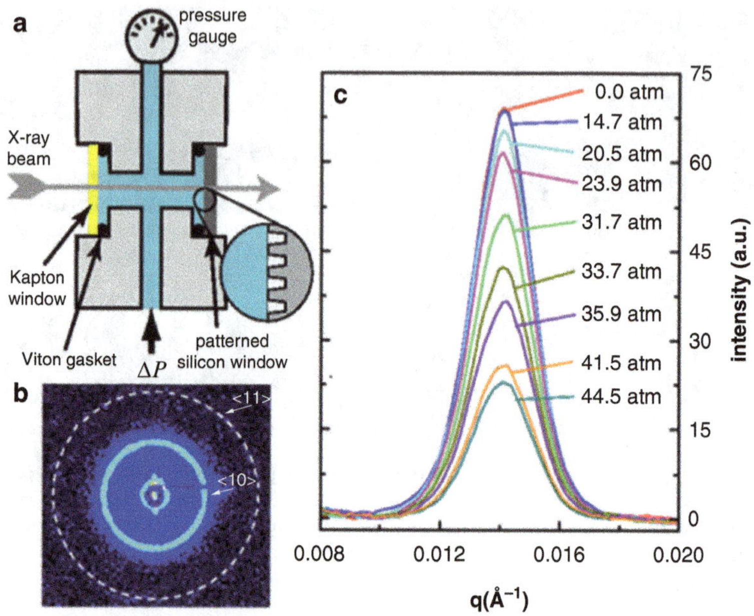

Fig. 9.18 (**a**) Schematics of the SAXS experiment. (**b**) SAXS pattern from the array of nanopillars at $\Delta P = 0$ atm, where the arrows denote the position of the (10) and (11) Bragg rings (the latter being highlighted by the *dashed line*). (**c**) Azimuthally integrated x-ray intensity scattered from the sample with nanopillars as a function of the wave vector transfer q at various pressures. Reprinted figure with permission from [35]. Copyright (2014) by the American Physical Society

$$\phi_V(\Delta P) = \frac{(\rho_X^*)_{Si}}{(\rho_X^*)_W}\left(1 - \sqrt{aR(\Delta P)}\right), \tag{9.10}$$

where $R(\Delta P) = I(\Delta P)/I(0)$ is the ratio between the (10) peak intensity at ΔP and at atmospheric pressure, and a is a normalization factor introduced to account for the observation that for all nanotextures there is some spontaneous infiltration ϕ_O, even at $\Delta P = 0$. ϕ_O was found through independent SAXS experiments to be 0.05 for the nanocones and nanolamellae and 0.08 for the nanopillars accounted for by setting a = 0.95 and 0.93, respectively, in Eq. (9.10).

The infiltration isotherms for nanocones and nanolamellae are shown in Fig. 9.19. Very little infiltration occurs up to a critical pressure ΔP_C ($\Delta Pc \approx 5$ and ≈ 25 atm for the nanocones and the nanolamellae, respectively), beyond which the isotherms increase strongly. Complete filling is reached at $\Delta P = 22$ atm for the nanocones and at 41 atm for the nanolamellae. Another quantitative difference between the infiltration of the two textures is that (up to $\phi_V = 0.5$) the infiltration curve for the nanolamellae is steeper than for the nanocones. When the pressure was progressively reduced to the atmospheric one (hollow symbols) no liquid extrusion was observed for the nanocones whereas ϕ_V decreased to 0.94 for the nanolamellae.

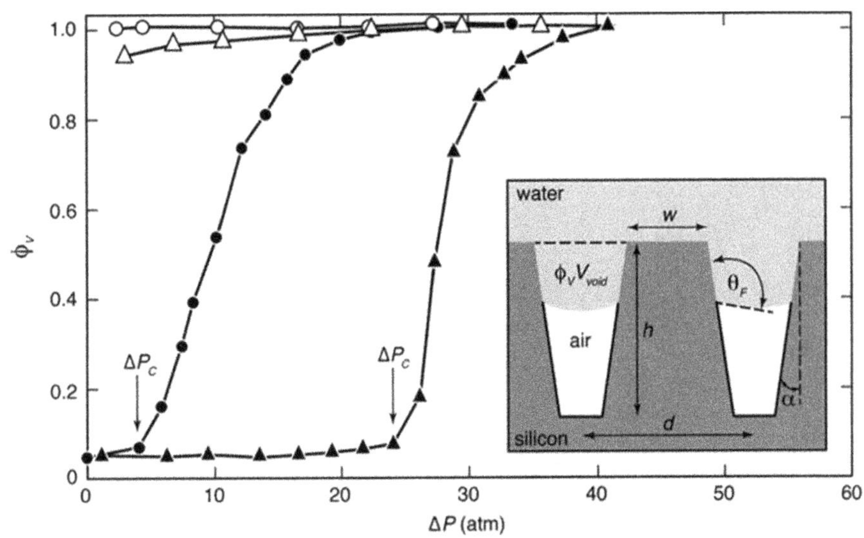

Fig. 9.19 Filling fraction ϕ_V versus hydrostatic pressure in excess of atmospheric pressure, ΔP, for the short nanocones (*filled circles*) with $h = 75$ nm, and the nanolamellae (*filled triangles*) with $h = 75$ nm). *Filled* and *hollow symbols* represent the intrusion and the extrusion curve, respectively. Reprinted figure with permission from [35]. Copyright (2014) by the American Physical Society

The isotherms for the taller nanocones and the nanopillars were qualitatively similar (Fig. 9.20a). For these textures, infiltration increased abruptly at $\Delta Pc \approx 5$ and 28 atm for the nanocones and the nanopillars, respectively. Complete filling was reached at $\Delta P \approx 35$ atm for the nanocones, whereas for the nanopillars the maximum filling of 0.95 was obtained at $\Delta P = 45$ atm.

Additional infiltration experiments were performed in which the pressure was repeatedly cycled to reach intermediate infiltrations before complete filling the textures. The detailed infiltration path is shown in Fig. 9.20b by arrows. These experiments demonstrated that both sorption and desorption isotherms depend not only on the texture geometry and the aspect ratio but also on the sample history.

Theoretical analysis of the data obtained indicated that macroscopic capillary theory might provide an adequate description of water infiltration into surface nanotextures with a variety of geometries. This suggested that line tension and thermal fluctuations of the gas-liquid interface did not play a predominant role at the 10 nm scale. As water infiltrates the texture, trapped gas dissolves irreversibly into the liquid and this mass transfer is not inhibited by the spatial confinement of the liquid-gas interface. Despite the complete gas dissolution observed at high pressure, the reappearance of the gas phase upon reducing the pressure may occur spontaneously owing to the texture geometry and favorable aspect ratios. This process appears to be affected by the kinetics of the gas-liquid exchange.

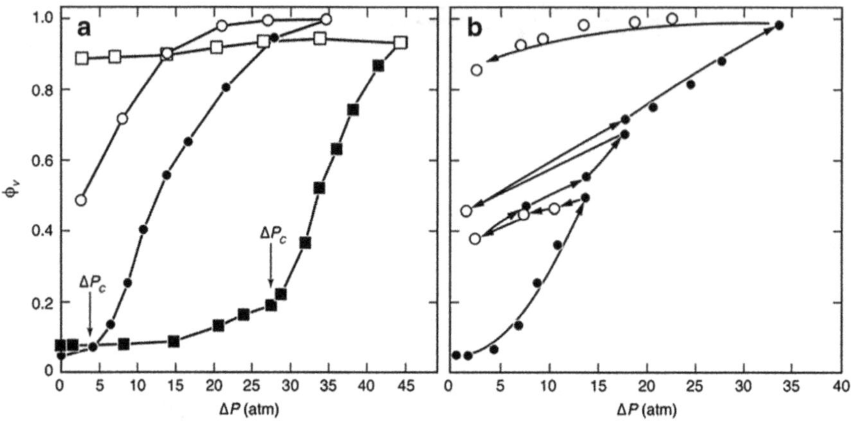

Fig. 9.20 (**a**) Infiltration isotherm for 170 nm tall nanocones (*filled circles*) and 180 nm tall nanopillars (*filled squares*). *Filled* and *hollow symbols* represent the intrusion and extrusion curve, respectively. (**b**) Isotherm for 170 nm tall nanocones obtained by varying the hydrostatic pressure cyclically as indicated by the *arrows*. Reprinted figure with permission from [35]. Copyright (2014) by the American Physical Society

The results of SAS studies may have implications for the understanding and the rational design of nanosized, multiphase systems for microchannel "laboratories on a chip", where channel wetting plays important role in device functionality.

9.4 Liquid–Liquid Demixing in Mesopores

Understanding the influence of confinement on the miscibility of liquid solutions has important practical implications as the preferential adsorption and the reduced miscibility window may affect the efficiency of separation technologies, oil recovery, lubrication and coating. From the fundamental standpoint, a detailed assessment of the role of disorder, confinement, and interfacial phenomena is required in order to quantify the behavior of confined solutions. Thermodynamic and dynamic behavior of fluids in disordered porous materials has been the subject of intense experimental and theoretical interest over the past decades [36]. A number of general trends observed so far include strong metastability, absence of the macroscopic phase separation, as well as considerable shift and shrinkage of the miscibility window on the phase diagram of confined solutions.

A suggestion that the porous medium can act as the "random" field on a liquid mixture because of its preference for one of the solution components over the other was first made in [37]. In this case system behavior is described by random-field Ising model (RFIM). This hypothesis should be relevant when the correlation length of the concentration fluctuations in the liquid mixture becomes longer than the mesh (or pore) size of the porous medium, which should occur near the critical temperature T_C of the liquid-liquid demixing. The validity of this argument should also depend on topological properties of the porous medium, and the relative strength of interaction between solution molecules and the pore surface [38].

The alternative "single pore" or "wetting" model predicted formation of frozen liquid microdomains below T_C in each individual pore independently of randomness [39]. According to this model, the behavior of binary solutions in small pores is predominantly controlled by the intermolecular interactions, which reduce the rate of domain growth near T_C and thus inhibit macroscopic phase separation. Depending on the relative strength of the molecule–molecule and molecule–surface interactions and the proximity to the critical demixing point, the solution components in a pore may form a "tube", in which one component is "plated" out onto the pore surface, a "plug", in which cylindrical plugs of both components coexist separated by hemispherical interfaces, or a "capsule", which represents plugs separated from the pore walls by a layer of the second solution component [39].

SAS provides spatial resolution required for investigating the preferential adsorption and demixing processes occurring in pores on the micro- and mesoscale. SANS was first employed to study phase demixing of water–lutidine mixtures in Vycor porous glass [40, 41] as well as in silica gels [42]. Scattering from the demixing solution in Vycor was described by the sum of a Lorentzian and a Lorentzian-squared term:

$$I(Q) = \frac{A_1}{1 + Q^2\xi_1^2} + \frac{A_2}{(1 + Q^2\xi_2^2)^2}. \tag{9.11}$$

In the single-phase region of the liquid solution $\xi_1 = \xi_2 = \xi$, and the Lorentzian describes the scattering from critical fluctuations of the correlation length ξ, and the Lorentzian-squared term describes the static response of the system to random field imposed by the porous medium, i.e. scattering from the concentration fluctuations that are "pinned" to the matrix. In the two-phase region, $\xi_1 \neq \xi_2$, and the Lorentzian-squared term describes scattering from the randomly distributed liquid domains of characteristic size ξ_2. A common observation made in [40–42] was that the correlation length never exceeded the pore size in Vycor even well inside the two-phase region of the bulk solution, in a qualitative agreement with the single-pore model. This observation was attributed to strong preferential adsorption of water on silica pore walls, which should shift the composition in the pore core far away from the critical concentration.

Contrast matching SANS was used in [43, 44] to study the structure of n-hexane + n-perfluorooctane solutions at the critical concentration of components in porous Vycor glass. Both liquids are characterized by a similar adhesion to silica dioxide, which should help to minimize the effect of preferential adsorption on the solution miscibility and the amplitude of critical concentration fluctuations in pores. The experiment was carried out at the PAXY diffractometer of the Laboratoire Leon Brillouin (Saclay-France). Slabs of Vycor ($32 \times 10 \times 1.84$ mm^3) were cleaned and placed in a quartz cell with the internal thickness 2 mm, close to the thickness of Vycor glass in order to minimize the scattering from the unconfined bulk fluid. The mixture was loaded at $T = 60\ ^\circ C$, i.e. well inside the single phase region of the bulk mixture (T_C of bulk solution 33 °C).

SANS scattering was measured in the temperature range $15 < T < 45\ ^\circ C$, which covers both the single-phase ($T > 33\ ^\circ C$) and the phase separation ($T < 33\ ^\circ C$) region of the phase diagram of bulk solution. A significant increase in the small-angle neutron scattering was observed in the limit of small $Q < 0.02\ \text{Å}^{-1}$. In this Q-range, the scattering data were fitted to the Ornstein-Zernike formula (first term in Eq. (9.11)) and the value of the correlation length ξ was extracted as a function of temperature. It was found that for solutions of fully deuterated and fully protonated components the correlation length of the concentration fluctuations may exceed (up to a factor of two) the pore diameter in Vycor ($D_P \cong 75$ Å). This effect was attributed to the negligible preferential adsorption in the studied system. It was concluded that, despite the cross sectional steric restriction imposed by the cylindrical pores, the critical fluctuations might extend along the pore axes, in a qualitative agreement with the RFIM.

SANS was used in [45] to study the local structure of a phase-separating mixture of isobutyric acid (iBA) + D_2O in a controlled porous glass (CPG-10), which constituted a random interconnected pore network similar to Vycor, but having larger pore diameter ~100 Å and total porosity ($\phi_P \cong 0.5$). Bulk solution, used in this study exhibits a lower miscibility gap with an upper critical solution temperature $T_C \cong 45\ ^\circ C$ and a critical volume fraction of iBA ≅ 0.37. The mesoscale structure of the liquid in the pore space was investigated as a function of temperature at two off-critical compositions of bulk mixture, corresponding to the D_2O-rich and the iBA-rich branch of the coexistence curve. Temperature range covered in the experiments included both the single-phase ($T < T_C$) and the phase separation region above T_C.

The SANS experiments were performed at the SANS-2 facility at GeNF, GKSS Geesthacht, Germany. Neutrons with mean wavelength $\lambda = 5.7$ Å ($\Delta\lambda/\lambda = 0.1$) were used. The range of the scattering vector $0.004 < Q < 0.22\ \text{Å}^{-1}$ was covered by using three detector-to-sample distances $L = 1$, 4, 14, and 19 m. The sealed sample cells with the CPG-10 disks were mounted in a container, the temperature of which was controlled to better than ±0.05 °C in the range $10 < T < 70\ ^\circ C$.

SANS from the D_2O-rich solution (weight fraction of $D_2O \cong 0.75$) did not reveal any variation with temperature. This was explained as due to the strong preferential adsorption of water, which displaced the minority component (iBA) from narrow pore regions into the sections with larger width, where iBA formed capsule-like domains with sizes independent of temperature. SANS data obtained from the iBA-rich sample (weight fraction of iBA ≅ 0.54) showed a measurable variation with temperature (Fig. 9.21). The data were fitted to the modified model function (Eq. 9.11), which accounted for the normalized scattering from the dry CPG matrix and an adsorbed liquid layer. The analysis of thus obtained parameters ξ_1 and ξ_2 suggested that transition from a single-phase to the microphase separated condition

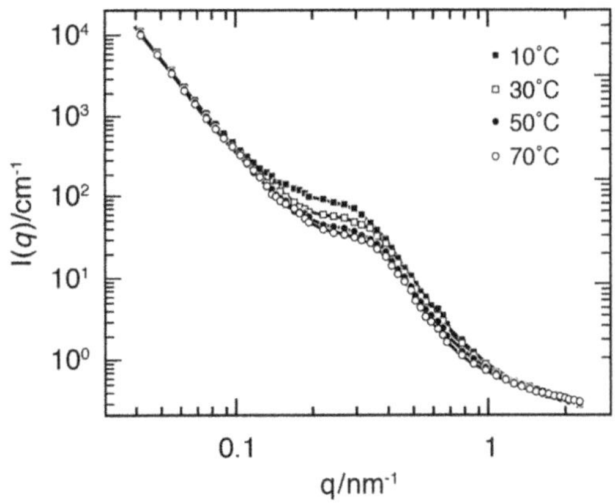

Fig. 9.21 SANS data for the iBA-rich mixture in CPG-10 at temperatures shown in the inset. Reprinted with permission from [45]. Copyright [2005], AIP Publishing LLC

lead to a formation of tube-like structures with water-rich layer at the pore walls and iBA component in the pore core.

Synchrotron SAXS was used to study a critical hexane/nitrobenzene solution permeated in a mesoporous resorcinol formaldehyde (RF) gel [46]. Calcinated RF aerogels are promising materials for catalysis applications and fuel cell electrodes. Their structure has been characterized thoroughly before using SAS methods [47, 48], which make them a suitable model of disordered mesoporous materials.

The RF gel was synthesized *via* polycondensation of resorcinol and formaldehyde in water and aged for three days at 343 K in a closed vial. Based on the composition of the starting solution, the solid fraction of the gel was estimated to be 25 vol.%. The mother saturating gel solvent (mostly water) was exchanged for a critical hexane/nitrobenzene solution by cutting gel into small cubes, 1 mm in size, washing them in a large excess of acetone, and subsequently immersing in a large excess of a critical hexane/nitrobenzene solution (critical concentration of nitrobenzene 42.2 mol.% and critical demixing temperature $T_C = 293.5$ K).

The SAXS measurements were conducted at the Dutch-Belgian station, BM26, of the European Synchrotron Radiation Facility (Grenoble, France). The SAXS patterns were measured on a 2D multiwire gas-filled detector placed at 5 m from the sample. A millimeter sized piece of gel was taken out of the supernatant solution and placed in hermetically closed aluminum cells. The temperature was controlled using a heating/freezing stage. The SAXS measurements were performed while the sample was cooled from 303 to 213 K at a rate of 5 K/min. The acquisition time was set to 12 s, which corresponded to one SAXS pattern per Kelvin.

Figure 9.22 shows the SAXS intensity measured from the RF gel saturated with the critical binary nitrobenzene/hexane solution as a function of temperature. Some extra scattering appears at low Q when the temperature is lowered below 288 K, which points to the structural changes in the system at the mesoscopic scale. In the limit of large values of Q, the intensity of x-rays follows the Porod law $I \sim Q^{-4}$ and

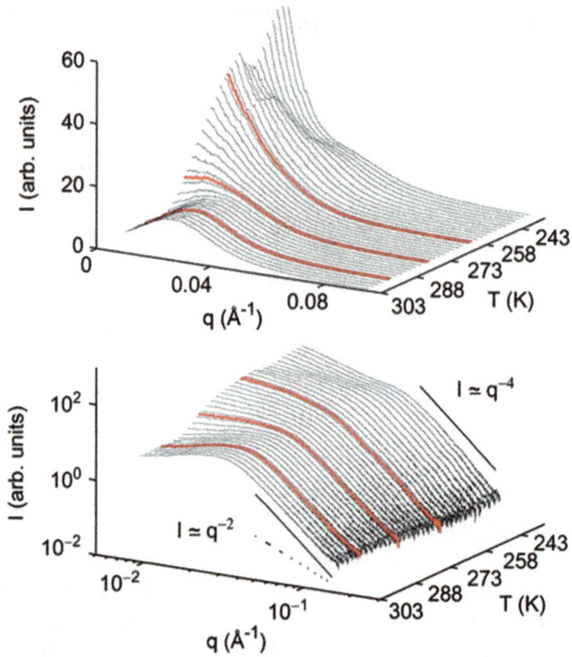

Fig. 9.22 Small-angle x-ray scattering patterns of the RF gel permeated with critical nitrobenzene/hexane solution, during cooling from 303 to 243 K. The same data are plotted in the *top* and *bottom graphs* on linear and logarithmic scales, respectively. The *two solid lines* in the *bottom graph* are Porod scattering of the type $I \sim Q^{-4}$, and the dashed line is the $I \sim Q^{-2}$ scattering relevant to concentration fluctuations. The three highlighted temperatures correspond to 293, 278 and 259 K. They are analyzed further in Fig. 9.23 [46]. Published with IUCr's copyright permission

not the Ornstein-Zernike function $I \sim Q^{-2}$ characteristic of the scattering from concentration fluctuations (Fig. 9.22, bottom). Analysis of the Porod invariant can deliver information about changes of the volume fraction of coexisting microphases (Sect. 6.3). In the present case, however the value of the invariant remained independent of temperature, which indicated that the observed evolution of the SAXS data was occurring mostly due to a reorganization of existing phases within the pores of the RF gel. These two observations lead to a conclusion that the scattering from the studied system cannot be adequately explained by neither RFIM nor the single pore model.

The scattering data from the mesoporous gel with two coexisting liquid phases in its pores were analyzed based on plurigaussian geometrical model, which was first introduced in the context of geological engineering [49] and could account for a variety of nanostructures of the pore filling phases. Figure 9.23 displays realizations of the plurigaussian model using the parameters derived from SAXS. At high temperature, the nitrobenzene phase forms a nanometer-thin layer that covers uniformly the surface of the solid. In this configuration there is no contact between

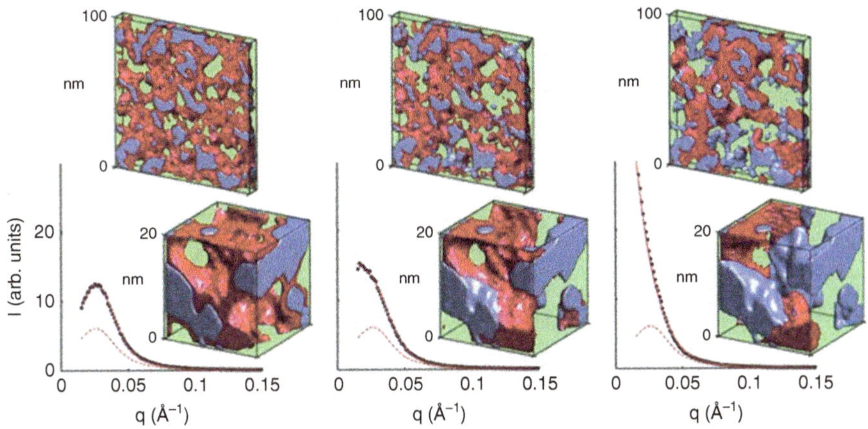

Fig. 9.23 Reconstructed nanostructure of the nitrobenzene/hexane phases inside the pores of the RF gel, at three different temperatures corresponding to the temperatures highlighted in Fig. 9.22. The solid phase is in *gray*, the nitrobenzene phase is in *red* and the hexane phase is shown as the *green background*. Two reconstructions are shown at each temperature, corresponding to two different scales. The experimental SAXS patterns are shown (*filled circles*) together with the SAXS curves of the reconstructions (*red line*). The theoretical SAXS curve of the solid filled with a homogeneous hexane/nitrobenzene solution is displayed for comparison (*dotted red line*) [46]. Published with UICr's copyright permission

the solid and hexane phases. A dewetting transition is observed when the temperature is decreased, through which the nanostructure of the nitrobenzene phase progressively changes from a layer to a collection of isolated plugs, which extend over the entire pore section. When the temperature is further decreased, a coarsening process is observed, when the number of plugs decreases and their size increases.

The SAXS data obtained and the plurigaussian reconstruction analysis of the solution morphology in the RF gel reveal existence of two distinct phases already at high temperature 303 K $> T_C = 293.5$ K. The observed morphology was explained by a difference in the dispersive forces exerted by the solid phase on nitrobenzene and hexane molecules. The nitrobenzene-rich phase can be viewed as a region of space where nitrobenzene molecules concentrate under the influence of dispersive forces exerted by the solid, rather than as a thermodynamic phase. At $T = 303$ K, bulk solution is miscible and there is no surface tension associated with the nitrobenzene/hexane interface. As the temperature is decreased below T_C, there is a thermodynamic penalty associated with formation of a nitrobenzene/hexane interface, which should be minimized by forming morphology with a smaller interface area. This is achieved *via* the nitrobenzene dewetting and the coarsening process, which reduces number of nitrobenzene plugs and increases their size. The final morphology is a result of a balance between the dispersive forces and the surface tension between liquid phases.

9.5 Supercooled Water in Confined Geometries

Abundance of water on the earth makes it an important participant of a variety of physical, chemical, biological and geological processes, as well as materials science and technological applications. The interfacial and confined water as well as aqueous solutions may be found in various geological systems, where they act as reaction media, reactants, and carriers of energy and matter in the natural environment [50]. Considerable amount of water in living organisms is confined in cells, where it hydrates membranes and large biomacromolecules. Confinement near hydrophobic and hydrophilic surfaces of nano- and mesopores may alter dramatically thermodynamic, dynamic, and structural properties of water molecules and result in the appearance of various surface transitions and property gradients [51]. Such alteration plays fundamental role in many processes such as protein folding, concrete curing, corrosion, as well as molecular and ionic transport [52].

Water properties have a set of unique physical, thermodynamic, structural, and density anomalies [53], which can be traced to the features of its structure that represents an extended hydrogen bonding network. One of the most well-known anomalies is the density maximum of H_2O occurring at about four degree above its melting temperature ~277 K. The density of bulk water below the density maximum and in the supercooled phase decreases rapidly with temperature before the onset of the ice nucleation precludes further supercooling. Since the density of ice increases with decreasing temperature, it was predicted that if it were possible to supercool bulk water below the currently achievable limit of ~240 K, a minimum of density should be reached at a certain temperature. Such density minimum was indeed observed in recent molecular dynamics experiments [54–56].

Ice nucleation at ~240 K can be suppressed by confining water in small pores, which may help to experimentally assess water properties within the normally inaccessible, low-temperature region. Such approach was applied in [57, 58] in which density of supercooled D_2O confined in cylindrical pores of MCM-41-S-15 with diameter 15 ± 1 Å was studied as a function of temperature using SANS.

The MCM-41-S-15 powder was made of micellar templated mesoporous silica matrices and had 1D cylindrical pores arranged in a 2D hexagonal lattice. It consisted of approximately spherical crystallites, or grains of the order of micrometer size. The sample was hydrated by exposing it to D_2O vapor in a closed chamber until it reached the full hydration level of 0.5 g of D_2O per 1 g of silica. SANS measurements were conducted at 14 temperatures between 160 and 290 K in steps of 10 K to monitor the variation of the density in the supercooled region.

SANS experiments were performed at NG3, a 30-m SANS spectrometer, in the National Institute of Standards and Technology Center for Neutron Research (NCNR). The incident monochromatic neutrons have an average wave length of $\lambda = 5$ Å with a fractional spread of $\Delta\lambda/\lambda = 10$ %. The sample to detector distance is fixed at 6 m, covering the range of magnitudes of neutron wave vector from 0.008 to 0.40 Å^{-1}. This Q range includes the interfacial scattering between different grains, as well as the Bragg peak due to the hexagonal array of silica pores within a grain.

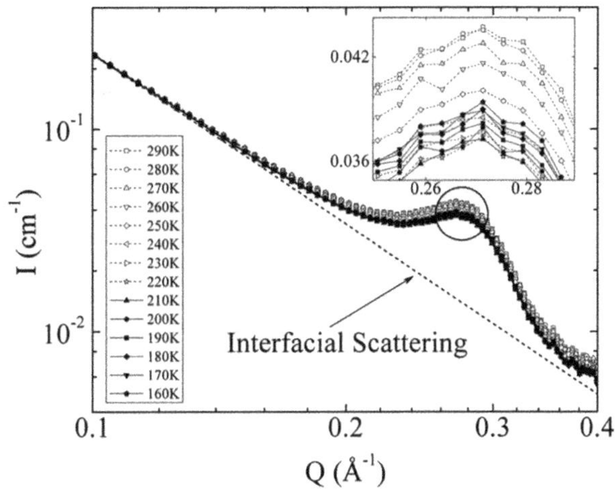

Fig. 9.24 SANS intensity as a function of Q, for MCM-41-S-15 hydrated with D_2O as a function of temperature (*inset*). Reprinted with permission from [57]. Copyright (2007) National Academy of Sciences, U.S.A.

The peak amplitude can be used to monitor the density of water in the fully hydrated sample as the SANS intensity is proportional to the square of the difference of the SLD of the confined liquid and that of the porous matrix. The SLD of a molecular liquid is proportional to the product of the scattering length of the nuclei and the physical (mass) density of the liquid (Eqs. 1.15 and 1.17). Assuming that the scattering lengths of D_2O and SiO_2 are temperature independent and the density of SiO_2 matrix does not change significantly with temperature, the variation of the peak amplitude may be treated as due to the variation of the density of confined D_2O.

SANS intensity as a function of Q at different temperatures is shown in Fig. 9.24. The dashed straight line represents the interfacial scattering coming from the surfaces of the grains of the silica crystallites. Each scattering curve shows a power law region and a first-order diffraction peak situated at $Q_{(10)} = 0.287$ Å^{-1}. The diffraction peak is due to Bragg diffraction from the 2D hexagonal lattice made of parallel D_2O-filled cylinders contained in grains of the silica crystallites. The height of the diffraction peak is proportional to the square of the difference of the SLD of D_2O inside the cylindrical pore, and the silica matrix. The peak amplitude decreases when the sample is cooled down in the temperature range from 280 to 220 K, which can be interpreted as due to the corresponding decrease of the D_2O density in pores. After the initial reduction, SANS intensity begins to increase with temperature in the range from 210 to 160 K.

Figure 9.25 compares the observed density minimum with the result of the computer simulations of supercooled water [55]. The SANS and computer simulation data agree well: $T_{max} - T_{min}$ calculated in [55] is ≈70 K, compared with 80 K reported in the SANS study [57]. The ratio of the maximum to minimum density from computer simulations (1.05), is close to 1.06 measured using SANS [57].

SANS experiments with D_2O- and H_2O-hydrated MCM helped to estimate the average density of confined water, which was found to be 8 % higher than the

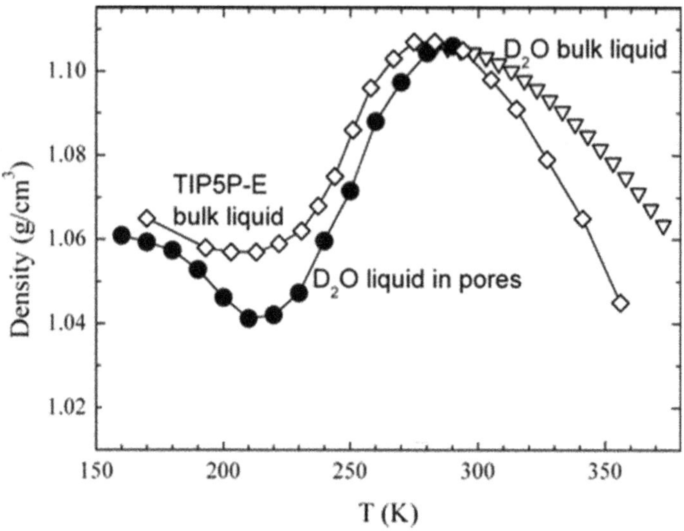

Fig. 9.25 Comparison of density vs. temperature curves at ambient pressure for bulk liquid D_2O (*open triangles*), confined liquid D_2O (*filled circles*), and molecular dynamics simulations of liquid TIP5P-E water (*open diamonds*). The density values for the TIP5P-E model (which is parameterized as model of H_2O) have been multiplied by 1.1 to facilitate comparison with the behavior of D_2O. Reprinted with permission from [57]. Copyright (2007) National Academy of Sciences, U.S.A.

density of bulk water inside 19 Å pores [58]. SANS measurements of the D_2O density profiles in MCM as a function of pressure revealed significant shift of the density minimum position from 210 to 170 K as the pressure was increased from ambient to 2900 bars [59].

SANS data analysis in [57–59] was performed based on the two-phase model, i.e. assuming a homogeneous distribution of water molecules in the mesopore volume. Later SAS studies of the confined water [60, 61] in MCM revealed the development of the inhomogeneous density profiles across the pore volume. The details of the inhomogeneities may be substrate- and temperature dependent, due to possible changes of the balance between water-water and water-surface interactions. As a consequence, the concept of average fluid density in pores may not be sufficient to fully understand the structure of confined supercooled water (Fig. 9.26).

9.6 Order-Disorder Transitions in Liquid Crystals

Orientational ordering, i.e. a process of alignment of molecular axes in a particular direction, is a key property of functional materials with switching capability, such as nematic liquid crystals (LCs), as well as ferromagnetic and ferroelectric

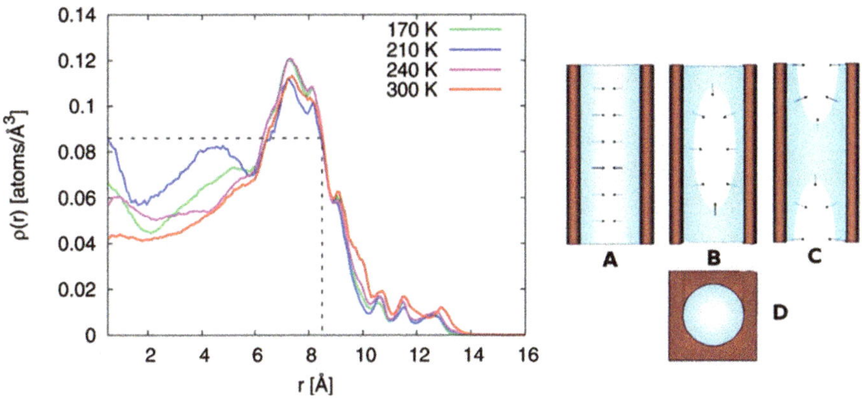

Fig. 9.26 *Left*: (**a**) The water density profile, $\rho(r)$ versus the distance from the pore axis, r calculated for MCM-41-S-15 at temperatures shown in the *inset*. The *black dashed lines* indicate the estimated wall position, and the average water density, according to the hydration level and total pore volume (TPV) of the sample. (**b**) schematic view of the three pore filling scenarios compatible with the density profile $\rho(r)$ shown in (**a**). *Right*: (**A**) there is a layer of water molecules wetting the pore surface and a depletion of density towards the pore center, resulting in density fluctuations along the r direction; (**B**) there is a cohesive failure between water molecules within the pore, giving density fluctuations along the radius and axis of the cylinder; (**C**) there is a meniscus at each end of the pore. When the temperature decreases, water molecules move in the direction shown by the arrows and the density fluctuations decrease in all three scenarios. (**D**) Top view of the cylinder, showing the density profile after averaging along the cylinder axis. Reprinted with permission from [61]. Copyright 2010 American Chemical Society

materials. LCs are extensively used in various applications, including displays, optical switches, and a variety of other electro-optical devices. They represent a class of soft materials with a weak long-range orientational and translational order. Therefore, their physical and structural properties are expected to be significantly modified by geometrical confinement, the chemistry of the pore surface, the pore dimensionality, and quenched disorder imposed by the pore wall morphology [62].

A number of studies were devoted to phase behavior and optical properties of LCs confined in various structures [63]. New properties of the confined LCs have been documented, including previously unknown phase transitions, extreme optical turbidity, and sluggish, glassy-like dynamic behavior. Two major areas of the SAS applications in the area of confined LCs may be outlined as investigation of the structure of LC—polymer composites [64, 65], and modification of the LC structure and phase behavior in porous media [66–70].

One of the most extensively studied LCs is 4-n-octyl-4′-cyanobiphenyl (8CB). In bulk, the sequence of phases formed with increasing temperature is crystal (K) to smectic A (A) to nematic (N) to isotropic (I), with the corresponding transition temperatures $T_{KA} = 294.5$ K, $T_{NA} = 306.3$ K, and $T_{NI} = 313.5$ K. A comprehensive study of the phase diagram of 8CB confined in porous silica was conducted in [67] using a combination of SANS, SAXS, and neutron diffraction techniques.

Porous matrix was made from a (100) oriented silicon substrate using an electrochemical anodization that results in formation of highly anisotropic, perpendicular to the wafer surface pores with an average diameter of 30 nm and high aspect ratio 1000:1.

SANS were carried out on the spectrometer PAXY of the Leon Brillouin Laboratory (Saclay, France), respectively. The experimental setup allowed covering a Q-range from 0.03 to 0.3 Å^{-1}. Neutron diffraction experiments were performed on the double axis spectrometer G6.1 of the same facility over the Q-range 0.1–1.8 Å^{-1}.

2D SANS patterns of 8CB confined in porous silica matrix measured during cooling the sample from 340 to 180 K are shown in Fig. 9.27. At two highest temperatures no measurable SANS signal was observed. At $T = 290$ K two

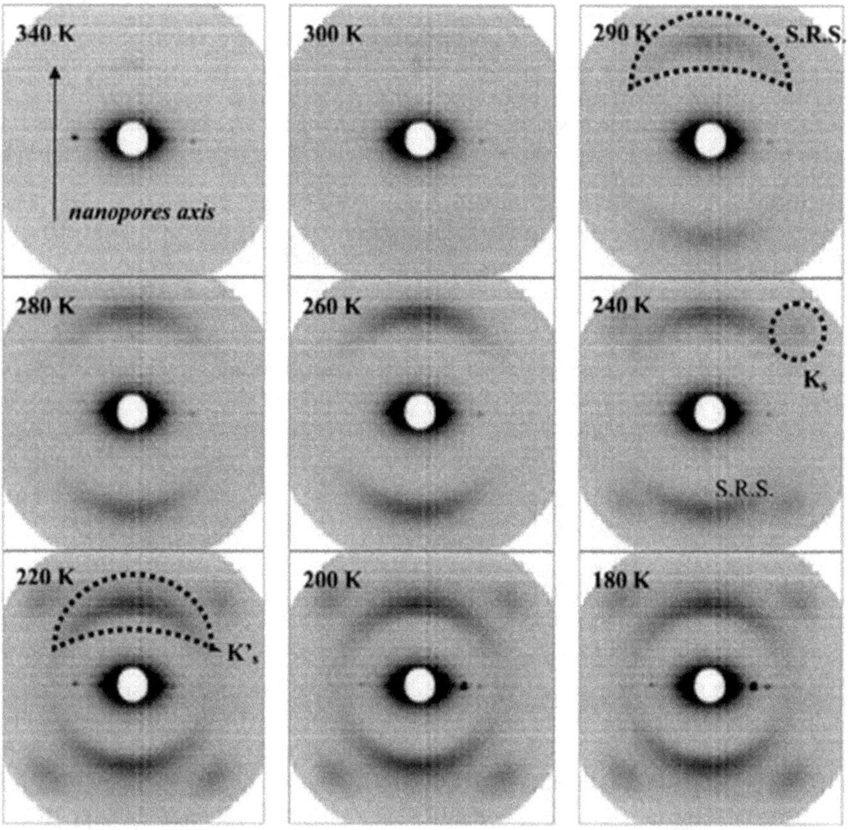

Fig. 9.27 2D SANS patterns of 8CB confined in porous silicon at different temperatures on cooling from 340 to 180 K. Appearance of Bragg peaks at 290 K (highlighted by *dashed lines*) indicates formation of the short-range smectic phase followed by the formation of K_s and K'_s phase at 240 and 220 K, respectively. Reprinted with kind permission from Springer Science + Business Media [67]

symmetric Bragg peaks appeared at $Q = 0.2$ Å^{-1}, which indicated the formation of a short-range smectic (SRS) phase. On further decreasing the temperature, four symmetric Bragg peaks appeared at 240 K ($Q = 0.25$ Å^{-1}) indicating the formation of the crystallization K_s phase coexisting with the SRS phase. At 220 K, the Bragg peak attributed to the SRS phase vanished and was replaced with two symmetric peaks located on the vertical axis at 0.15 Å^{-1}. This additional phase has never been observed in bulk 8CB, but could possibly be related to the so-called K'_s crystallization phase first observed in [68] for 8CB confined in porous silica glass with a pore size diameter of 10 nm. This phase coexists with the K_s phase at 220 K and no change in the neutron scattering intensity was observed on further decreasing the temperature down to 180 K.

Figure 9.28 shows the phase diagram of confined 8CB reconstructed based on the SANS and neutron diffraction data [67]. The smectic peak intensity of bulk 8CB is also shown for comparison. The nematic to smectic transition was not observed in confined 8CB probably due to quenched disorder effects induced by confinement in narrow pores. Crystallization onset was shifted to extremely low temperature as compared to bulk 8CB. The formation of crystallization phase K_s was observed as the occurrence of a unique peak at $Q = 0.25$ Å^{-1} at 240 K. At the same time, the

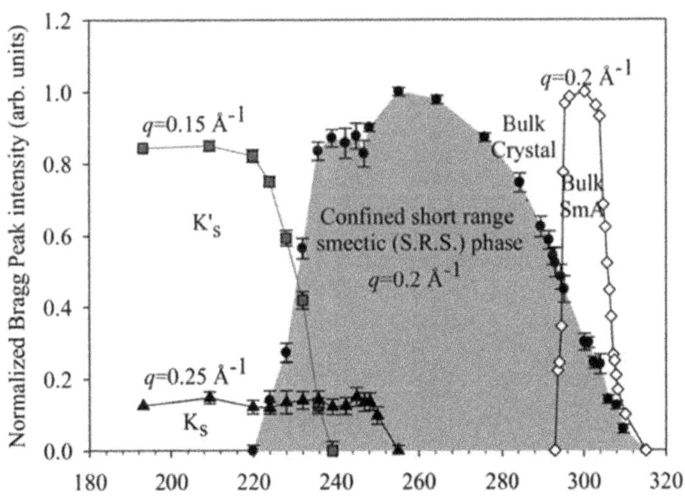

Fig. 9.28 Temperature variation of the maximum intensity of a selection of characteristic Bragg peaks of the smectic ($Q = 0.2$ Å^{-1}), K'_s ($Q = 0.15$ Å^{-1}) and K_s ($Q = 0.25$ Å^{-1}) phases of 8CB confined in porous silicon (*filled symbols*) during cooling. The smectic region of bulk 8CB is added for comparison (*open diamonds*). Each smectic Bragg peak intensity is rescaled with respect to its largest value. Reprinted with kind permission from Springer Science + Business Media [67]

intensity of the SRS peak decreased by ~15 % of its maximum value, which could be used as an estimate of the volume fraction of 8CB converted from the SRS to the K_s phase. On further cooling, SRS and K_s phases coexisted down to 220 K, where the remaining 85 % of SRS phase finally crystallized forming the K'_s phase.

A complete suppression of the smectic transition in 8CB confined in 10 nm diameter silicon pores was also observed in [68]. Below the bulk smectic transition $T^0_{NA} = 306.7$ K, a Bragg peak located around $Q = 0.2$ Å^{-1} was detected (Fig. 9.29).

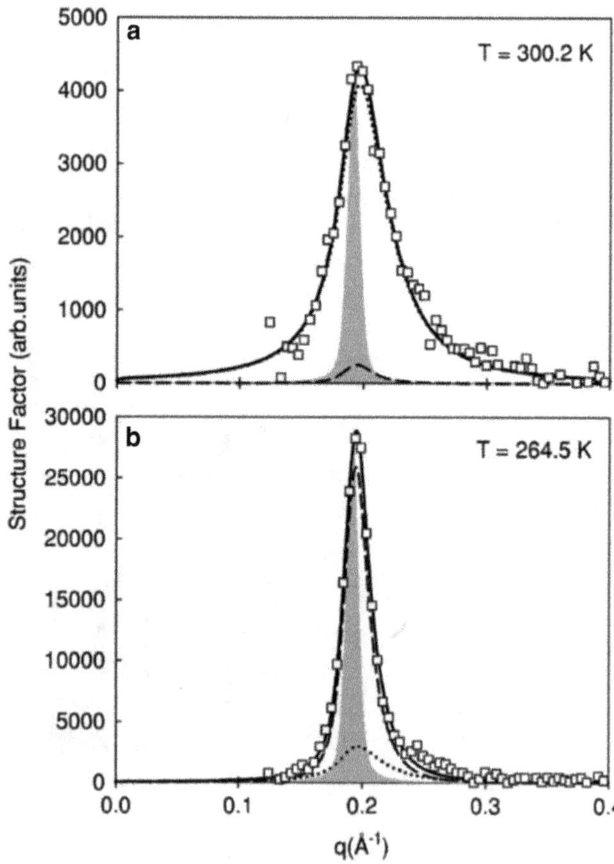

Fig. 9.29 Neutron structure factor of 8CB confined in porous silicon in the region of the smectic Bragg peak after subtraction of the diffuse scattering background. The spectrometer resolution (*shaded area*) highlights the broadening of the peak. The *full line* results from the analysis with a two components model including a thermal term (*dotted line*) and a disorder term (*dashed line*) in (Eq. 9.12). The intensity arises essentially from thermal fluctuations at 300.2 K, whereas it is dominated by the random-field contribution at 264.5 K. Note the difference in the vertical scales between the two graphs. Reprinted figure with permission from [68]. Copyright (2006) by the American Physical Society

The Bragg peak related to the smectic translational order emerged gradually starting from about the bulk transition temperature T^0_{NA} down to the lowest temperature accessible with no sign of crystallization. Even at the lowest temperatures measured, the peak intensity did not saturate, indicating that the smectic order evolved continuously and reversibly over more than 50 K.

The extreme depression of the crystallization point of 8CB in porous silicon offers an opportunity to study this short-ranged smectic phase at equilibrium over an extended temperature range. The measured smectic peak remains significantly broader than the experimental resolution, even at the lowest temperatures (Fig. 9.29). This demonstrates that confinement leads to a finite extension of the smectic correlations.

The shape of Bragg peak may be analyzed using formalism derived from random field theories. After subtraction of a diffuse background, the scattered intensity in the region of the Bragg peak may be represented as a sum of two contributions:

$$I(Q) = \frac{\mathrm{A}_{therm}}{1 + \left(Q_{\|} - Q_0\right)^2 \xi_{\|}^2 + Q_{\perp}^2 \xi_{\perp}^2 + cQ_{\perp}^4 \xi_{\perp}^4} + \frac{\mathrm{A}_{disorder}}{\left[1 + \left(Q_{\|} - Q_0\right)^2 \xi_{\|}^2 + Q_{\perp}^2 \xi_{\perp}^2 + cQ_{\perp}^4 \xi_{\perp}^4\right]^2}. \tag{9.12}$$

The first term in Eq. (9.12) describes contribution of thermal fluctuations, whereas the second term—the effects of quenched disorder. $\xi_{\|}$ and $\xi_{\perp}$ is the smectic correlation length in the direction parallel and perpendicular to the smectic wave vector Q_O, respectively.

Fitting the experimental data to Eq. (9.12) allowed determining of both correlation lengths as a function of temperature. As is seen in Fig. 9.30, above T^0_{NA} the smectic correlation length $\xi_{\|}$ of confined 8CB is about 30 Å. In bulk 8CB, similar values were observed in the smectic thermal fluctuations a few degrees below the transition point. In contrast with the bulk, $\xi_{\|}$ of 8CB confined in porous silicon does not reveal any divergent behavior on approaching the transition temperature T^0_{NA} from above. $\xi_{\|}$ starts increasing moderately below T^0_{NA} and reaches a saturating value of 120 Å at 255 K. The observed moderate variation of $\xi_{\|}$ and $\xi_{\perp}$ does not allow for extracting any critical indices. In general, the behavior of both parameters is similar to the variation of the correlation lengths reported for 8CB in aerogels and aerosils [69–71], although the stronger confinement induced by porous silicon reduces their absolute values considerably.

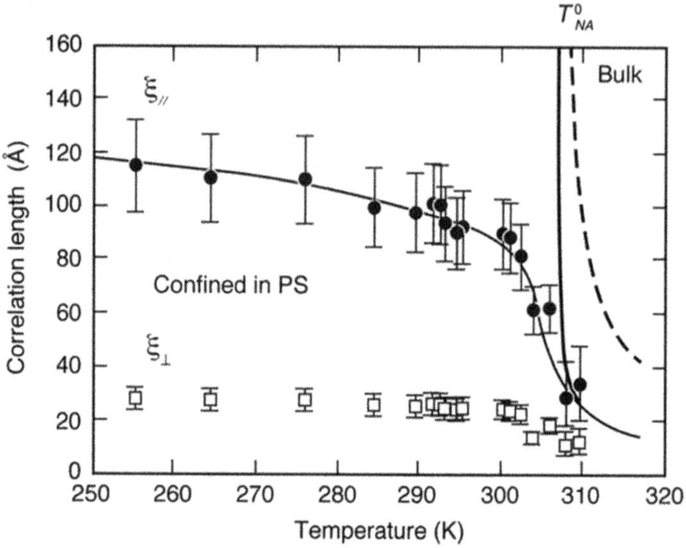

Fig. 9.30 The temperature variation of the smectic correlation length $\xi_{\parallel}$ (*solid circles*) and $\xi_{\perp}$ (*open squares*) for 8CB confined in porous silicon. The *gray solid line* is a guide to the eye. The same quantities for bulk 8CB are plotted in *dashed* and *solid lines* respectively. Reprinted figure with permission from [68]. Copyright (2006) by the American Physical Society

References

1. Simon, P., Gogotsi, Y.: Capacitive energy storage in nanostructured carbon – electrolyte systems. Acc. Chem. Res. **46**, 1094 (2013)
2. Gu, W., Yushin, G.: Review of nanostructured carbon materials for electrochemical capacitor applications: advantages and limitations of activated carbon, carbide-derived carbon, zeolite-templated carbon, carbon aerogels, carbon nanotubes, onion-like carbon, and graphene. WIREs Energ. Environ. **3**, 424 (2014)
3. Oren, Y.: Capacitive delonization (CDI) for desalination and water treatment – past, present and future (a review). Desalination **228**, 10 (2008)
4. Portet, C., Yushin, G., Gogotsi, Y.: Electrochemical performance of carbon onions, nanodiamonds, carbon black and multiwalled nanotubes in electrical double layer capacitors. Carbon **45**, 2511 (2007)
5. Raymundo-Pinero, E., Kierzek, K., Machnikowski, J., Beguin, F.: Relationship between the nanoporous texture of activated carbons and their capacitance properties in different electrolytes. Carbon **44**, 2498 (2006)
6. Boukhalfa, S., He, L., Melnichenko, Y.B., Yushin, G.: Small-angle neutron scattering for in situ probing of ion adsorption inside micropores. Angew. Chem. Int. Ed. **52**, 4618 (2013)
7. Boukhalfa, S., Gordon, D., He, L., Melnichenko, Y.B., Nitta, N., Magasinski, A., Yushin, G.: In situ small angle neutron scattering revealing ion adsorption in microporous carbon electrical double layer capacitors. ACS Nano **8**, 2495 (2014)
8. Chmiola, J., Yushin, G., Gogotsi, Y., Portet, C., Simon, P.: Anomalous increase in carbon capacitance at pore size below 1 nm. Science **313**, 1760 (2006)

9. Kaneko, K., Ohiba, T., Ohkubo, T.: Nanospace molecular science and adsorption. Adsorption **11**, 21 (2005)
10. Haumann, M., Rusager, A.: Hydroformylation in room temperature ionic liquids (RTILs): catalyst and process developments. Chem. Rev. **108**, 1474 (2008)
11. Stefanopoulos, K.L., Romanos, G.E., Vangeli, O.C., Mergia, K., Kanellopoulos, N.K., Koutsioubis, A., Larez, D.: Investigation of confined ionic liquid in nanostructured materials by a combination of SANS, contrast-matching SANS, and nitrogen adsorption. Langmuir **27**, 7980 (2011)
12. Romanos, G.E., Stefanopoulos, K.L., Vangeli, O.C., Mergia, K., Beltsios, K.G., Kanellopoulos, N.K., Lairez, D.: Investigation of physically and chemically ionicf liquid confinement in nanoporous materials by a combination of SANS, contrast-matching SANS, XRD and nitrogen adsorption. J. Phys. Conf. Ser. **340**, 012087 (2012)
13. Romanos, G.E., Vangeli, O.C., Stefanopoulos, K.L., Kouvelos, E.P., Papageorgiou, S.K., Favvas, E.P., Kanellopoulos, N.K.: Methods of evaluating pore morphology in hybrid organic-inorganic porous materials. Micropor. Mesopor. Mater. **120**, 53 (2009)
14. Impéror-Clerc, M., Davidson, P., Davidson, A.: Existence of a microporous corona around the mesopores of silica-based SBA-15 materials templated by triblock copolymers. J. Am. Chem. Soc. **122**, 11925 (2000)
15. Shin, T., Findenegg, G.H., Brandt, A.: Surfactant adsorption in ordered mesoporous silica studied by SANS. Prog. Colloid. Polym. Sci. **133**, 116 (2006)
16. Perdikaki, A.V., Vangeli, O.C., Karanikolos, G.N., Stefanopoulos, K.L., Beltsios, K.G., Alexandridis, P., Kanellopoulos, N.K., Romanos, G.E.: Ionic liquid-modified porous materials for gas separation and heterogeneous catalysis. J. Phys. Chem. C **116**, 16398 (2012)
17. Radlinski, A.P.: Small-angle neutron scattering and the microstructure of rocks. Rev. Miner. Geochem. **63**, 363 (2006)
18. Radlinski, A.P., Boreham, C.J., Wignall, G.D., Lin, J.S.: Microstructural evolution of source rocks during hydrocarbon generation: a small-angle scattering study. Phys. Rev. B **53**, 14152 (1996)
19. Radlinski, A.P., Boreham, C.J., Lindner, P., Randl, O., Wignall, G.D., Hinde, A., Hope, J.M.: Small angle neutrom scattering signature of oil generation in artificially and naturally matured hydrocarbon source rocks. Org. Geochem. **31**(1) (2000)
20. Radlinski, A.P., Kennard, J.M., Edwards, D.S., Hinde, A.L., Davenport, R.: Hydrocarbon generation and expulsion from early cretaceous source rocks in the Browse Basin, North West Shelf, Australia: a SANS study. Aust. Pet. Prod. Explor. Assoc. J. **151**, 2014 (2014)
21. Blossey, R.: Self-cleaning surfaces – virtual realities. Nat. Mater. **2**, 301 (2003)
22. Tuteja, A., Choi, W., Ma, M., Mabry, J.M., Mazzella, S.A., Rutledge, G.C., McKinley, G.H., Cohen, R.E.: Designing superoleophobic surfaces. Science **318**, 1618 (2007)
23. Cao, L., Jones, A.K., Sikka, V.K., Wu, J., Gao, D.: Anti-icing superhydrophobic coatings. Langmuir **25**, 12444 (2009)
24. Theofanous, T.G., Tu, J.P., Dinh, A.T., Dinh, T.N.: The boiling crisis phenomenon – part I: nucleation and nucleate boiling heat transfer. Exp. Therm. Fluid. Sci. **26**, 775 (2002)
25. Bocquet, L., Lauga, E.: A smooth future? Nat. Mater. **10**, 334 (2011)
26. White, E.R., Mecklenburg, M., Singer, S.B., Aloni, S., Regan, B.C.: Imaging nanobubbles in water with scanning transmission electron microscopy. Appl. Phys. Express **4**, 055201 (2011)
27. Seddon, J.R.T., Stefan Kooij, E., Poelsema, B., Zandvlet, H.J.W., Lohse, D.: Surface bubble nucleation stability. Phys. Rev. Lett. **106**, 056101 (2011)
28. Zhang, H., Lamb, R.N., Cookson, D.J.: Nanowetting of rough superhydrophobic surfaces. Appl. Phys. Lett. **92**, 254106 (2007)
29. Checco, A., Hofmann, T., DiMasi, E., Black, C.T., Ocko, B.M.: Morphoogy of air nanobubbles trapped at hydrophobic nanopatterned surfaces. Nano Lett. **10**, 1354 (2010)
30. Melnichenko, Y.B., Lavrik, N.V., Popov, E., Bahadur, J., He, L., Kravchenko, I.I., Smith, G., Pipich, V., Szekely, N.K.: Cavitation on deterministically nanostructured surfaces in contact with an aqueous phase: a small-angle neutron scattering study. Langmuir **30**, 9985 (2014)

31. Engel, M., Stuhn, B., Schneider, J.J., Cornelius, T., Naumann, M.: Small-angle X-ray scattering (SAXS) off parallel, cylindrical, well-defined nanopores: from random pore distribution to highly ordered samples. Appl. Phys. **97**, 99 (2009)
32. Zickler, G.A., Jahnert, S., Wagermaier, W., Funari, S.S., Findenegg, G.H., Paris, O.: Physisorbed films in periodic mesoporous silica studied by in situ synchrotron small-angle diffraction. Phys. Rev. B **73**, 184109 (2006)
33. Reyssat, M., Pepin, A., Marty, F., Chen, Y., Quéré, D.: Bouncing transitions on microtextured materials. Europhys. Lett. **74**, 306 (2006)
34. Deng, T., Varanasi, K.K., Hsu, M., Bhate, N., Keimel, C., Stein, J., Blohm, M.: Nonwetting of impinging droplets on textured surfaces. Appl. Phys. Lett. **94**, 133109 (2009)
35. Checco, A., Ocko, B.M., Rahman, A., Black, C.T., Tasinkevych, M., Giacomello, A., Dietrich, S.: Collapse and reversibility of the superhydrophobic state on nanostructured surfaces. Phys. Rev. Lett. **112**, 216101 (2014)
36. Gelb, L.D., Gubbins, K.E., Radhakrishnan, R., Sliwinska-Bartkowiak, M.: Phase separation in confined systems. Rep. Prog. Phys. **62**, 1573 (1999)
37. de Gennes, P.G.: Liquid-liquid demixing inside a rigid network – qualitative features. J. Phys. Chem. **88**, 6469 (1984)
38. Andelman, D., Joanny, J.F.: Scaling phenomena in disordered systems. In: Pynn, R., Skjeltorp, A. (eds.) Metastability and Landau Theory for Random Fields and Demixing in Porous Media, p. 1991. Plenum, New York, NY (1991)
39. Liu, A.J., Durian, D.J., Herbolzheimer, E., Safran, S.A.: Wetting transition in a cylindrical pore. Phys. Rev. Lett. **65**, 1897 (1990)
40. Dierker, S.B., Wiltzius, P.: Statics and dynamics of a critical binary fluid in a porous medium. Phys. Rev. Lett. **66**, 1185 (1991)
41. Lin, M.Y., Sinha, S.K., Drake, J.M., Wu, X.I., Thiyagarajan, P., Stanley, H.B.: Study of separation of a binary fluid mixture in confined geometry. Phys. Rev. Lett. **72**(2207) (1994)
42. Frisken, B.J., Cannell, D.S., Lin, M.Y., Sinha, S.K.: Neutron scattering studies of binary mixtures in silica gels. Phys. Rev. E **51**, 5866 (1995)
43. Formisano, F., Teixeira, J.: Appearance of critical fluctuations in a binary fluid mixture confined in Vycor glass. J. Phys. Condens. Matter **12**, A351 (2000)
44. Formisano, F., Teixeira, J.: Critical fluctuations of a binary fluid mixture confined in a porous medium. Eur. Phys. J. E. **1**, 1 (2000)
45. Schemmel, S., Rother, G., Eckerlebe, H., Findenegg, G.: Local structure of a phase-separating binary mixture in a mesoporous glass matrix studied by small-angle neutron scattering. J. Chem. Phys. **122**, 244718 (2005)
46. Gommes, C.J.: Three-dimensional reconstruction of liquid phases in disordered mesopores using in situ small-angle scattering. J. Appl. Crystallogr. **46**, 493 (2013)
47. Pekala, R.W., Schaefer, D.W.: Structure of organic aerogels. 1. Morphology and scaling. Macromolecules **26**, 5487 (1993)
48. Gommes, C.J., Roberts, A.P.: Structure development of resorcinol-formaldehyde gels: microphase separation or colloid aggregation. Phys. Rev. E **77**, 041409 (2008)
49. Armstrong, M., Galli, A.G., Beucher, H., Le Loc'h, G., Renard, D., Doligez, B., Eschard, R., Geffroy, F.: Plurigaussian Simulations in Geosciences. Springer, Berlin (2003)
50. Cole, D.R., Herwig, K.W., Mamontov, E., Larese, J.Z.: Neutron scattering and diffraction studies of fluids and fluid-solid interactions. Rev. Miner. Geochem. **63**, 313 (2006)
51. Brovchenko, I., Oleinikova, A.: Interfacial and Confined Water. Elsevier, Amsterdam, Boston, Heidelberg (2008)
52. Fayer, M.D., Levinger, N.: Analysis of water in confined geometries and at interfaces. Annu. Rev. Anal. Chem. **3**, 89 (2010)
53. http://www1.lsbu.ac.uk/water/water_anomalies.html
54. Brovchenko, I., Geiger, A., Oleinikova, A.: A multiple liquid-liquid transitions in supercooled water. J. Chem. Phys. **118**, 9473 (2003)

55. Paschek, D.: How the liquid-liquid transition affects hydrophobic hydration in deeply supercooled water. Phys. Rev. Lett. **94**, 217802 (2005)
56. Brovchenko, I., Geiger, A., Oleinikova, A.: A liquid-liquid phase transitions in supercooled water studied by computer simulations of various water models. J. Chem. Phys. **123**, 044515 (2005)
57. Liu, D., Zhang, Y., Chen, C.C., Mou, C.Y., Poole, P.H., Chen, S.H.: Observation of the density minimum in deeply supercooled confined water. Proc. Natl. Acad. Sci. U. S. A. **104**, 9570 (2007)
58. Liu, D., Zhang, Y., Liu, Y., Wu, J., Chen, C.C., Mou, C.Y., Chen, S.H.: Density measurement of 1-D confined water by smal-angle neutron scattering method: pore size and hydration level dependencies. J. Phys. Chem. B **112**, 4309 (2008)
59. Zhang, Y., Faraone, A., Kamitahara, W.A., Liu, K.H., Mou, C.Y., Leao, J.B., Chang, C., Chen, S.H.: Density hysteresis of heavy water confined in a nanoporous silica matrix. Proc. Natl. Acad. Sci. U. S. A. **108**, 12206 (2011)
60. Erko, M., Wallacher, D., Hoell, A., Haus, T., Zizak, I., Paris, O.: Density minimum of confined water at low temperatures: a combined study by small-angle scattering of X-rays and neutrons. Phys. Chem. Chem. Phys. **14**, 3852 (2012)
61. Mancinelli, R., Bruni, F., Ricci, M.A.: Controversial evidence on the point of minimum density in deeply supercooled confined water. J. Phys. Chem. Lett. **1**, 1277 (2010)
62. Zanotti, J.M., Morineau, D.: Dynamics of soft matter: neutron applications. In: Sakai, V.G., Alba-Simionesco, C., Chen, S.H. (eds.) Surface and Confinement Effects in Nano/Mesoporous Materials. Springer, New York, NY (2012)
63. Crawford, G.P., Zumer, S.: Liquid Crystals in Complex Geometries Formed by Polymer and Porous Networks. CRC Press, Boca Raton, FL (1996)
64. Pesce de Silveira, N., Ehrburger-Dolle, F., Rochas, C., Rigacci, A., Pereira, F.V., Westfahl Jr., H.: Smectic ordering in polymer liquid crystal – silica aerogel nanocomposites. J. Therm. Anal. Calorim. **79**, 579 (2005)
65. Zhang, R., Zeng, X., Prehm, M., Liu, F., Grimm, S., Geuss, M., Steinhart, M., Tschierske, C., Ungar, G.: Honeycombs in honeycombs: complex liquid crystal alumina composite mesostructures. ACS Nano **8**, 4500 (2014)
66. Zidansek, A., Lahajnar, G., Kralj, S.: Phase transitions in 8CB liquid crystal confined to a controlled-pore glass: deuteron NMR and small angle X-ray scattering studies. Appl. Magn. Reson. **27**, 311 (2004)
67. Guegan, R., Morineau, D., Lefort, R., Beziel, W., Guendouz, M., Noirez, L., Henschel, A., Huber, P.: Rich polymorphism of a rod-like crystal (8CB) confined in two types of unidirectional nanopores. Eur. Phys. J. E. **26**, 261 (2008)
68. Guegan, R., Morineau, D., Loverdo, C., Beziel, W.: Evidence of anisotropic quenched disorder effects on a smectic liquid crystal confined in porous silicon. Phys. Rev. E **73**, 011707 (2006)
69. Bellini, T., Radzihovsky, L., Toner, J., Clark, N.A.: Universality and scaling in the disordering of a smectic liquid crystal. Science **294**, 1074 (2001)
70. Clegg, P.S., Stock, C., Birgeneau, R.J., Garland, C.W., Roshi, A., Iannacchione, G.S.: Effect of a quenched random field on a continuous symmetry breaking transition: nematic to smectic-A transition in octyloxycyanobiphenyl-aerosil dispersions. Phys. Rev. E **67**, 02173 (2003)
71. Ocko, B.M., Birgeneau, R.J., Litster, J.D.: Crossover to tricritical behavior at the nematic to smectic-A transition: an X-ray scattering study. Z. Phys. B. Condens. Matter. **62**, 487 (1986)

Chapter 10
Supercritical Fluids in Confined Geometries

Understanding the influence of confinement on the structure and thermodynamic properties of supercritical fluids in pores of different surface chemistry and topology is fundamental to the successful development and optimization of the variety of technologies involving fluid–solid interactions. Adsorption behavior of supercritical fluids (SCFs) is fundamentally different from that of subcritical vapors and this chapter begins with a general description of the specifics of the supercritical adsorption. Presented examples illustrate the capability of SAS methods to deliver unique, pore-size-dependent information on the properties of supercritical CO_2, hydrogen, methane, and other fluids confined in pores of different engineered and natural porous solids. Applications of SAS for studying pore interconnectivity and structural stability of porous materials under pressure are also discussed.

10.1 Specifics of the Supercritical Fluid Adsorption

A supercritical fluid (SCF) is a substance at a pressure (P) and temperature (T) above its liquid–gas critical point, where the coexisting (subcritical) liquid and vapor phases become indistinguishable [1]. Unlike subcritical vapor, which can be condensed into liquid at a certain temperature-dependent saturated vapor pressure, the density of SFC can be varied continuously and reach high, liquid-like values without condensation. Even at high density, the viscosity of SFC remains an order of magnitude lower than that of the corresponding liquid at the same (P,T). Hence, the diffusivity of solutes in SCFs is much higher than in liquids, which facilitates their mass transfer. The combination of tunable solvent power, practically unlimited miscibility with gases and many small solutes, as well as advantageous mass transfer properties make SCFs an attractive alternative to liquid solvents for a variety of technological applications such as heterogeneous catalysis, adsorptive separations, regeneration of adsorbents, storage of fuel gases, and SCF chromatography [2].

Y.B. Melnichenko, *Small-Angle Scattering from Confined and Interfacial Fluids*, DOI 10.1007/978-3-319-01104-2_10

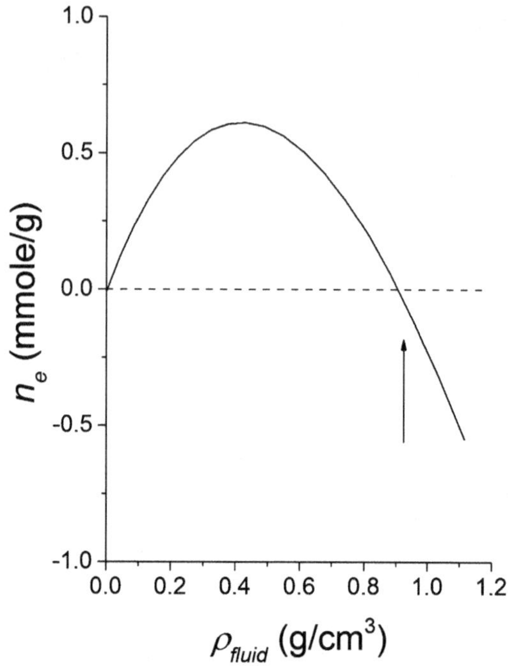

Fig. 10.1 Generic excess adsorption isotherm of a supercritical fluid in a porous medium as a function of the fluid density. The *arrow* shows the value of ρ_{fluid} at which $n_e = 0$ and thus $\rho_{fluid} = \rho_{ads} = 0.9$ g/cm^3

The adsorption behavior of SCFs is fundamentally different from that of subcritical fluids, the adsorption isotherms of which are well defined by the standard IUPAC classification (Sect. 8.1). One of the most noticeable features that distinguish adsorption isotherms of SCFs and subcritical fluids is a maximum in the excess adsorption (n_e) of SCFs, which is observed with increasing pressure (or fluid density, Fig. 10.1).

Consider a system with the total volume V_{total} which consists from a solid adsorbent, unadsorbed fluid and adsorbed phase with the volumes (V_{solid}), (V_{fluid}), and (V_{ads}), respectively:

$$V_{total} = V_{solid} + V_{fluid} + V_{ads} = V_{solid} + V_{void}, \tag{10.1}$$

where V_{void} is the void volume accessible to the supercritical fluid. By definition, the *excess* adsorption n_e (also known as Gibbs adsorption) is the excess material present in the pores of the adsorbent over that corresponding to the density of the fluid in the bulk phase at the same temperature and pressure. Thus, in terms of the relative mass (n, g/g), the *excess* adsorption n_e can be represented as [3]:

$$n_e = n_{total} - V_{void}\rho_{fluid}, \tag{10.2}$$

where n_{total} is the total amount (g/g) of supercritical fluid in the volume occupied by the adsorbent, V_{void} is the pore volume in cm^3/g and ρ_{fluid} is the fluid density in g/cm^3. Equation (10.2) illustrates the physical meaning of the excess adsorption, which is calculated neglecting the volume occupied by the adsorbed phase i.e. as if the entire void volume V_{void} was accessible to the unadsorbed (bulk) fluid.

Conversely, the *absolute* adsorption (n^{abs}), is calculated taking into account the reduction of the void volume accessible to the fluid due to formation of the adsorbed phase:

$$n^{abs} = n_{total} - V_{fluid}\rho_{fluid}. \tag{10.3}$$

Equations (10.2) and (10.3) can be rearranged to give:

$$n_e = V_{ads}\left(\rho_{ads} - \rho_{fluid}\right) \tag{10.4}$$

and

$$n^{abs} = n_e\left(\frac{\rho_{ads}}{\rho_{ads} - \rho_{fluid}}\right). \tag{10.5}$$

Under supercritical conditions, the density of the fluid phase increases monotonically with pressure, whereas the density of adsorbed phase initially increases and then levels off at a value corresponding to the maximal amount of adsorbate, which can be packed into the pores. Consequently, according to Eq. (10.4), the *excess* adsorption should initially increase, go through the maximum, become zero at $\rho_{ads} = \rho_{fluid}$ and finally become negative at higher pressures (Fig. 10.1). The value of $\rho_{fluid}(n_e = 0)$ may be used as an estimate of the density of the adsorbed phase in a porous medium.

All conventional methods (gravimetric, volumetric, total desorption, isotope exchange, etc.) measure n_e of both SCF and subcritical fluids as a function of temperature and pressure. The difference between n^{abs} and n_e is usually very small in the case of *subcritical* adsorption due to the low bulk fluid density $\rho_{fluid} \to 0$, which gives $n^{abs} \cong n_e$ (Eq. 10.5). In the *supercritical* region and at high pressure this difference becomes essential. Equations (10.4) and (10.5) show that the calculation of the *absolute* adsorption from n_e requires information on the density (or volume) of the adsorbed phase, which is not readily accessible by existing experimental methods [4].

The maximum in the supercritical fluid adsorption isotherms has been well documented experimentally [5], and a progress in the description of the characteristics of supercritical fluids in confined spaces has been made based on different theoretical approaches [4, 6–9], as well as computer simulations [9–11]. It is now generally recognized that when supercritical fluid is in contact with a solid adsorbent, the adsorption potential between the adsorbate molecules and the surface atoms should create a region near the solid surface with the average density higher than that in the homogeneous bulk phase. In large meso- and macropores, the adsorption force is similar to that of an open surface, which results in formation of an adsorbed phase region extending to a distance of few molecular diameters.

Due to a high compressibility of supercritical fluids, significant density gradients are anticipated to form in the interfacial region, and such gradients are expected to become even more long- ranged as the liquid–gas critical point is approached from above. Complete micropore filling is possible even far away from the critical point due to an enhancement of the adsorption energy caused by the overlapping of the pore wall adsorption potentials.

Understanding of the structure and thermodynamic properties of SCFs in pores of different surface chemistry and topology is fundamental to the successful development and optimization of the variety of technologies involving SCF–solid interactions. Examples given in this Chapter will illustrate the capability of SAS methods to deliver unique, pore-size-dependent information on the properties of supercritical CO_2, hydrogen, methane and other fluids confined in pores of different engineered and natural porous solids.

10.2 Density Fluctuations Near the Liquid–Gas Critical Point of Confined Fluids

Phase behavior of fluids can be drastically modified by confinement and disorder. The liquid–vapor critical behavior in porous systems with quenched disorder, such as silica aerogels, is of particular interest since it may provide an experimental realization of the Random Field Ising Model (RFIM) [12]. Theoretical considerations suggest that a new set of critical indices may appear in the fluid with quenched correlated impurities implying possible transition to a new universality class for confined systems. Experiments on the phase diagrams reported the bulk (Ising model) values of the critical index β of the coexistence curves of confined ^4He and N_2 [13, 14]. At the same time, studies of the superfluid transitions and heat capacity of ^4He in aerogels demonstrated substantial narrowing of the phase diagram and a much lower critical temperature compare to bulk fluid [15]. For ^3He–^4He mixtures confined in aerogels with porosity of 98 or 99.5 %, the presence of 2 % or even 0.5 % silica has dramatic effects on entire phase diagram causing the phase separation curve to detach from the lambda line and stabilizing a region of dilute ^4He superfluid inaccessible in bulk mixtures [16]. Liquid–gas transitions for neon and helium are accompanied by hysteresis and sluggish dynamics persisting up to close to the critical temperature [17].

SANS is an ideal tool for investigating the critical behavior of confined liquid solutions near the critical demixing point (Sect. 9.4) as well as individual fluids near the liquid–gas critical point as it can probe the local structure and thermodynamic properties of fluids in small pores with sizes of the order on 10–1000 Å. SANS was applied in [18, 19] to study the temperature variation of the correlation length (ξ) of the density fluctuations as well as susceptibility (χ) of CO_2 confined in a random porous matrix of a silica aerogel with ~96 % porosity, a surface area of 400 m^2/g, a density of 0.1 g/cm^3, and pore size distribution peaked around the nominal pore

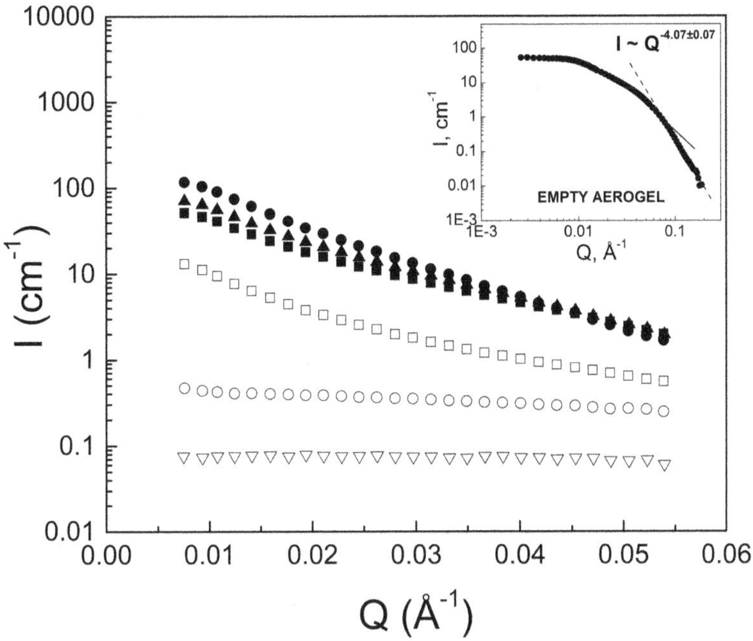

Fig. 10.2 Absolute neutron scattering cross section per unit volume measured from bulk (*open symbols*) and confined CO_2 (*solid symbols*) in the homogeneous region. (*Open inverted triangle, filled square*) $T = 80$ °C; (*open circle, filled triangle*) $T = 40$ °C; (*open square, filled circles*) $T = 31.4$ °C. The average density of CO_2 is $\rho_C = 0.468$ g/cm^3. The *inset* shows scattering from empty aerogel at ambient conditions. *Solid line* is described by Eq. (10.6), whereas *dash line* corresponds to the Porod law $I \sim Q^{-4}$. Reprinted figure with permission from [18]. Copyright (2004) by the American Physical Society

diameters $2R_P \sim 60$–70 Å. Aerogel was obtained from Oscellus Technologies, Livermore, CA and shaped into a cylinder (17 mm OD, 10 mm length), which fit tightly into the SANS high-pressure cell (Sect. 4.2, Fig. 4.2a). Temperature scans were performed in the range 80 °C $\geq T \geq$ 25 °C that included the liquid–gas critical temperature of bulk CO_2 ($T_C = 31.06$ °C [20]) at five different constant average densities of CO_2 ($\rho_{CO2} = 0.3$, 0.4, 0.468, 0.55 and 0.65 g/cm^3), i.e. within ±35 % deviation from the critical density of the bulk $CO_2 \rho_{C,BULK} = 0.468$ g/cm^3 [20].

SANS experiments were performed on the KWS-2 SANS facility at the FRJ reactor in Julich, Germany. The neutron wavelength was $\lambda = 6.3$ Å ($\Delta\lambda/\lambda = 10$ %). Three sample-detector distances of 14, 4, and 1.4 m were used for measuring scattering from empty aerogel to give an overall range of momentum transfer $0.003 < Q < 0.2$ Å^{-1}. Measurements of empty aerogel were repeated after each temperature scan and revealed no changes in the aerogel structure after exposure to CO_2. Scattering from bulk and confined CO_2 was measured in the Q-range $0.007 < Q < 0.07$ Å^{-1}. The SANS data sets were corrected for instrumental backgrounds and normalized to an absolute (±5 %) differential cross section per unit sample volume [$I(Q)$ in units of cm^{-1}] by means of pre-calibrated secondary standards.

The intensity of scattering from empty aerogel is shown in the inset of Fig. 10.2. $I(Q)$ was fitted to the function [21]:

$$I(Q) = \frac{I(0)}{\left[1 + Q^2 \xi_G^2\right]^{(D_f - 1)/2}} \times \frac{\sin\left[(D_f - 1)\tan^{-1}(Q\xi_G)\right]}{(D_f - 1)Q\xi_G}, \tag{10.6}$$

where ξ_G is the gel correlation length, which represents the upper length scale limit of the fractal correlations, D_f is the fractal dimension, and $I(0)$ is the extrapolated intensity of scattering at $Q = 0$. The solid line is the result of fitting (10.6) to the data, which provided an excellent fit in the Q range up to ~0.1 Å^{-1}, after which a crossover to the Porod law was observed. The values of fit parameters $\xi_G = 85 \pm 9$ Å and $D_f = 2.5 \pm 0.1$ were found to be in a good agreement with ξ_G and D_f of base catalyzed aerogels of similar density [22].

SANS from bulk CO_2 as a function of pressure and temperature was measured in a separate experiment. Several representative scattering curves from bulk and confined CO_2 as a function of temperature in the one phase region are shown in Fig. 10.2. The values of ξ_{BULK} for bulk CO_2 as well as $I(0)$ proportional to the fluid compressibility were obtained by fitting $I(Q)$ to the Ornstein–Zernike formula [23]:

$$I(Q) = \frac{I(0)}{\left(1 + Q^2 \xi^2\right)}. \tag{10.7}$$

The variation of ξ_{BULK} as a function of T is shown in Fig. 10.3. As may be seen, ξ_{BULK} tends to diverge as $T \to T_C$, and decreases rapidly below T_C at which point the formation of the meniscus is observed. On the log-log scale, the slope of $\xi_{BULK}(T - T_C/T_C)$ gives the critical index of the bulk correlation length

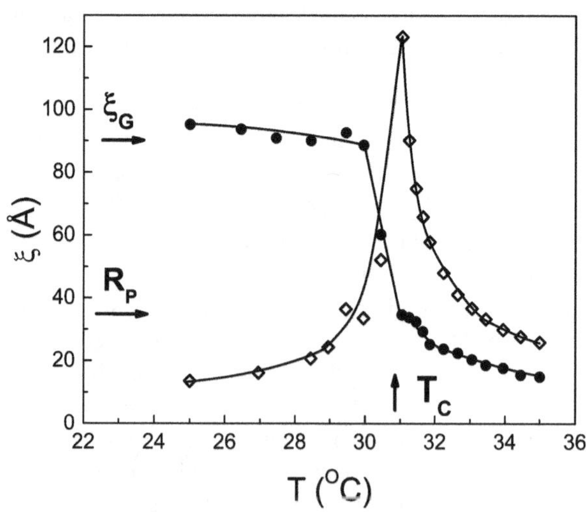

Fig. 10.3 Correlation length of the density fluctuations in bulk CO_2 (open diamonds) and confined (solid circles) CO_2 as a function of temperature. Arrow shows the critical liquid–gas temperature of bulk CO_2. Reprinted figure with permission from [18]. Copyright (2004) by the American Physical Society

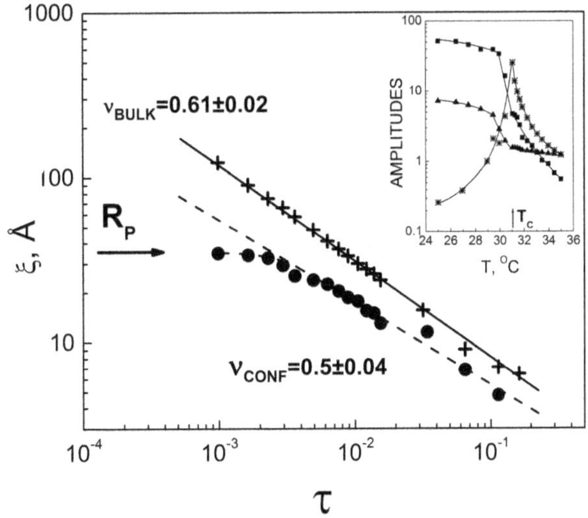

Fig. 10.4 ξ_{BULK} (crosses) and ξ_{CONF} (*circles*) vs. reduced temperature $\tau = (T - T_C)/T_C$ ($T_C = 31.06$ °C). The *inset* shows the temperature variation of amplitudes in Eqs (10.7) and (10.8). Bulk CO_2: (*asterisk*) [$I(0)$, cm^{-1}]. Confined CO_2: (*filled square*) [I_{LOR}], (*filled triangle*) [I_{LSQ}]. Reprinted figure with permission from [18]. Copyright (2004) by the American Physical Society

$\nu_{BULK} = 0.61 \pm 0.02$, which agrees within experimental error with the Ising model exponent $\nu = 0.630 \pm 0.001$ [23] (Fig. 10.4).

The scattering for confined CO_2 was interpreted quantitatively using an equation similar to that utilized previously to interpret SANS results on the influence of confinement on the liquid-liquid critical phenomena in partially miscible liquid solutions [24]:

$$I(Q) = \frac{I_{LOR}}{(1 + Q^2\xi^2)} + \left[\frac{I_{LSQ}}{(1 + Q^2\xi^2)}\right]^2 I_{aerog}(Q) + B_{inc}. \tag{10.8}$$

The angle-independent background ($B \sim 0.5$ cm^{-1}) in Eq. (10.8) arises mainly from hydrogen atoms in the methoxy groups on the aerogel surface. The first (Lorentzian) term is identical to Eq. (10.7) and describes the genuine scattering from critical fluctuations inside pores, whereas the second (Lorentzian-squared) term arises from the response of the order parameter (fluid density) to the "field" induced by aerogel and is related to the critical absorption [25]. The scattering from aerogel itself is taken into account in Eq. (10.8) via the "background" term $I_{aerog}(Q)$, which represents scattering from CO_2-saturated aerogel at $T = 80$ °C $\gg T_C$, when contribution of the density fluctuations is negligible.

In the one-phase region ($T > 31$ °C) $I(Q)$ is described by (10.8) with the same value of ξ in both Lorentzian and Lorentzian squared terms, as theoretically predicted [26]. This result is in a qualitative agreement with results obtained for liquid solutions in rigid porous matrices of Vycor glass and silica gels (Sect. 9.4). The temperature variation of the correlation length of bulk and confined CO_2 (ξ_{CONF}) at average density $\rho = \rho_{C,BULK} = 0.468$ g/cm^3 is compared in Fig. 10.4.

At $T > T_C = 31$ °C, in the one-phase region, ξ_{CONF} is systematically smaller than ξ_{BULK} and the difference between the two parameters increases as $T \to T_C$. These trends arise because ξ_{BULK} diverges and ξ_{CONF} saturates at the value corresponding to the pore radius $R_P \sim 30$–35 Å. The log-log slope of $\xi_{CONF}(\tau)$ gives $\nu_{CONF} = 0.5 \pm 0.04$ in the range of the reduced temperatures $\tau = (T - T_C/T_C) \geq 6 \times 10^{-3}$, which is close to the mean field value of $\nu = 0.5$ [23]. At $\tau \to 0$, ξ_{CONF} tends to round off at $\xi_{CONF} \sim R_P$. Below $T = 31$ °C, ξ_{CONF} in Eq. (10.7) shows increases abruptly and tends to saturate at the value roughly corresponding the correlation length of aerogel $\xi_G \sim 90$ Å. At this point aerogel becomes strongly opaque, but exhibits no evidence of macroscopic phase separation (i.e. formation of the liquid phase).

The lack of critical divergence of ξ_{CONF} was interpreted in [18, 19] as the result of the competition between the genuine critical fluctuations in the "core" of aerogel pores and critical absorption of CO_2 molecules on the silica strands of the aerogel, as both processes are characterized by a similar correlation length. The dilute quenched disorder induced by the host matrix suppressed density fluctuations in the critical region at all average densities of CO_2 studied. The results provided strong experimental evidence for supercritical fluids in aerogels to fall in the category of phase separation under confinement with critical absorption effects, in qualitative agreement with the single pore model [27].

10.3 Adsorption of Supercritical CO_2 in Porous Silica

10.3.1 Silica Aerogels

The neutron transmission and x-ray absorption coefficients are parameters routinely measured in SAS experiments. Their values are required for converting the intensity of scattering in counts per second into a neutron (or x-ray) cross section in units of cm^{-1} [28, 29]. Neutron transmission and x-ray absorption data can provide information on the SCF adsorption at a variety of different conditions. The attenuation of a neutron beam passing through a porous adsorbent (e.g. silica aerogel) is due to scattering (coherent and incoherent) and absorption of neutrons by silicon and oxygen nuclei. The beam can be furthermore attenuated if aerogel contains adsorbing and or scattering impurities such as hydrogen containing alkyl groups. The transmission T_1 of blank aerogel is described by Beer's law [30]:

$$T_1 \equiv \frac{I}{I_0} = T_0 \exp(-N_{SiO2}\sigma_{SiO2}L), \tag{10.9}$$

where I_0 and I are the incident and attenuated neutron intensities, respectively, N_{SiO2} is the number of silica molecules per unit volume (number density of SiO_2), σ_{SiO2} is the total neutron cross section of SiO_2 and L is the sample thickness. The pre-factor $T_0 \leq 1$ accounts for additional attenuation of neutron beam due to impurities.

The saturation of an aerogel with a fluid (e.g. SC CO_2) should decrease the total transmission due to additional fluid density-dependent attenuation of the beam caused by scattering and absorption of neutrons by CO_2. The transmission of the two-phase system representing aerogel matrix (phase 1) homogeneously saturated with a non-adsorbing fluid (phase 2) is given by:

$$T_2 = \exp(-N_{SiO2}\sigma_{SiO2}L - N_{CO2}\sigma_{CO2}L) = T_1\exp(-N_{CO2}\sigma_{CO2}L). \tag{10.10}$$

The number density (N_{CO2}) and physical density (ρ_{CO2}) of CO_2 are related via

$$N_{CO2} = \rho_{CO2}\frac{N_A}{M_{CO2}}, \tag{10.11}$$

where N_A is Avogadro's number (6.017×10^{23}), $M_{CO2} \cong 44.01$ is the molecular weight and $\sigma_{CO2} = 14.01 \times 10^{-24}$ cm^2 is total neutron cross section of a CO_2 molecule. Combining Eqs. (10.9) and (10.11), we obtain for the two-phase system (with no adsorption):

$$T_2 = T_1\exp(-C\rho_{CO2}L), \tag{10.12}$$

where $C \cong 0.19$ cm^2/g for ρ_{CO2} in g/cm^3 and a total sample thickness $L = 1$ cm.

In the presence of an adsorbed (third) phase, the transmission will further decrease in comparison to that described by (10.12), and the transmission of such a three-phase system is given by:

$$T_3 = T_1\exp - \left[(N_{unads,CO2} + N_{ads,CO2})\sigma_{CO2}L\right]. \tag{10.13}$$

Equation (10.13) has two unknown parameters: the number density of unadsorbed CO_2 ($N_{unads,CO2}$) and the number density of adsorbed CO_2 ($N_{ads,CO2}$),which cannot be independently determined from the transmission measurements because both the volume of the adsorbent phase and its density are not known. Following the Gibbs model [3], let us neglect the volume occupied by the adsorbed phase ($V_{ads} = 0$) which is equivalent to assumption that $N_{unads,CO2} = N_{CO2}$. In this case the first term in the right-hand side of Eq. (10.13) can be approximated by Eq. (10.11), and the number density of CO_2 molecules in the adsorbed phase should be substituted for the number density of CO_2 (N_e) corresponding to the *excess* adsorption. Thus, Eq. (10.13) becomes:

$$T_3 = T_1\exp - (N_{CO2}\sigma_{CO2}L + N_e\sigma_{CO2}L) = T_2(\exp - N_e\sigma_{CO2}L) \tag{10.14}$$

and the value of N_e can be determined through the difference between the transmission of the two-phase and the three-phase systems:

$$N_e = \frac{\ln T_2 - \ln T_3}{\sigma_{CO2}L}. \tag{10.15}$$

The *excess* adsorption n_e can be calculated for an aerogel sample with a known aerogel density $\rho_{aerogel}$:

$$n_e = \frac{M_{CO2} N_e}{\rho_{aerogel} N_A}. \quad (10.16)$$

The described approach was used in [31, 32] to measure excess adsorption of SC CO_2 in silica aerogels as a function of pressure and temperature. Base-catalyzed silica aerogels with a nominal density $\rho_{aerogel} = 0.1$ g/cm^3, corresponding to ~96 % porosity, were obtained from Oscellus Technologies, Livermore, CA. The aerogels are composed of thin silica strands with the mesh size 60–70 Å and a highly developed surface area ~700 m^2/g (Sect. 7.2.2). The surface of native (nonoxidized) aerogels is covered by abundant methoxy groups (=Si–O–CH_3) formed during supercritical drying of the precursor gel in supercritical methanol. Scattering and transmission data were taken along two isotherms ($T = 35$ and 80 °C) as a function of pressure in the range $0 \leq P \leq 25$ MPa, which corresponds to the bulk fluid densities in the range $0 \leq \rho_{CO2} \leq 0.9$ g/cm^3.

SANS experiments were performed on the KWS-2 SANS facility at the FRJ2 reactor in Jülich, Germany. The neutron wavelength was $\lambda = 6.3$ Å ($\Delta\lambda/\lambda = 0.10$ %). Two sample-detector distances (SDD) of 4 and 1.4 m were used for measuring scattering from the empty aerogel to give an overall range of momentum transfer $0.01 < Q < 0.2$ Å^{-1}. The transmission at each pressure and temperature was measured continuously *in situ* using He^3 straight beam monitor positioned in the center of the beam stop. The SANS data sets were corrected for instrumental backgrounds and normalized to an absolute (±5 %) differential cross section per unit sample volume [$I(Q)$ in units of cm^{-1}] by means of pre-calibrated secondary standards.

The neutron transmission of CO_2 in silica aerogels in the strong adsorption regime, corresponding to the bulk fluid density $\rho_{CO2} < 0.5$ g/cm^3 is shown in Fig. 10.5 along with the transmission of the bulk CO_2. The strong negative deviation of the transmission of CO_2-saturated aerogel from that of bulk CO_2 indicates formation of a dense adsorbed fluid phase with density higher than that of bulk CO_2 at same thermodynamic conditions.

As discussed above, the difference between the transmission of two- and three-phase systems can be used to determine the *excess* adsorption parameter n_e (Eqs. 10.15 and 10.16), the variation of which as a function of the density of bulk fluid ρ_{CO2} is shown in Fig. 10.6. As expected, n_e (ρ_{CO2}) at $T = 35$ °C initially increases, goes through a maximum and then decreases between $0.3 < \rho_{CO2} < 0.4$ g/cm^3. The excess adsorption is maximal at $\rho_{CO2} \sim 0.374$ g/cm^3, i.e. at the fluid densities at which minima in transmission are observed (Fig. 10.5). The variation of n_e at $T = 80$ °C $\gg T_C$ of bulk CO_2 is much more moderate although qualitatively similar to that near the liquid–gas critical temperature, where strong critical adsorption is anticipated.

Analysis of the available literature on adsorption of SC CO_2 shows that the maximal values of the *excess* adsorption in the near-critical conditions vary

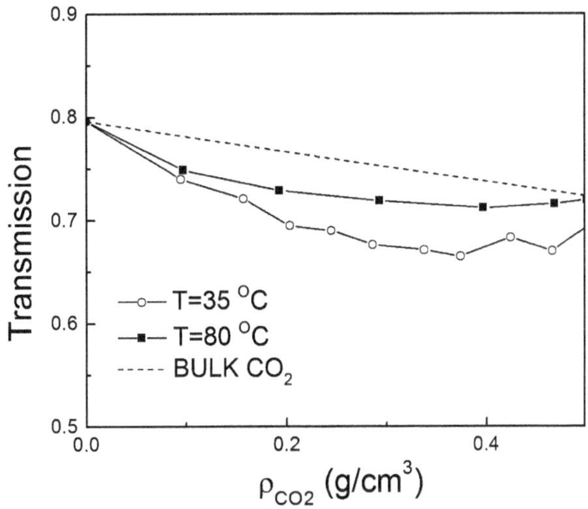

Fig. 10.5 Transmission of silica aerogel vs. density of CO_2 at two temperatures specified in the *inset*. *Dashed lines* represent transmission of bulk CO_2. Reprinted from [32] with kind permission from Springer Science + Business Media

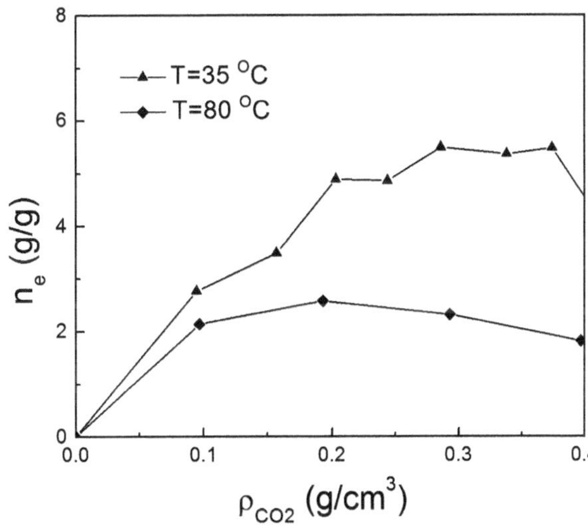

Fig. 10.6 Excess adsorption of CO_2 in silica aerogel as a function of the fluid density at two temperatures shown in the *inset*. Reprinted with permission from [31]. Copyright [2006], AIP Publishing LLC

generally in the range between 5 mmol/g (zeolites) and 20 mmol/g (superactivated carbon), which translates into $0.2 \leq n_e$ (g/g) ≤ 0.9. It follows from the data shown in Fig. 10.6, that the maximal *excess* adsorption of SC CO_2 in aerogels at $T = 35$ °C ($n_e \sim 5.74$ g/g) far exceeds n_e for CO_2 in other adsorbers, which is related to extremely high porosity (large values of V_{void}) and high surface area of silica aerogels.

The coherent cross section $I(Q)$ in units of cm^{-1} of a two-phase system (e.g. a porous silica matrix with homogeneously distributed SC CO_2), is proportional to the neutron contrast between SiO_2 and CO_2 molecules [33]:

$$I(Q) \sim \left(\rho_1^* - \rho_2^*\right)^2 = \left(\frac{b_{SiO2}}{M_{SiO2}}\rho_{SiO2} - \frac{b_{CO2}}{M_{CO2}}\rho_{CO2}\right)^2. \quad (10.17)$$

Here ρ_1^* and ρ_2^* are the scattering length densities of silica and bulk (unadsorbed) CO_2, respectively, $b_{SiO2} = 1.58 \times 10^{-12}$ cm and $b_{CO2} = 1.825 \times 10^{-12}$ cm are the scattering lengths of silica and CO_2, respectively. $M_{SiO2} = 60.08$ is the molecular weight and ρ_{SiO2} is the density of SiO_2. Equation (10.17) predicts that for a two-phase system (e.g. porous silica matrix with homogeneously distributed CO_2 molecules), the square root of *I(Q)* should be proportional to the fluid density $I(Q)^{1/2} \sim \rho_{CO2}$ and thus the formation of a third phase corresponding to the adsorbed fluid will result in deviations from the linear variation of $I(Q)^{1/2}$ vs. ρ_{CO2}. Furthermore, Eq. (10.17) can be used to calculate the zero average contrast (ZAC) condition at which the neutron contrast between the matrix and the fluid becomes zero and thus $I(Q) = 0$ (Sect. 5.4). The ZAC condition can be reached at $\rho_{CO2} = 1.39$ g/cm^3 if the density of silicon matrix of the porous material is taken to be equal to the density of amorphous silica $\rho_{SiO2} = 2.2$ g/cm^3. The skeletal density of silica in aerogel strands is smaller than that of the amorphous silica ($\rho_{SiO2} \cong 2.0$ g/cm^3 [34]) and the calculated ZAC condition in this case corresponds to $\rho_{CO2} = 1.27$ g/cm^3.

Representative SANS curves obtained from aerogel at different CO_2 densities are shown in Fig. 10.7. The variation of *I(Q)* vs. pressure (or fluid density) is non-monotonic, i.e. the cross section first increases with *P* and ρ_{CO2}, reaches maximum and then decreases at higher density. The values of the extrapolated scattering at zero scattering angle *I*(0) at different pressures as well as the variation of $I(0)^{1/2}$ vs. ρ_{CO2} is shown in Fig. 10.8. As is seen in this Figure, a significant *positive* deviation from the straight line corresponding to the two-phase model (Eq. 10.17) is observed.

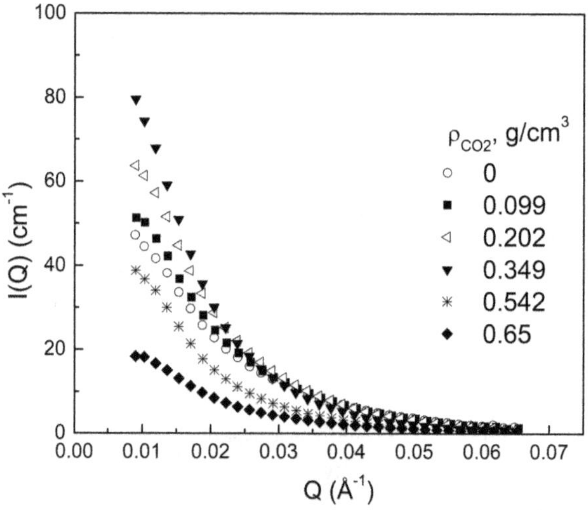

Fig. 10.7 Representative scattering curves from CO_2 saturated silica aerogel at $T = 35$ °C as a function of the fluid density. Reprinted with permission from [31]. Copyright [2006], AIP Publishing LLC

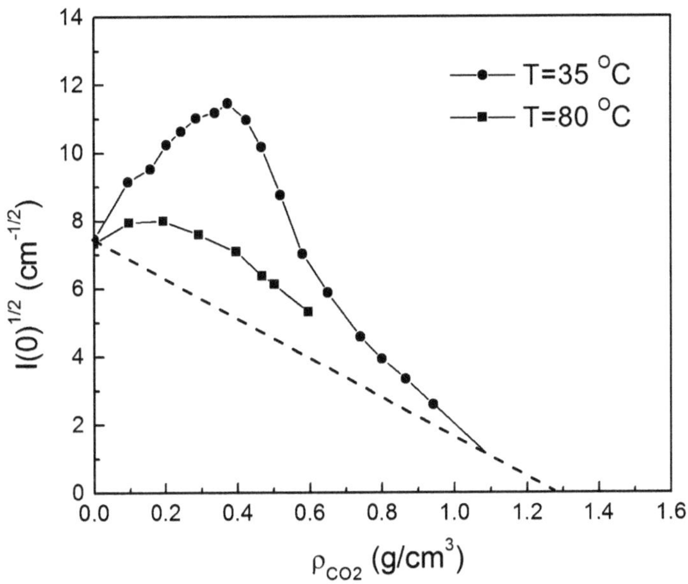

Fig. 10.8 Representative scattering curves from CO_2 saturated aerogel at T = 35 and 80 °C as a function of the fluid density shown in the *inset*. *Dashed line* is the variation of $I(Q)^{1/2}$ vs. ρ_{CO2} calculated using Eq. (10.17) for the aerogel-CO_2 system with no fluid adsorption. Reprinted with permission from [31]. Copyright [2006], AIP Publishing LLC

The positive deviation from a two-phase model indicates the presence of the third phase (adsorbed CO_2) with an extremely high average scattering length density. The adsorption of CO_2 begins by filling the smallest pores with densified fluid, which increases the average SLD of the porous matrix and hence the neutron contrast between the average SLD of the matrix and meso- as well as macropores filled with low density CO_2. The resulting higher contrast leads to an increase in the scattering cross section. The maximum deviation from a two phase model is observed when all nanopores become filled with the densified fluid, after which the scattering starts to decrease due to lowering of the contrast between the average SLD of the matrix and the SLD of the compressed fluid in meso- and macropores pores. The trajectory of $I(0)^{1/2}$ vs. ρ_{CO2} tends to cross the straight line at an extrapolated fluid density of $\rho_{CO2} \cong 1.07$ g/cm^3. At this point the densities of the adsorbed and unadsorbed fluid in larger pores become equal and therefore $\rho_{CO2} \cong 1.07$ g/cm^3 can be taken as an estimate of the density of the adsorbed phase (Sect. 10.1).

This average density of the adsorbed phase is higher than the saturated liquid density of CO_2 at 7 °C (0.68 g/cm^3) and is similar to the density of close-packed molecules with a diameter corresponding to the van der Waals volume ($\rho_{vdW} \sim 1.03$ g/cm^3 [35, 36]). This result is also consistent with high densities of the adsorbed phase reported for CO_2 adsorbed in activated carbon with the average pore size 17 Å ($\rho_{ads} \sim 1.03$ g/cm^3) [35] as well as in NaY zeolite with the pore sizes

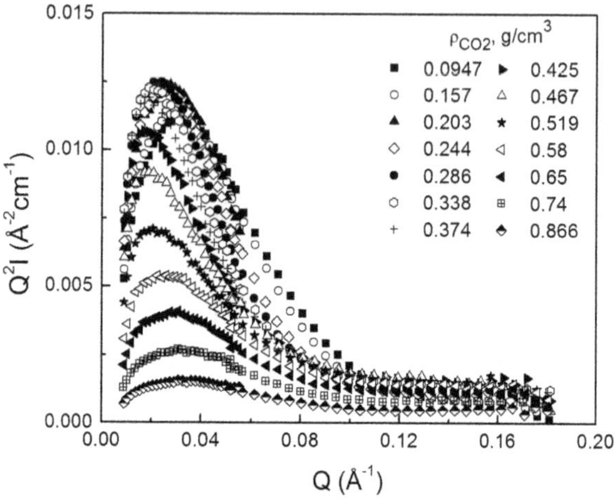

Fig. 10.9 Typical Kratky plots for CO_2 saturated aerogel at T = 35 °C and different fluid densities specified in the inset. Reprinted with permission from [31]. Copyright [2006], AIP Publishing LLC

in the range 25 Å ($\rho_{ads} \sim 1.27$ g/cm^3) [36]. The formation of an adsorbed CO_2 phase with a density far exceeding that of the bulk fluid ($\rho_{ads} \geq 1.1$ g/cm^3) was documented in porous silica with pore diameters in the range 20–150 Å using FTIR [37]. Evidence for the formation of densified CO_2 layers formed in microporous carbon at near-critical conditions was provided using neutron powder diffraction [38–40].

As discussed in Sect. 6.9.2, the volume fraction (ϕ_3) of the adsorbed phase at different temperature and pressure can be determined provided the scattering invariant measured using SAS is supplemented by additional information on the average density of the adsorbed phase (ρ_3) at the same (*T*,*P*) [31, 32, 41]. Typical Kratky plots of $Q^2I(Q)$ vs. Q, based on SANS data [31] from CO_2 saturated aerogel at $T = 35$ °C and different fluid densities are shown in Fig. 10.9. The evaluation of the scattering invariant Ξ from the three-phase system corresponding to the area under each curve in Fig. 10.9 was carried out by the numerical integration and the results are shown in Fig. 10.10. The average density of the adsorbed CO_2 in aerogel as a function of (*T*,*P*) was calculated based on the transmission data shown in Fig. 10.5 using methodology described in [32] (Fig. 10.11a). The values of Ξ obtained from SANS experiments, $\phi_1 = 0.04$, as well as the SLD of silica $\rho_1^* = 3.16 \times 10^{10}$ cm^{-2}, the SLD of unadsorbed $CO_2 \rho_2^*$ calculated using the equation of state of CO_2 [42], and the SLD of the adsorbed $CO_2 \rho_3^*$ determined through the transmission measurements were substituted in Eqs. (6.78) and (6.79), which were numerically solved at each pressure and temperature. The values of the volume fraction of the adsorbed phase ϕ_3 as a function of (*T*,*P*) are shown in Fig. 10.11b.

The data confirm the results of SAS as well as computer modelling studies [18, 31, 43] that indicate the formation of the adsorbed phase in SC CO_2 saturated silica aerogels with density much exceeding that of the unadsorbed (bulk) fluid at similar thermodynamic conditions. The value of the density ($\rho_3 = 1.12$ g/cm^3) and

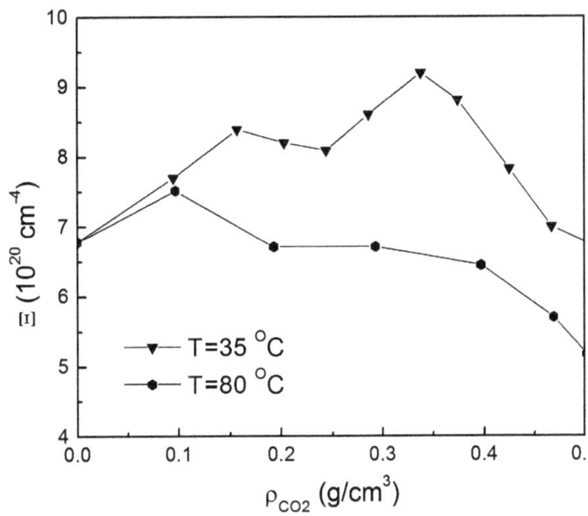

Fig. 10.10 Scattering invariant of a silica aerogel having both unadsorbed and adsorbed CO_2 phases (three phase system described by Eq. (6.77)). Reprinted with permission from [31]. Copyright [2006], AIP Publishing LLC

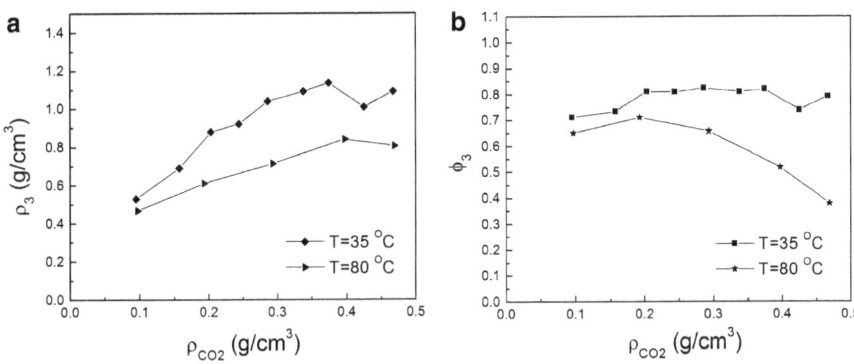

Fig. 10.11 The variation of density (**a**) and the volume fraction (**b**) of the adsorbed phase in pores of silica aerogel with a nominal density of 0.1 g/cm^3 at two temperatures specified in the *inset*. Reprinted from [32] with kind permission from Springer Science + Business Media

volume fraction ($\phi_3 = 0.82$) of the adsorbed phase calculated based on the SAS invariant—transmission tandem approach are in a good agreement with the estimates based on the analysis of difference in scattering between the two- and three-phase models ($\rho_3 = 1.07$ g/cm^3, $\phi_3 = 0.78$, respectively) [31]. The adsorbed phase of CO_2 in aerogels is formed not only in the proximity of the gas-liquid critical point but also at temperatures exceeding $T_C = 31.06$ °C by ~50 °C. At $T = 80$ °C, the values of ρ_3 and ϕ_3 are by 20–30 % lower than corresponding values at $T = 35$ °C, which underlines importance of the critical adsorption in the formation of the adsorbed fluid phase near T_C.

As was discussed in Sect. 6.9.3, detailed structural information on the properties of the adsorbed phase in porous solids may be extracted based on the analysis of the oscillatory deviations from the Porod law in the limit of high Q. Such approach was

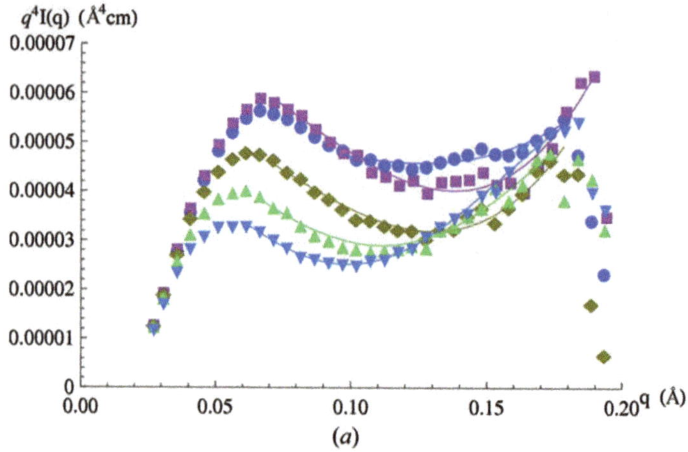

Fig. 10.12 Representative Porod plots of the scattering intensities from CO_2 saturated silica aerogel measured at $T = 35$ °C at different pressures: (*filled circle*) $P = 0$ MPa; (*filled square*) $P = 4.61$ MPa; (*filled diamond*) $P = 6.63$ MPa; (*filled triangle*) $P = 7.6$ MPa, (*filled inverted triangle*) $P = 7.92$ MPa. Reprinted from [44] with IUCR's copyright permission

applied in [44, 45] to analyze the phase behavior of supercritical CO_2 in silica aerogel with volume fraction $\phi_1 = 0.08$ of silica (porosity 92 %), surface area of 400 m^2 g^{-1} and nominal density of 0.2 g/cm^3 using SANS.

The Porod plots of the SANS scattering measured from CO_2 saturated aerogel at $T = 35$ °C are shown in Fig. 10.12. The solid lines represent fits to Eq. (6.80). The fitting parameters A, R, and δ allowed to calculate the dimensionless SLDs $\hat{\rho}_2^*$ (6.92) and $\hat{\rho}_3^*$ (6.93), and hence to determine corresponding physical densities of the adsorbed and unadsorbed phases using Eqs. (1.15) and (1.17). Volume fractions of the adsorbed ϕ_3 phase at different temperatures and pressures were obtained using $\phi_3 = \widehat{S}\delta$, where δ is the thickness of the adsorbed phase on the pore wall.

The variation of the density and volume fraction of adsorbed CO_2 in silica aerogel with a nominal density of 0.2 g/cm^3 as a function of fluid density is shown in Fig. 10.13. The observed patterns are qualitatively similar to those obtained in [32] for a factor of two less dense silica aerogel with a nominal density of 0.1 g/cm^3 (Fig. 10.11). The density of the adsorbed CO_2 far exceeds the density of bulk CO_2 at similar (T,P) and is of the order of 1–1.1 g/cm^3 in both aerogel samples. At high pressure and $T = 35$ °C the volume fraction of the adsorbed phase reaches value of ~0.6 and remains approximately independent of pressure. At $T = 80$ °C, i.e. far above the liquid–gas critical temperature, ϕ_3 increases initially, goes through the maximum and starts to decrease at higher pressures/fluid densities. This effect correlates with the variation of the thickness of the adsorbed phase, which at $T = 35$ °C stays approximately constant within the range of 30–35 Å at $\rho_{CO2} > 0.4$ g/cm^3 and decreases from ~30 to ~12 Å at $T = 80$ °C in the same range of the fluid density (Fig. 10.14). The observed difference in the variation of ϕ_3 and δ

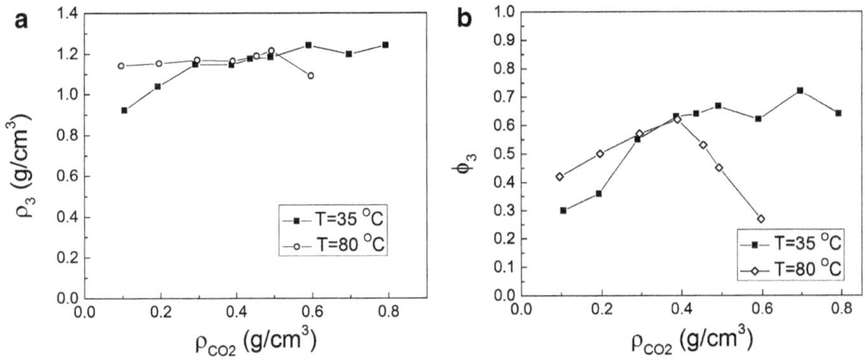

Fig. 10.13 The variation of density (**a**) and the volume fraction (**b**) of the adsorbed phase in pores of silica aerogel with a nominal density of 0.2 g/cm^3 at two temperatures shown in the *inset*. Reprinted from [44] with IUCR's copyright permission

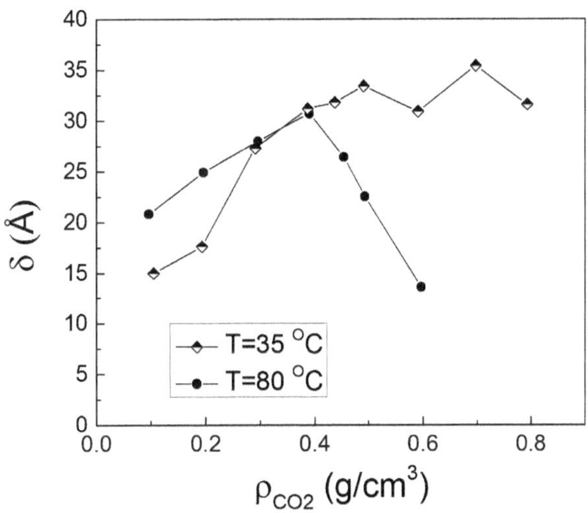

Fig. 10.14 Thickness of the adsorbed layer δ of CO_2 in silica aerogel as a function of the fluid density. The variation of δ is similar to that of ϕ_3 as both parameters are proportional to each other: $\phi_3 = \widehat{S}\delta$. Reprinted from [44] with IUCR's copyright permission

at $T \cong T_C$ and $T \gg T_C$ can be explained by the effect of critical adsorption, which diminishes at high temperature, when intermolecular interactions become dominant compare to the molecule-surface interactions that prevail near T_C [46].

10.3.2 Porous Fractal Silica

The results of computer simulations [47], suggest the shape and size of the pores are key parameters that may impact the behavior of fluids under confinement. Experimental investigation of the influence of the pore size (R) on the adsorption capacity is usually complicated as it requires synthesizing a series of samples with relatively monodisperse pores of the same shape and different size in each sample. At the same time, if the sample structure is characterized by a wide pore size distribution, SAS and USAS techniques allow to look inside pores of different sizes simultaneously, and monitor phase behaviour as a function of pore size in a single sample on the length scale between few Angstrom and micrometers.

SANS was applied to investigate the phase behaviour of supercritical CO_2 in porous fractal silica (PFS, Figs. 7.4 and 7.10) with volume fraction of silica $\phi_s = 0.15$ (porosity 85 %) at different pressure and temperature [48]. The availability of a fractal pore size distribution provided opportunity to monitor different stages of adsorption (micropore filling, formation of adsorbed film) in meso- and micropores in a single experiment.

PFS sample was prepared by a template method [49]. SANS experiments were conducted at ORNL on the General Purpose SANS instrument [50] with a neutron wavelength of $\lambda = 4.8\text{Å}(\Delta\lambda/\lambda \sim 0.13)$. Sample-detector distance was chosen to cover an overall range of scattering vectors (Q) $0.0086 < Q\,\text{Å}^{-1}$, where $Q = 4\pi\lambda^{-1} sin\theta$, in which 2θ is the scattering angle. The upper cut-off of the Q-range $Q = 0.2$ Å^{-1} was defined by the limited acceptance angle imposed by the design of the high-pressure cell used in these experiments. The data were corrected for instrumental background as well as detector efficiency and put on absolute scale [cross section $I(Q)$ in units of cm^{-1}] by means of pre-calibrated secondary standards.

Representative SANS data from the studied PFS with $\phi_s = 0.15$ at two temperatures T = 35 °C, and 60 °C as a function of pressures are shown in Fig. 10.15. The upper axis in each Figure shows the pore radius R calculated using $R = 2.5/Q$ (Sect. 6.10). As may be seen in Fig. 10.15, the variation of $I(Q)$ at constant temperature is different in various Q-ranges, or equivalently in pores of different sizes. In mesopores with $R > 40$ Å the cross section initially increases and then decreases as a function of pressure. In smaller pores with $R < 40$ Å the cross section is continuously decreasing as a function of pressure (fluid density).

The non-monotonic behavior of the cross section at low Q is illustrated in Fig. 10.16, which shows the variation of $I(Q)$ at $Q = 8.62 \times 10^{-3}$ Å^{-1} (mesopores) and at $Q = 0.1568 \times 10^{-3}$ Å^{-1} (micropores). The *positive* deviation from the two-phase model in the low Q region is qualitatively similar to the scattering behavior of CO_2 saturated silica aerogel (Fig. 10.8). As in aerogel, the adsorption of CO_2 in PFS begins by filling the smallest pores with densified fluid, which increases the average SLD of the porous matrix. The resulting higher neutron contrast leads to an increase in the scattering cross section at low Q as in this case scattering is averaged over a wider length scale. The maximum is reached when all nanopores become filled with the adsorbed fluid, after which the scattering starts to decrease due to a lower contrast between the matrix and the compressed fluid in larger pores. The trajectory of $I(0)^{1/2}$ vs. ρ_{CO2} tends to merge the straight line a fluid density of $\rho_{CO2} \cong 0.8$–0.9 g/cm^3, which can be taken as an estimate of the density of the adsorbed phase (Sect. 5.5).

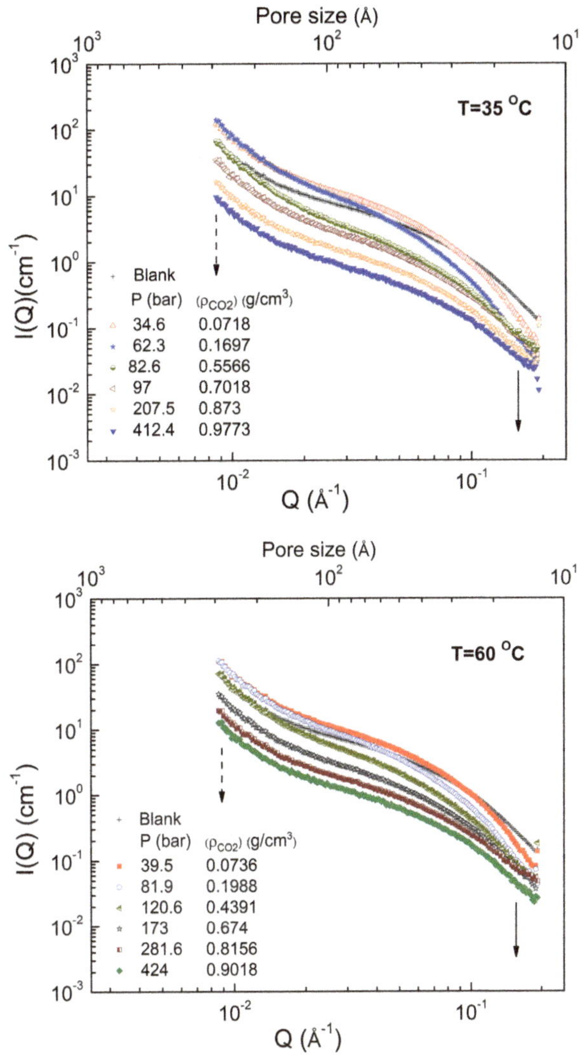

Fig. 10.15 Representative SANS data taken at T = 35 °C (*top panel*) and at T = 60 °C (*bottom panel*) as a function of CO_2 pressure and density specified in the *insets*. SANS data are shown for every third measured pressure for the sake of clarity. *Vertical lines* indicate representative Q values used to analyze the variation of $I(Q)$ in two characteristic pore size domains: mesopores ($Q = 8.62 \times 10^{-3}$ Å^{-1}, pore size $R = 290$ Å), and micropores ($Q = 0.1568$ Å^{-1}, pore size $R = 16$ Å). Reprinted with permission from [48]. Copyright 2010 American Chemical Society

The variation of $I(Q)$ in micropores is radically different: $I(Q)$ decreases continuously with pressure over a Q-range $Q > \sim 0.07$ Å^{-1}, which corresponds to pore sizes $R < 40$ Å (Fig. 10.16b). As may be seen in the Figure, $I(Q)$ decreases more rapidly than predicted by the two-phase model. The observed *negative* deviation from the two-phase model indicates that the average density of CO_2 in small pores is higher than that of the bulk fluid at the same thermodynamic conditions. Assuming that small pores are completely and completely filled with the adsorbent, the fluid density in small pores $(\rho_{CO2})_{pore}$ can be calculated using [47]:

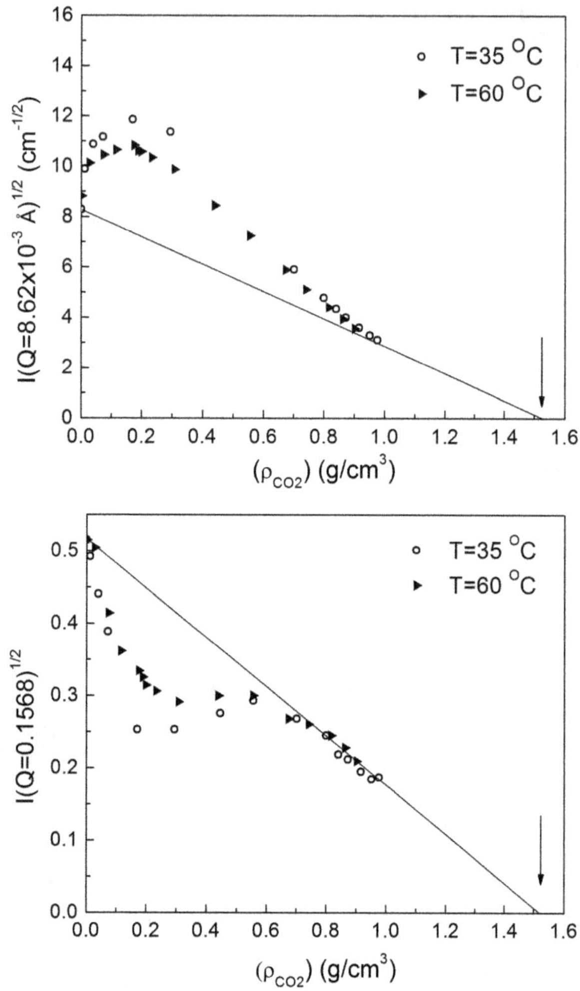

Fig. 10.16 Variation of SANS from mesopores of PFS with $R = 290$ Å (*top panel*) and $R = 16$ Å as a function of CO_2 density at two temperatures shown in the *inset*. *Solid lines* were calculated for the two-phase model (CO_2 in PFS with no adsorption). The arrows indicate zero average contrast density $\rho_{CO2} = 1.52$ g/cm^3 calculated using Eq. (10.17) for the silica density 2.4 g/cm^3. Reprinted with permission from [48]. Copyright 2010 American Chemical Society

$$\frac{I[Q,(\rho_{CO2})]}{I[Q,(\rho_{CO2})=0]} = \frac{\left[3.79 - 2.49(\rho_{CO2})_{pore}\right]^2}{(3.79)^2} = \left[1 - 0.66(\rho_{CO2})_{pore}\right]^2. \quad (10.18)$$

Thus calculated fluid density in pores with $R = 16$ Å as a function of bulk fluid density is shown in Fig. 10.17. As may be seen, a significant fluid densification occurs at both studied temperatures T = 35 and 60 °C. The "enrichment" factor $(\rho_{CO2})_{pore}/(\rho_{CO2})$ at $(\rho_{CO2}) \sim 0.2$ g/cm^3 is as high as ~5 at T = 35 °C and ~3.5 at T = 60 °C. At higher pressure, the difference between the density of the confined and bulk CO_2 gradually decreases and completely diminishes at $(\rho_{CO2})_{bulk} \sim 0.7$–0.8 g/cm^3. At this and higher densities the properties of bulk and confined CO_2 become identical: pores are evenly filled with dense CO_2 and scattering varies with pressure according to the two-phase model.

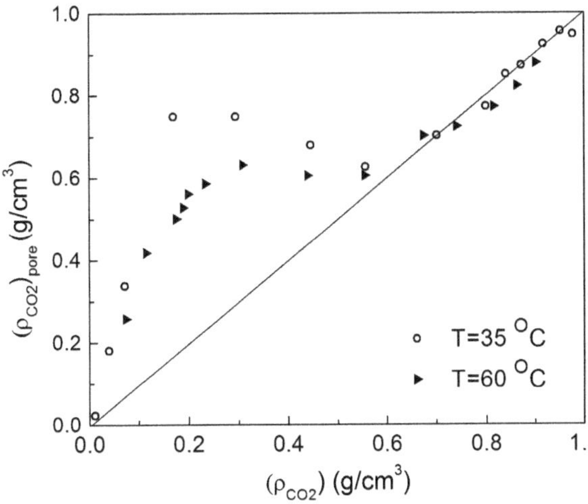

Fig. 10.17 Variation of the average CO_2 density in micropores of PFS with $R = 16$ Å as a function of the bulk fluid density at temperatures specified in the *inset*. *Solid line* is described by equation $(\rho_{CO2})_{pore} = (\rho_{CO2})$. Reprinted with permission from [48]. Copyright 2010 American Chemical Society

10.4 Methane in Porous Carbons

CO_2 emissions from anthropogenic sources are directly linked to the rising level of atmospheric CO_2 and to global warming. Carbon capture and sequestration in geological formations has been proposed as a measure for arresting the rising concentration of atmospheric carbon dioxide. Coal seam sequestration accompanied by the enhanced coalbed methane (ECBM) recovery is a particularly promising climate-change mitigation option as ECBM may help to offset the cost of compression, transportation, and storage of CO_2 by producing methane.

Fluid sorption and diffusivity are primary parameters, which control CO_2 sequestration as well as ECBM. The mechanism of gas transport in coal may be due to molecular diffusion through the micropores, bulk diffusion through the macro- or mesopores, or permeation through the cleats [51]. The gas mobility in pores may be studied by macroscopic techniques, such as chromatography, permeance and uptake measurements, which measure non-equilibrium transport diffusion coefficients dependent on a gradient in pressure or composition. One of the few experimental techniques, which may deliver information on the fluid mobility on a microscopic scale is quasi-elastic neutron scattering (QENS) technique [52].

QENS and SANS were used in [53] to study the diffusion of methane in nanoporous carbon aerogel with a pore size distribution similar to that of coal studied in [54]. Carbon aerogel samples were obtained from Oscellus Technologies, Livermore, CA. According to the manufacturer, carbon aerogels were made by preparing an organic phenolic-like aerogel followed by conversion to carbon aerogel by pyrolysis. The BET surface area of the samples studied was 480 m^2/g, the BJH adsorption average pore diameter was 48 Å, and the density of the aerogels was $\rho = 1.06$ g/cm^3 which gave aerogel porosity of $\approx$60 % based on the density of amorphous carbon 1.8 g/cm^3. Aerogel samples for QENS experiments were shaped into cylinders of 5 mm in diameter and 50 mm in length. The surface of cylinders was polished to ensure a tight fit in the QENS sapphire high-pressure cell to avoid free space which could be occupied with bulk methane.

QENS experiments were carried out on the time-of-flight spectrometer back-scattering spectrometer (BASIS) at the SNS, ORNL [55]. The selected dynamic range was from −100 to 100 μeV, and the instrument energy resolution function had full width at half maximum (FWHM) of 3.0 μeV. During the first series of the experiments the carbon-aerogel was used as obtained from the manufacturer. As a result molecular water was present in the sample due to atmospheric exposure. The measurements were done at 0.137, 0.206, 0.482, 0.965, 1.378, 5.515 and 8.273 MPa. For the next set of experiments, the pressure cell with the sample inside was heated overnight under high vacuum at 473 K. Following water out-gassing, the pressure cell was sealed and transferred to the spectrometer. The experiments were conducted at 0.482, 0.965, 1.378, 2.757, 3.447, 4.826 and 6.17 MPa. The data acquisition time was 5–6 h at each pressure to obtain good statistics. The neutron transmission of the methane filled aerogel was higher than 90 % which minimized the contribution of multiple scattering.

The spectra obtained at various pressures are shown in the Fig. 10.18 at a Q value of 0.9 Å^{-1}. The data shows significant elastic scattering and a broad QENS signal.

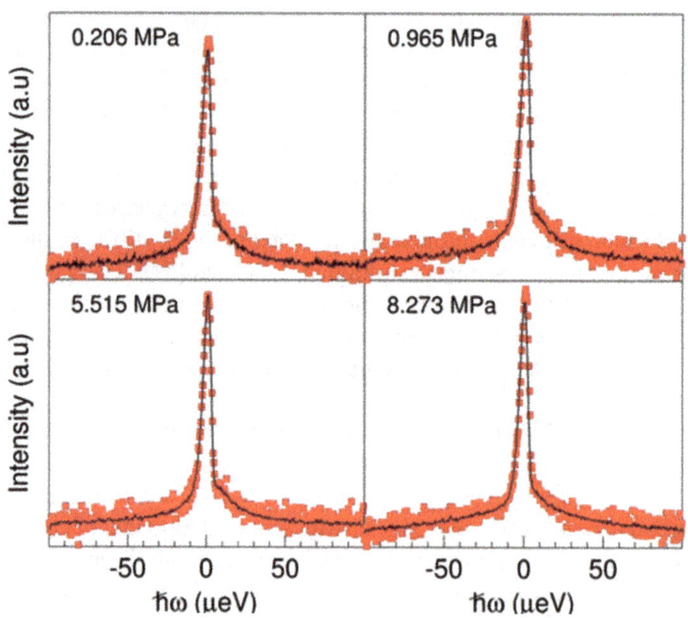

Fig. 10.18 The scattering intensities measured at $Q = 0.9$ Å^{-1}. The spectra are obtained for methane in carbon aerogel with water absorbed due to the atmospheric exposure at the pressures shown in the *insets*. The *solid line* through the data points are the Lorentzian fits. Reprinted from [53]. Copyright (2010), with permission from Elsevier

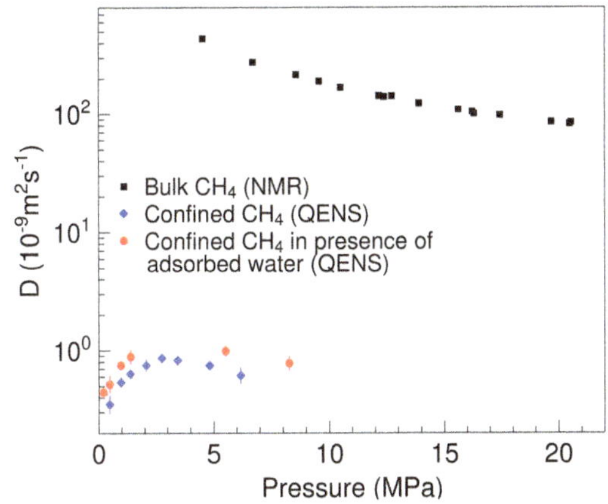

Fig. 10.19 Self-diffusivity of methane in nano-porous carbon aerogel at room temperature as a function of CH_4 pressure. *Circles* represent the data measured from carbon aerogel pores, which contain traces of adsorbed water from atmospheric exposure. *Diamonds* represent the diffusivities in pores of dried aerogel. For comparison, the diffusivity of bulk methane measured by NMR at room temperature is also shown. Reprinted from [53]. Copyright (2010), with permission from Elsevier

The elastic intensity is due to the scattering from carbon aerogel and the adsorbed molecules immobile on the time scale of the experiment. Since hydrogen atoms have a large incoherent scattering cross section the scattering intensity mainly originated from this source. In the incoherent scattering process, one measures the self-intermediate scattering function and the quasileastic broadening is due to the single particle motions. In this case, at low Q the HWHM, Γ, is related to the diffusivity, D, as $\Gamma = \hbar DQ^2$, where $\hbar$ is the reduced Planck's constant.

The diffusivity of confined methane from both sets of measurements and the diffusivity of bulk methane [56] as a function of pressure are shown in Fig. 10.19. In the case of the carbon aerogel with traces of the adsorbed moisture, the diffusivity of methane is somewhat higher than that in the dried sample. The higher diffusivity of methane in the "wet" sample may be explained by the hydrophobicity of methane which enhances its density fluctuations when it is in contact with water. In the dry aerogel, the value of D obtained at the lowest pressure measured (0.482 MPa) was $3.47 \pm 0.41 \times 10^{-10}$m^{2}s^{-1}. The highest value for $D = 8.55 \pm 0.33 \times 10^{-10}$m^{2}s^{-1} was obtained at a pressure of 2.75 MPa and at the highest pressure measured (6.17 MPa), D decreased to $6.07 \pm 0.62 \times 10^{-10}$m^{2}s^{-1}. These values are about two orders of magnitude smaller than diffusivities of bulk methane gas [56].

The described QENS study of the mobility of confined supercritical methane in carbon aerogel was complimented by the SANS study of the d-methane (CD_4) adsorption in same aerogel sample [53]. Deuterated methane was used in this study to minimize incoherent scattering from hydrogen and SANS curves measured as a function of CD_4 pressure (CD_4 density ρ_{CD4}) are shown in Fig. 10.20. The variation of $I(Q)^{1/2}$ vs. ρ_{CD4} in meso- and micropore domains (Fig. 10.21) is qualitatively

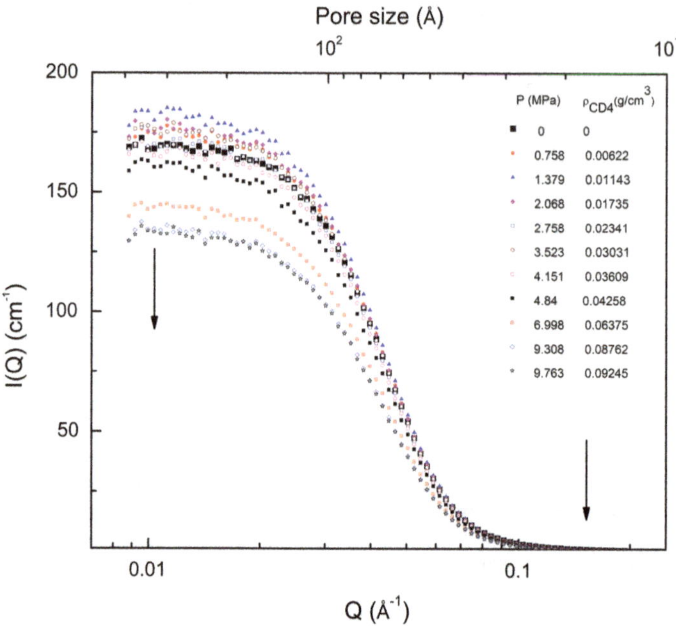

Fig. 10.20 SANS from d-methane-saturated carbon aerogel as a function of pressure (fluid density) shown in the *inset*. Pressure scan conducted at $T = 298$ K. The variation of scattering is different in different regions of the scattering vector Q, or equivalently pore size R shown in the upper layer. *Arrows* indicate representative Q values used to analyze the variation of $I(Q)$ in two characteristic domains: mesopores ($Q = 0.01033 \times 10^{-3}$ Å^{-1}, $R \approx 240$ Å), and micropores ($Q = 0.1527$ Å^{-1}, $R \approx 16$ Å). Reprinted from [53]. Copyright (2010), with permission from Elsevier

similar to that observed for CO_2 saturated porous silica (Figs. 10.16 and 10.17). SLD of the carbon matrix (ρ_1^*), can be calculated using physical density of carbon 1.8 g/cm^3 as $\rho_s^* = 6 \times 10^{12}$ cm^{-2} and SLD of CD_4 can be calculated at each density of bulk methane ρ_{CD4} using $\rho_{CD4}^* = [(\rho_{CD4}) \times 10]\, 10^{12}$ cm^{-2} (Eq. 5.29). Therefore, for the two-phase sample (no fluid adsorption) we obtain:

$$I(Q) \sim k_n^2 \sim \left[6 - 10(\rho_{\mathrm{CD4}})_{\mathrm{bulk}}\right]^2 \times 10^{24}\,\mathrm{cm}^{-4}. \tag{10.19}$$

It follows from Eq. (10.19) that ZAC should be reached at $(\rho_{CD4}) \approx 0.6$ g/cm^3 and the straight lines in Fig. 10.21 represent the variation of the cross section of the two phase system. Similarly to (10.18), the density of confined CD_4 in micropores of carbon aerogel can be calculated using:

$$\frac{I[Q, (\rho_{\mathrm{CD4}})]}{I[Q, (\rho_{\mathrm{CD4}}) = 0]} = \frac{\left[6 - 10(\rho_{\mathrm{CD4}})_{\mathrm{pore}}\right]^2}{(6)^2} = \left[1 - 1.66(\rho_{\mathrm{CD4}})_{\mathrm{pore}}\right]^2. \tag{10.20}$$

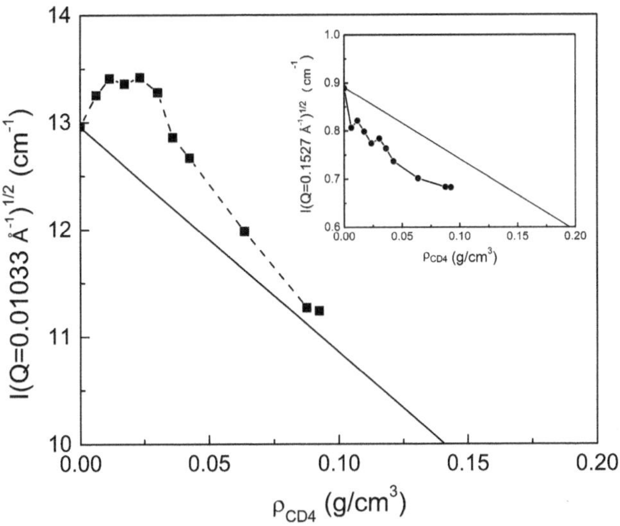

Fig. 10.21 Variation of SANS from mesopores with $R = 240$ Å ($Q = 0.01033 \times 10^{-3}$ Å^{-1}) and $R = 16$ Å ($Q = 0.1527$ Å^{-1}) (*inset*) as a function of CD_4 density. *Straight lines* were calculated using Eq. (10.19) for CD_4 in carbon aerogel with no adsorption (the two-phase model). Reprinted from [53]. Copyright (2010), with permission from Elsevier

An example of such calculation for micropores with $R = 16$ Å is shown in the inset in Fig. 10.22. As may be seen, the average methane density in small pores is constantly higher than the density of bulk methane at same thermodynamic conditions. This result agrees with the results of previous SAS studies, which demonstrated significant methane densification in small pores of activated carbons [57, 58]. The dramatic fluid densification occurs already at ~1 MPa and the "enrichment" factor $(\rho_{CD4})_{pore}/(\rho_{CD4})$ at this pressure is as high as ~9. At higher pressures ($P \sim 10$ MPa) it decreases down to ~1.5, which indicates a trend of the density of unadsorbed phase to approach the density of the adsorbed methane. Extrapolation of last four points in Fig. 10.21 shows that $(\rho_{CD4})_{pore}$ should approach (ρ_{CD4}) at the density of d-methane ~0.12 g/cm^3, which can be taken as an estimate of the density of the adsorbed CD_4.

SANS data can be used to explain apparently counterintuitive non-monotonic variation of the methane diffusivity vs. pressure revealed by QENS. At low pressure, the adsorption process begins in small, sub-nanometer pores, which become filled with the adsorbed phase characterized by the liquid-like density and diffusivity. As the pressure rises up to ~2.76 MPa, pores with progressively bigger sizes become filled with methane. The contribution of the sluggish mobility in the near-surface adsorbed phase decreases and the total diffusivity increases in larger pores due to the growing fraction of mobile molecules in the internal pore volume. At $P > 2.76$ MPa, the effect of intermolecular collisions becomes important, which results in a progressively lower total mobility, as is observed in the bulk state.

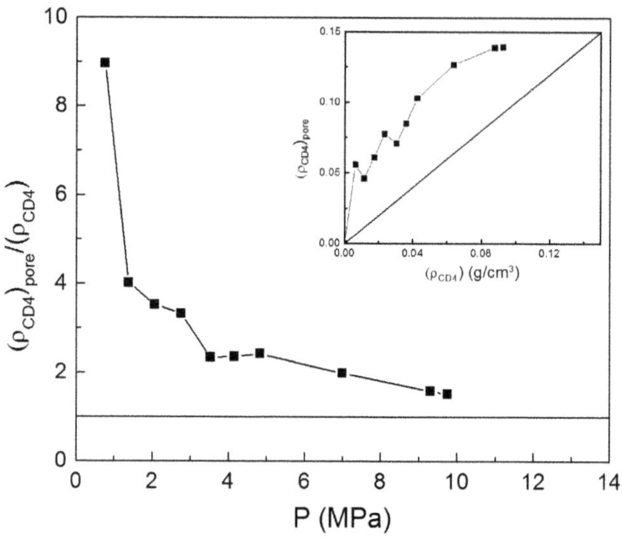

Fig. 10.22 Ratio of densities of confined and bulk CD_4 in micropores with the radius $R = 16$ Å. *Dashed line* was used to estimate the density of the adsorbed phase $\rho_a \approx 0.125$ g/cm^3. *Inset* shows the variation of the average CO_2 density in micropores with $R = 16$ Å as a function of the bulk density of d-methane. *Solid line* is described by equation $(\rho_{CD4})_{pore} = (\rho_{CD4})$. Reprinted from [53]. Copyright (2010), with permission from Elsevier

10.5 Hydrogen Storage in Activated Carbons

Entrapping hydrogen within the nanopores of solid adsorbents represents an attractive option for on-board storage of hydrogen for transportation applications. The key advantage of the physisorption process for the storage of hydrogen is the higher density values that can be achieved with the adsorbed gas, compared to that of the compressed phase, translating into higher storage capacities at lower pressures. The densification of hydrogen, or any adsorbed gas in general, is the result of the higher "internal" pressures experienced by the gas molecules confined within the pores. These pressures can be several times higher than of the applied external pressure.

Detailed information on the adsorbed phase density within pores of various sizes is a critical prerequisite for designing and developing the next generation of hydrogen storage materials as controlling the size and number of pores may lead to improved storage capacity and heat of adsorption. However, acquiring quantitative data on the density of adsorbed H_2 remains an outstanding challenge because of numerous experimental complications. As was demonstrated in earlier Chapters, SANS can provide unique pore-size-specific information on the phase behavior of confined fluids and may be used to evaluate density of the adsorbed fluid in pores of various sizes, including subnanometer length scales.

SANS was applied in [59] to evaluate the density of adsorbed H_2 gas in the nanopores and mesopores of ployfurfuryl alcohol-derived activated carbon (PFAC)

obtained by physical activation of the carbonized polymer. *In situ* SANS is well suited to study H_2 adsorbed in porous carbons because the negative neutron scattering length of hydrogen provides increases contrast between the adsorbed phase and the solid matrix at elevated pressure. The PFAC material had the BET surface area of 1530 m^2/g and a pore volume of 0.68 cm^3/g, of which 90 % was contained in micropores with width of <20 Å.

In-situ high-pressure SANS studies were conducted using the GP-SANS instrument at ORNL, with a neutron wavelength (λ) of 4.8 Å ($\Delta\lambda/\lambda \approx 0.13$). The sample–detector distance was chosen to cover an overall range of scattering vectors (Q) of 0.01 $Å^{-1} < Q < 0.8$ $Å^{-1}$. A powder PFAC sample was outgassed overnight at 300 °C, loaded into a thin-walled aluminum container with an internal thickness of 1 mm, and subsequently mounted inside the high-pressure cell. The H_2 pressure was varied from vacuum up to ~200 bar, and all of the measurements were conducted at room temperature (~23 °C). The scattering patterns were corrected for detector efficiency, electronic noise, and scattering from the empty pressure cell. The two-dimensional scattering patterns were radially averaged and put on an absolute scale [cross section $I(Q)$ in cm^{-1}] by means of precalibrated secondary standards.

The scattering patterns from the PFAC under vacuum and at different H_2 pressures (Fig. 10.23) show that $I(Q)$ increases continuously with pressure for all

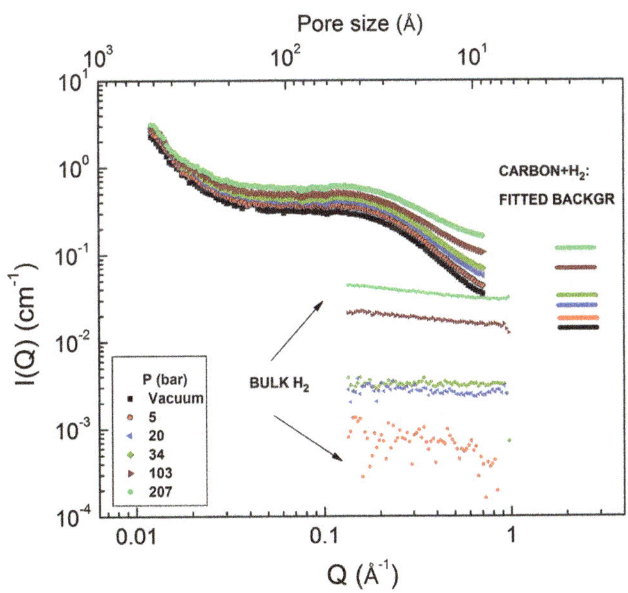

Fig. 10.23 SANS patterns from PFAC under vacuum and at elevated pressures of H_2 at room temperature (raw data before correction for incoherent background). Scattering from bulk H_2 measured at the same pressures as the confined H_2 is also shown. Fitted values of the incoherent background at each pressure are shown at the *right*. The observed significant differences between the scattering from bulk H_2 and H_2-loaded carbon are due to the extreme hydrogen densification in small pores. Reprinted with permission from [59]. Copyright 2010 American Chemical Society

values of the scattering vector. The observed variation of $I(Q)$ is due to the cumulative effect of the densification of the adsorbed H_2 in pores and the increasing incoherent background from H_2 with pressure. The pressure-dependent contribution of the incoherent background (I_{inc}) was determined by fitting the experimentally measured functions $I(Q)$ at each pressure to the Debye equation [60]:

$$I(Q) = \frac{B}{\left(a^{-2} + Q^2\right)^2} + I_{inc}, \tag{10.21}$$

where B is a pressure-dependent variable that accounts for the strong variation of the scattering-length density of H_2 with pressure and a is the Debye correlation length in the autocorrelation function of the scattering-length density fluctuations $\gamma(r) = \exp(-r/a)$ (Sect. 6.1); a is independent of pressure for completely saturated pores. The fits were performed over the range $Q > 0.1$ Å^{-1}, and the obtained values of I_{inc} were subtracted from the experimentally measured intensities to provide the coherent cross sections $I(Q)$. The density of H_2 in the pores $(\rho_{H2})_{pore}$ at different values of Q (or, equivalently, at different pore sizes $R \approx 2\pi/Q$) was evaluated using the ratios of the coherent intensities measured for H_2-loaded and blank carbons [$I(Q, P)$ and $I(Q, P = 0)$, respectively]:

$$\frac{I(Q,P)}{I[Q,(P=0)]} = \left[1 + 5.6(\rho_{H2})_{pore}\right]^2. \tag{10.22}$$

The densities of H_2 in carbon pores calculated using Eq. (10.22) are plotted in Fig. 10.24a as functions of pressure for the selected pore sizes. The densities of liquid H_2 at the triple point and gaseous H_2 at the corresponding pressures at room temperature are also shown for comparison. It is evident that at room temperature, for any given values of the external pressure and pore size, the density of *confined* H_2 is considerably higher than that of the bulk gas and approaches the density of liquid H_2 at the highest pressure for the narrowest pore studied. The effect of pore size on the density at constant pressure is illustrated in Fig. 10.24b: the density is systematically higher in the narrow pores and decreases with increasing pore size, reaching almost constant values in the mesoporous range (>20 Å).

The interrelation between the structure and adsorption capacity of H_2 in a different activated carbon (ultramicroporous carbon, UMC), obtained through chemical activation of wood, was studied in [61] using a combination of SANS, scanning transmission electron microscopy (STEM), and gas adsorption methods. Characterization using N_2 adsorption at 77 K and CO_2 adsorption 273 K confirmed that UMC had the surface area of 2450 m^2/g and a pore volume of 1.19 cm^3/g, of which 70 % was contained in micropores with width of <20 Å and ~53 % in narrow micropores with width <7 Å. Scattering from UMC as a function of H_2 pressure was compared with that from PFAC in the same pressure region between vacuum and 34.5 bar.

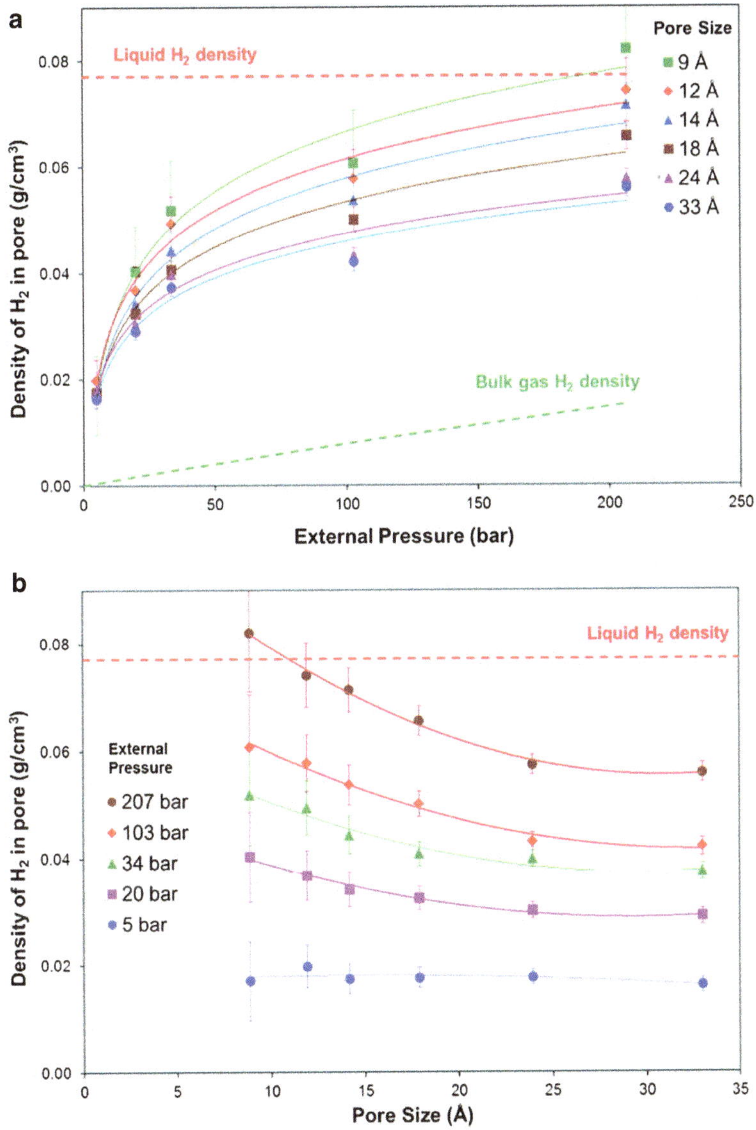

Fig. 10.24 Room-temperature densities of adsorbed H_2 as functions of (**a**) pressure for selected pore sizes and (**b**) pore size at constant pressures. Reprinted with permission from [59]. Copyright 2010 American Chemical Society

In situ SANS measurements were conducted at the GP SANS instrument of the High Flux Isotope Reactor, at ORNL [50] with a neutron wavelength of 4.72 Å ($\Delta\lambda/\lambda \sim 0.13$). The sample-to-distance was chosen to cover an overall range of scattering vectors $0.005 < Q < 0.65$ Å^{-1}. The maximum Q-limit was determined by the acceptance angle of the high pressure cell. This limit allowed the study of sub-nanometer pores. Powder carbon samples were outgassed for ~5 h at 300 °C, loaded into a thin-wall aluminum container with 1 mm internal thickness, and subsequently mounted inside a high-pressure cell. The samples were studied under vacuum (pressure less than 10^{-3} Tor) to explore the texture and pore structure, and then hydrogen was injected to study its phase behavior under confinement. The hydrogen pressure was varied from vacuum up to ~34.5 bar, and all measurements were conducted at room temperature (~23 °C). Average acquisition time for each scattering curve was ~30 min. All data were corrected for instrumental background, transmission and detector efficiency. The corrected data was converted on absolute scale with the cross section $I(Q)$ in units of cm^{-1} using pre-calibrated secondary standards.

As may be seen in Fig. 10.25, the scattering intensity from both UMC and PFAC carbons increases continuously with H_2 pressure. The increase of $I(Q)$ is especially pronounced in the high Q range. The observed variation of $I(Q)$ may be attributed to the combined effects of the increasing neutron contrast caused by the densification of adsorbed H_2 in micropores, and the increasing incoherent scattering. The total scattering intensity can be decomposed into the fractal scattering as well as scattering from H_2 adsorbed in meso- and micropores. The effect of H_2 pressure on

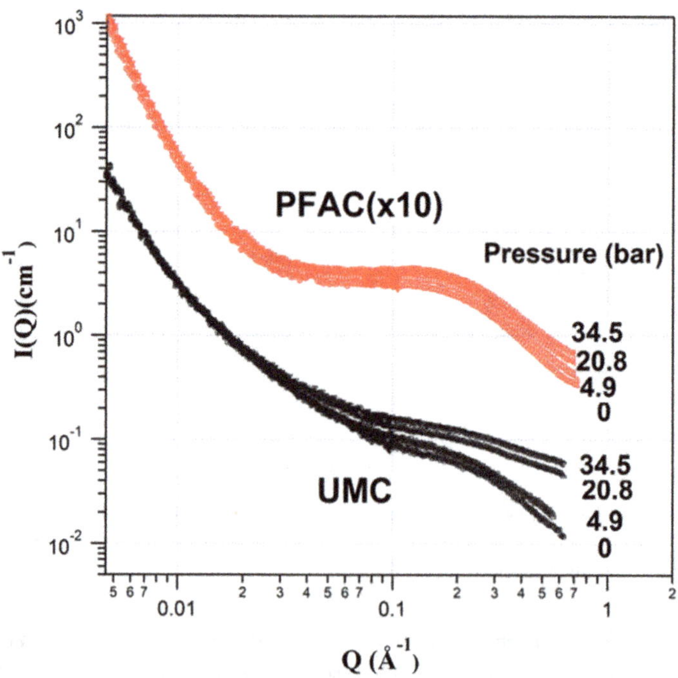

Fig. 10.25 SANS profiles of carbons under vacuum and as a function of H_2 pressure at room temperature. The scattering curves for PFAC were shifted up by a factor of 10 for clarity [61]. Reprinted from [61]. Copyright (2014), with permission from Elsevier

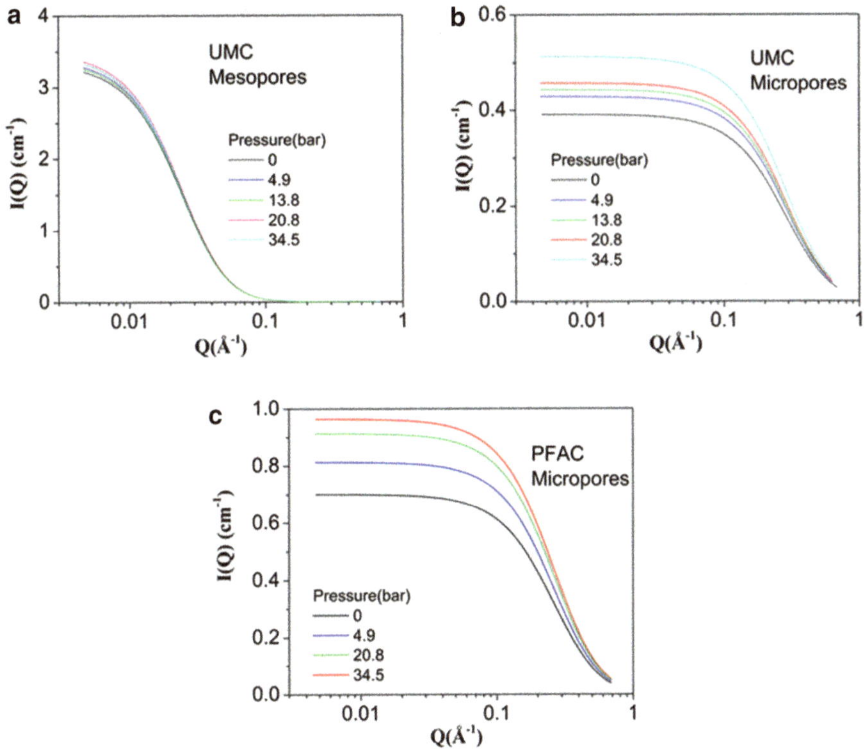

Fig. 10.26 Decomposed scattering profiles for mesopores (**a**) and micropores (**b**) of UMC, as well as micropores of PFAC (**c**) as a function of H_2 pressure. Reprinted from [61]. Copyright (2014), with permission from Elsevier

scattering intensity contribution from UMC mesopores and micropores is presented in Fig. 10.26a, b. As expected, the scattering intensity in the high Q range (micropores) increases with pressure for both UMC and PFAC (Fig. 10.26b, c, respectively) indicating the increasing contrast between adsorbed hydrogen and the carbon matrix. In contrast, only a subtle variation of scattering from UMC mesopores is observed in the low Q range (Fig. 10.26a).

Scattering components (Fig. 10.26) were used to extract the average H_2 density in micro- and mesopores in both carbons as a function of external pressure based on the Porod invariant analysis (Sect. 6.2). The results are shown in Fig. 10.27, which also gives the density of bulk H_2 calculated using the equation of state for H_2 [62]. Hydrogen adsorbed in micropores has higher density than the unadsorbed gas. The density of hydrogen confined in the *micropores* of PFAC is larger than that in the micropores of UMC, suggesting stronger interactions with the pore walls in PFAC due to smaller pore size in this carbon. The density of hydrogen in UMC *mesopores* is much lower and increases only moderately with pressure, which suggests absence of strong densification in these pores even at elevated pressures. The density of H_2 gas at the maximum pressure studied (~35 bar) is 0.0026 g/cm^3.

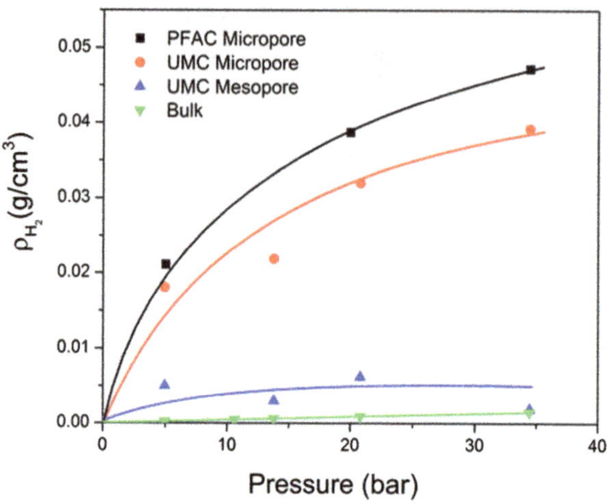

Fig. 10.27 Densities of hydrogen confined in micropores of PFAC and UMC, as well as in mesopores of UMC. For the reference, the density of bulk hydrogen as a function of pressure is shown. The *solid lines* are guide for eyes. Reprinted from [61]. Copyright (2014), with permission from Elsevier

The H_2 density in micropores in both carbons is about 20 times higher, which correlates with the results of the previous SANS study [59]. SANS measurements indicate that adsorbed H_2 fills predominantly the narrowest pores in both UMC and PFAC, and the density of adsorbed H_2 in micropores increases with the external pressure. At the same time, H_2 density in UMC *mesopores* is relatively low, close to the density of bulk gas, and undergoes only subtle changes with pressure.

These findings were supported by theoretical calculations, using first-principle methods of treating the van der Waals interactions within density functional theory, which confirmed that H_2 density in subnanometer pores indeed could reach a liquid-like density even at 50 bar and 298 K [63, 64]. The described SAS results demonstrate the advantage of adsorptive storage over compressed gas storage and emphasize the greater efficiency of micropores over mesopores in the adsorption process [65].

10.6 CO_2 Sequestration in Coal

CO_2 emissions from anthropogenic sources are frequently directly linked to the rising level of atmospheric CO_2 and to global warming. Carbon capture and sequestration in underground geological formations (saline aquifers, depleted oil and gas reservoirs as well as unmineable coal seams) is one of the proposed measures for arresting the rising concentration of atmospheric carbon dioxide. Sequestration of CO_2 in unminable deep coal seams is particularly attractive due to the relatively high sorption capacity exhibited by coal as well as the proximity of geological sites potentially suitable for geological storage to coal-burning power plants. In addition, injected CO_2 displaces methane present in the coal, which contributes to improved methane recovery in the basins with coalbed methane.

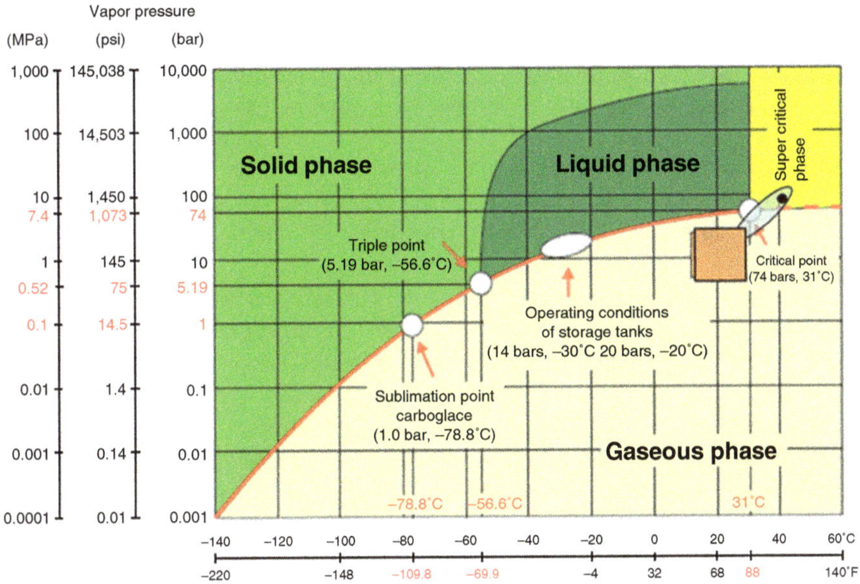

Fig. 10.28 CO_2 phase diagram and underground conditions in potential storage sites in unminable coal seams in Australia (*ellipse*) and USA (*square*). Image courtesy of M. Mastalerz

The amount of CO_2 that is expected to be adsorbed by a specific coal is one of the critical parameters for sequestration technology. In addition to the petrographic composition, the sorption capacity of coal depends on the seam depth, which defines temperature (T) and pressure (P) at which CO_2 will be adsorbed after injection in a particular geological formation. As soon as both T and P increase with depth, below ~800 m both pressure and temperature exceed liquid–gas critical parameters of CO_2 ($T_C \sim 31.1$ °C, $P_C \sim 7.2$ MPa), and the fluid will be injected and adsorbed at supercritical conditions [66]. Many of the proposed sequestration sites in USA, Europe, and Australia correspond to conditions of CO_2 supercriticality (Fig. 10.28). At the same time, due to a much lower critical temperature ($T_C \sim -83$ °C), methane will remain supercritical at any depth. Understanding the interplay between the seam depth as well as the phase state of CO_2 and methane is important to CO_2 sequestration and enhanced coalbed methane recovery technologies.

Numerous laboratory studies of CO_2 adsorption into coal have been conducted over the last decade providing information about rates and volumes of adsorbed CO_2 as well as selectivity of CO_2 sorption over methane under specific conditions [67]. Even though there have been numerous studies conducted, important questions about the structural stability as well as adsorption mechanisms and controls on sub- and supercritical CO_2 adsorption in coal remain open. Because the understanding of the process of CO_2 adsorption into the coal beds is necessary to ensure successful CO_2 injection operations on an industrial scale, the details of this process that remain unknown need to be resolved.

SANS and USANS techniques were applied in [68] to study CO_2 adsorption process in-situ in coals under simulated hydrostatic reservoir conditions. Three coal seams were sampled to represent a range of the hydrostatic reservoir conditions (pressure, P, and temperature, T) and the corresponding CO_2 phase properties: the Seelyville from the Illinois Basin (USA), Bulli 4 from the Sydney Basin (Australia), and Baralaba from the Bowen Basin (Australia). Based on the temperature and pressure conditions, the sampled coals span a range from subcritical (Seelyville and Bulli 4) to supercritical (Baralaba) thermodynamic conditions of *in situ* carbon dioxide.

SANS experiments were performed at the NIST Center for Neutron Research, Gaithersburg, Maryland, USA. A NG7 30 m SANS instrument was used with the neutron wavelength tuned to $\lambda = 6$ Å. For each sample, two sample-to-detector distances (1.4 and 15.3 m) were used which corresponds to the overall range of the scattering vectors $0.002 \leq Q \leq 0.4$ Å^{-1}. However, the real experimental high-Q cutoff was $Q \sim 0.2$ Å^{-1} due to the limited acceptance angle imposed by the design of the high-pressure cell used in our experiments. Scattering from the Seelyville coal in the high-Q domain $0.08 < Q < 1$ Å^{-1} was measured at ambient conditions using the General Purpose SANS instrument at Oak Ridge National Laboratory, Oak Ridge, Tennessee, USA. USANS experiments were performed at NIST, using the BT5 perfect crystal SANS instrument. Application of these instruments allowed a broad range of pore sizes, from approximately 35 Å (3.5 nm) to 40,000 Å (4 μm) to be probed by neutrons.

The coal samples (approximately 1.0 mm thick, 16 mm diameter) were dried overnight under vacuum at 60 °C and subsequently mounted inside a SANS/USANS pressure cell. The neutron beam size used was about 1 cm^2 in area and acquisition times were of the order of 30 min for SANS and several hours for USANS. SANS and USANS scattering profiles of each coal sample were acquired at T = 16 °C in the pressure range from 1 to 50 bar of CO_2 (for Seelyville), at T = 27 °C and pressure range 1–100 bar (for Bulli 4), and at T = 35 °C and pressure range 1–200 bar (for Baralaba). Such a selection of temperatures and pressures was made to simulate hydrostatic reservoir conditions of these coals in their respective subsurface (P, T) environments. Figure 10.29 shows the coexistence curve of bulk CO_2 calculated using equation of state [42] with superimposed thermodynamic paths studied for each coal. The thermodynamic conditions for the Seelyville coal were subcritical and corresponded to a gas phase of the bulk fluid. Temperature and pressure range covered for Bulli 4 also began at low pressures, which corresponded to a gaseous phase of carbon dioxide and ended at $P = 100$ bar, which corresponded to a single phase liquid. Finally, the thermodynamic path followed in the experiments with Baralaba coal laid entirely in the supercritical region.

SANS patterns from the Seelyville coal as a function of pressure are shown in Fig. 10.30. The variation of the scattering curve with pressure is different in different regions of the scattering vector and corresponding pore sizes. In the region (1), where $3 \times 10^{-5} \leq Q \leq 3 \times 10^{-4}$ Å^{-1} (which corresponds to pore size $R = \pi/Q$ $1 \times 10^5 \geq R \geq 1 \times 10^4$ Å), the scattering curves move down with increased pressure

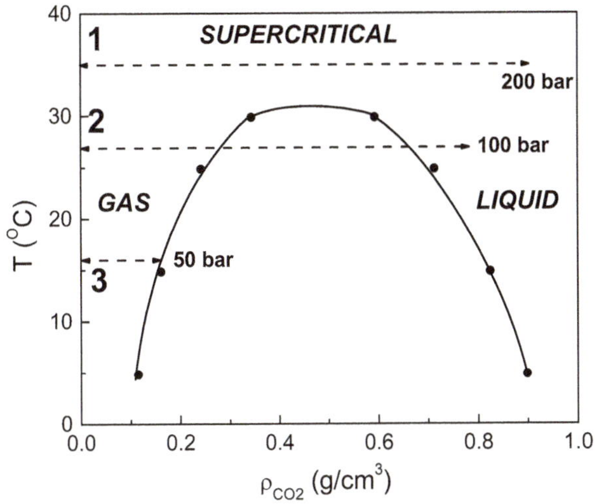

Fig. 10.29 Thermodynamic paths covered during SANS and USANS experiments for (*1*) Baralaba coal (35 °C, 1–200 bar); (*2*) Bulli 4 coal (27 °C, 1–100 bar); (*3*) Seelylville coal (16 °C, 1–50 bar). *Solid line* represents the gas-liquid coexistence curve for the bulk CO_2 calculated using the equation of state. 1 bar = 10^5 Pa. Reprinted from [68]. Copyright (2009), with permission from Elsevier

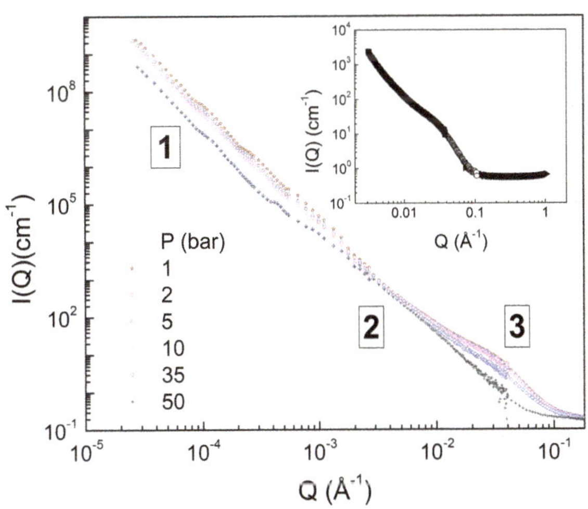

Fig. 10.30 Combined SANS/USANS data for Seelyville coal as a function of pressure specified in the *inset*. The variation of scattering with pressure is different in different regions of the scattering vector Q marked 1, 2, and 3. The *inset* highlights the flat scattering in the high-Q region ($Q > 0.1$ Å^{-1}), which demonstrates the absence of small pores with sizes below 35 Å. 1 bar = 10^5 Pa. Reprinted from [68]. Copyright (2009), with permission from Elsevier

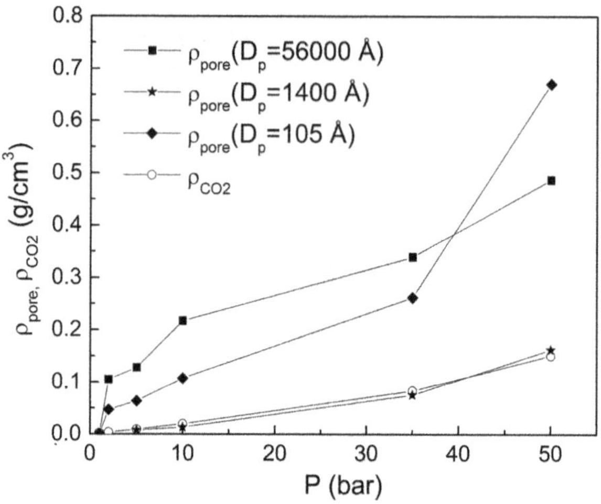

Fig. 10.31 Variation of bulk and average density of CO_2 adsorbed in pores of Seelyville coal as a function of pressure. Reprinted from [68]. Copyright (2009), with permission from Elsevier

in parallel on the log-log scale retaining the same slope of -3.30 ± 0.02. In region (2), where $0.001 \leq Q \leq 0.01$ Å^{-1} (which corresponds to R between 3000 and 300 Å), the variation of $I(Q)$ with pressure is more subdued. Finally, in the region (3), where $0.01 \leq Q \leq 0.1$ Å^{-1}, (which corresponds to R between about 300 and 30 Å), the variation of $I(Q)$ with pressure is generally more pronounced than in region (2), and it is Q-dependent. Such a rich pressure dependent variation of scattering is probably related to the relatively high total porosity of this (~7.8 %) and related interconnectivity of pores in this coal.

The SLD of Seelyville coal $\rho_1^* = 2.2 \times 10^{10}$ cm^{-2} was calculated based on its petrochemical composition. This value was substituted into (10.18) to calculate the average density of CO_2 in pores of different sizes (Fig. 10.31). The average CO_2 density in micrometer and nanometer size pores is systematically and significantly higher than the density of bulk CO_2 calculated for similar temperature and pressure conditions using the equation of state [42]. At the subsurface conditions ($T = 16$ °C, $P = 50$ bar), the average fluid density in *micrometer size pores* of the Seelyville coal (Q-region-1 in Fig. 10.30) is $\rho_{pore} = 0.489$ g/cm^3, i.e. by a factor of about three higher than the expected density of unadsorbed (gaseous) CO_2 at the same thermodynamic conditions ($\rho_{CO2} \approx 0.15$ g/cm^3). In *small pores* with $R \sim 105$ Å, the average density of the confined fluid is even higher ($\rho_{pore} = 0.671$ g/cm^3) and $\rho_{pore}/\rho_{CO2} \approx 4.4$. In *intermediate size pores* with $R \approx 1000$ Å the average fluid density in pores is similar to the density of bulk fluid, which indicates that adsorption (fluid densification) does not occur in these pores.

The ratio of the average CO_2 density in coal pores to the density of bulk CO_2 at subsurface conditions of Seelyville coal (16 °C, 50 bar) calculated for the entire range of the studied pore sizes is shown in Fig. 10.32. Both Figs. 10.31 and 10.32 illustrate that the dense condensed phase starts to form in Seelyville coal immediately after the introduction of CO_2, which indicates that the porous coal matrix

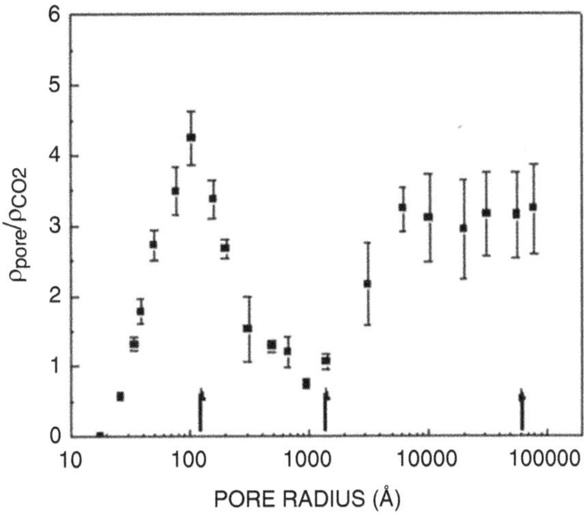

Fig. 10.32 The ratio of the average CO_2 density in pores of Seelyville coal and the density of the bulk fluid as a function of the pore radius. *Arrows* mark the values of the pore sizes at which the densities of ρ_{pore} plots of $I(Q)^{1/2}$ versus ρ_{CO2} shown in Fig. 10.31 were calculated. Reprinted from [68]. Copyright (2009), with permission from Elsevier

helps to condense the gaseous CO_2 and form coexisting gas and liquid phases. Calculations based on the equation of state suggest that pumping CO_2 into Seelyville coal at 16 °C should saturate coal pores at the gas density of ≈0.15 g/cm^3 at 16 °C (see Fig. 10.29). In reality, the CO_2 confined in coal is subject to adsorption and condensation, which leads to the formation of the dense adsorbed phase with a much higher average fluid density (0.489 g/cm^3). This indicates that approximately half of the micrometer size pores of the Seelyville coal are filled with liquid CO_2 at (P,T) conditions at which the bulk fluid should remain in a gaseous state. For the nanometer size pores, the average fluid density much exceeds the average density of coexisting gas and liquid, which indicates that most of the small pore volume is occupied by liquid CO_2.

SANS patterns from Baralaba coal with w = total porosity ~5.4 % as a function of pressure are shown in Fig. 10.33. For this coal, only two characteristic Q-regions of the scattering vector and corresponding pore sizes can be identified. These regions are marked region 1 ($2 \times 10^{-5} < Q < 0.01$ Å^{-1}), which corresponds to the pore size $1.6 \times 10^5 > R > 300$ Å) and region 2 ($0.01 < Q < 0.1$ Å^{-1}), which corresponds to the pore size $300 > R > 30$ Å).

SANS patterns from Bulli 4 coal as a function of pressure are shown in Fig. 10.34. The scattering from this high rank and relatively low total porosity (~1 vol.%) coal does not change within experimental error over the range of scattering vectors explored in this paper $2 \times 10^{-5} < Q < 0.2$ Å^{-1}. Negligibly small variation of scattering with the density of the invading CO2 is indicative of very low pore interconnectivity and accessibility to fluids in this coal (Sect. 10.7).

The obtained in [68] results demonstrate the unique capability of the neutron scattering techniques to monitor the evolution of microstructure and adsorption capacity of coal exposed to CO_2 environment at subsurface-like temperature and pressure conditions. The observed phase behavior of the injected CO_2 under confinement, particularly the densification from a gaseous phase to a liquid state,

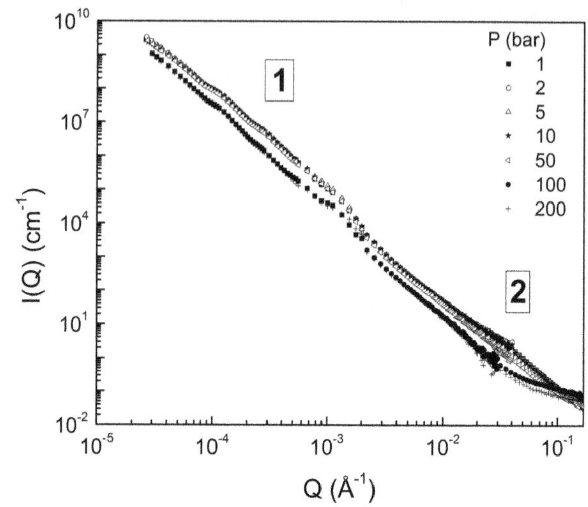

Fig. 10.33 Combined SANS/USANS data for Baralaba coal as a function of pressure specified in the *inset*. The variation of the scattering pattern with pressure is different in the two in different regions of the scattering vector Q marked *1* and *2*. Reprinted from [68]. Copyright (2009), with permission from Elsevier

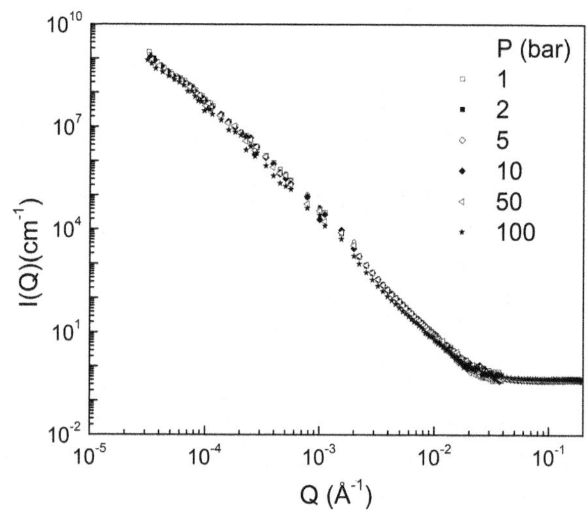

Fig. 10.34 Combined SANS and USANS data for the Bulli 4 coal as a function of pressure specified in the *inset*. Reprinted from [68]. Copyright (2009), with permission from Elsevier

have very significant operational and reservoir capacity ramifications when assessing the suitability of unminable coal seams for use as carbon dioxide sequestration reservoirs. The results show that the sorption capacity of coal is sample-dependent and it is also strongly affected by the phase state (subcrititcal or supercritical) of the injected fluid in addition to its structural and petrographic characteristics.

10.7 Pore Interconnectivity and Accessibility to Fluids in Coal and Shales

Fluids containing inorganic and organic solutes (including hydrocarbons) and gaseous species (e.g. carbon dioxide, CO_2, and methane, CH_4) can occupy the pores or fractures of numerous types of complex heterogeneous solids. These solid materials include such practical systems as supported catalysts, ceramics and composites, membranes, rock, minerals, soil, and bone. A number of factors dictate how fluids migrate into and through these micro- and mesoporous media, wet and ultimately adsorb and react with the solid surfaces. These include the size, shape, distribution and interconnectivity of pores, as well as the chemistry and physical properties of the solids and fluid molecules.

Pore accessibility and its variation with pore size is not defined solely by the structure of a specific porous solid. Accessibility may vary considerably depending on temperature and pressure, which determine the phase of the invading medium (i.e. gas, liquid, or supercritical fluid) as well as on the chemistry-driven specifics of the molecule-surface interaction potentials. Even for a particular solid/fluid combination, the accessible porosity and the variation of $c_{AC}(R)$ may depend on the proximity of the fluid phase state to its critical point at which the critical adsorption effects may become dominant. In the case of coal and other organic porous materials, the measured $c_{AC}(R)$ may also depend on the time scale of the experiment, as the molecular diffusion in such solids might occur quickly through an interconnected network of pores having access to the external surface as well as slowly through the solid matrix.

Coal is a porous material with pore sizes that span wide length scales including macro-, meso- and microporous regimes. The porosity plays a key role in all aspects of coal utilization, such as extraction of methane from coal seams, gasification, combustions, liquefaction, production of metallurgical coke and activated carbon as well as geological sequestration of CO_2. The debate about the nature and structure of the pores in coal is ongoing [67, 69]. According to a widely accepted consensus, coal is a solid that contains slit-like pores interconnected by narrow capillary constrictions and connected to the surface [70]. Some experimental results suggest however that a significant proportion of pores in coal may not be open to the external surface [71, 72]. It is not known what fraction of pores in coal is inaccessible to greenhouse gases such as methane and carbon dioxide, and the issue of selectivity of access to pores of different sizes is even more obscure. Such information is particularly important for the practice of ECBM as experimental data on pore accessibility and adsorption selectivity could help to understand the fundamental limits to the ability of CO_2 to displace methane in subsurface conditions during sequestration of CO_2 in coal seams.

A number of experimental methods have been used to characterize porosity in coal, including gas adsorption, mercury intrusion porosimetry, transmission electron microscopy (TEM), as well as SAS. Each of the methods has its limitations, e.g. gas adsorption and mercury porosimetry can only provide information about "open" porosity and TEM can only be used to assess pore connectivity inside a very limited sample volume. Thus far, SAS represents the only non-invasive technique that can be used for evaluating the *total porosity* (i.e. sum of the inaccessible and accessible pore volumes) over the range of pore sizes 0.4 nm to 5 μm.

In order to quantify the volume fraction of inaccessible pores, the porous solid may be saturated with "contrast matching" fluid, i.e. fluid with the SLD value close to that of the solid matrix. In this case, the scattering from open pores is eliminated and the residual scattering provides a "fingerprint" of the inaccessible porosity. SANS has been used before to evaluate the fraction of inaccessible pores in coal using *liquid mixtures* of protonated and deuterated solvents as a contrast matching medium. Mixtures of H_2O and D_2O were used in [73–76] to create contrast matching in coals. It was later realized that in coals many of the functional groups can exchange hydrogen with pore saturating water on time scales varying from seconds to weeks. Therefore, the isotope exchange may alter the H_2O/D_2O ratio in pores and significantly shift the local contrast matching condition, which was not fully recognized in early studies.

An alternative method of obtaining contrast matching in porous media is to use non-adsorbing or weakly adsorbing supercritical fluids or gases, such as CO_2 or deuterated methane (d-methane) and measure the scattering patterns as a function of pressure (Sect. 5.10). Using d-methane (CD_4) rather than "normal" methane (CH_4) helps to minimize the contribution of incoherent scattering from hydrogen to the SANS data. Furthermore, the SLD for methane can be varied with pressure and, unlike CH_4, d-methane has a positive SLD. An important advantage of utilizing gases or supercritical fluids is their excellent penetrability into porous structure due to their order(s) of magnitude lower viscosity than their corresponding liquids.

The scattering from a two-phase system with randomly distributed interconnected pores (e.g. coal with pores filled with air or weakly adsorbing fluid) is described by Eq. (6.23). In this Equation, ρ_1^* is the SLD of the solid matrix and ρ_2^* is proportional to the fluid density, which depends on pressure and temperature. At small pressure $\rho_1^* \gg \rho_2^*$, and therefore neutron contrast and $d\Sigma/d\Omega(Q)$ initially decreases with increasing P. If all pores are accessible to the fluid, $d\Sigma/d\Omega(Q)$ should virtually vanish at a certain P, corresponding to the zero average contrast (ZAC) (or contrast matched) pressure (P_{ZAC}) at which $\rho_1^* = \rho_2^*$. At $P > P_{ZAC}$, as ρ_2^* becomes greater than ρ_1^*, the cross section $d\Sigma/d\Omega(Q)$ will start increasing again. In the two-phase approximation, any residual scattering observed at $P = P_{ZAC}$ is attributed to the scattering from inaccessible pores, which do not belong to the interconnected porous channels having access to the external surface (Fig. 10.35).

First SAS studies of the interconnectivity of pores in different coals using supercritical fluids and pressurized gases were conducted in [77] using a combination of SANS and USANS techniques. Three coal samples were investigated in this study: one from the Illinois Basin in the USA, collected from 167 m depth (Seelyville). Two other coals were samples of commercial coals from the Bowen Basin, Queensland, Australia. The selected coal samples had different total porosities and thus potentially different volume of accessible pores.

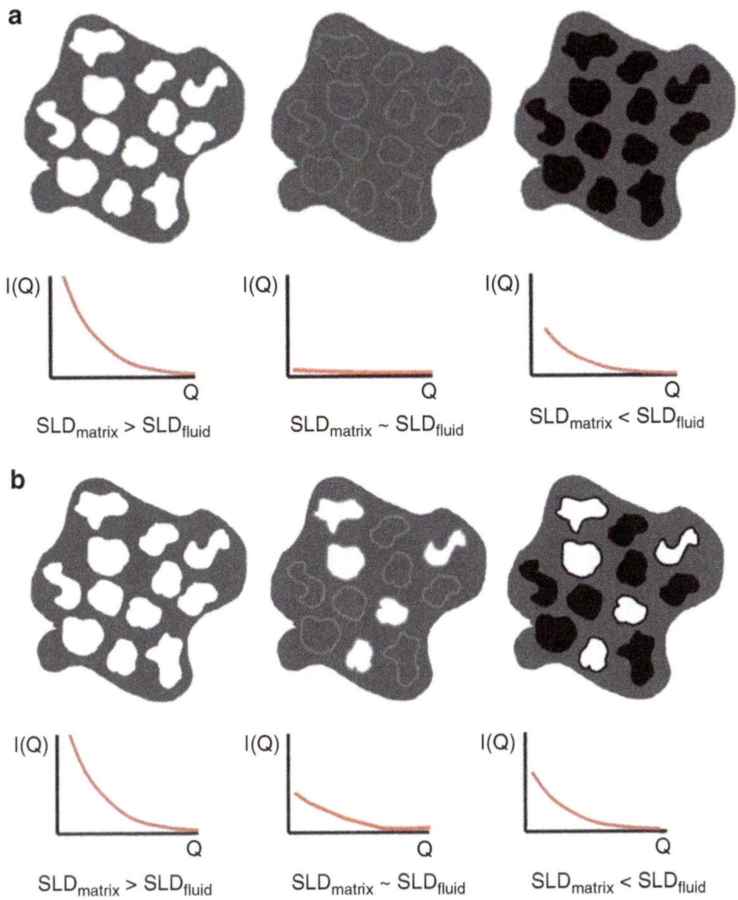

Fig. 10.35 Qualitative presentation of contrast-matching experiments with fluid saturated porous systems. (**a**) All pores are accessible to fluid molecules; (**b**) pores are partially accessible to fluid molecules. Note that even in case of completely accessible pores, certain residual scattering may be observed in contrast matched point due to finite fluctuations of the SLD of the solid matrix. Reprinted from [77]. Copyright (2012), with permission from Elsevier

All coals were prepared in the form of coarse powder (particle size 1–0.5 mm). Results of the performed petrochemical analysis were used to calculate SLD of the studied coals shown in Table 10.1. This Table also gives the pressure and density of CD_4 at T = 23 °C and supercritical CO_2 at T = 60 °C at which the SLD of each porous matrix is contrast matched by the fluid. Note that densities of a deuterated fluid calculated using equation of state of the "normal" fluid should be normalized by the ratio of atomic weights of deuterated and protonated molecules. For methane CD_4 such normalization factor is 1.25.

SANS experiments were conducted at ORNL on the General Purpose SANS instrument with neutron wavelengths of $\lambda = 12$ Å and $\lambda = 4.8$ Å ($\Delta\lambda/\lambda \sim 0.13$).

Table 10.1 Scattering length densities of coal [77]

Coal	SLD (10^{10} cm^{-2})	P_{ZAC} (bar)/ρ_{CD4} (g/cm^3, 23 °C)	P_{ZAC} (bar)/ρ_{CO2} (g/cm^3, 60 °C)
Seelyville	2.20	224.7/0.220	380.1/0.88
Coal 1	3.45	504.7/0.345	–
Coal 2	2.34	246.7/0.234	–

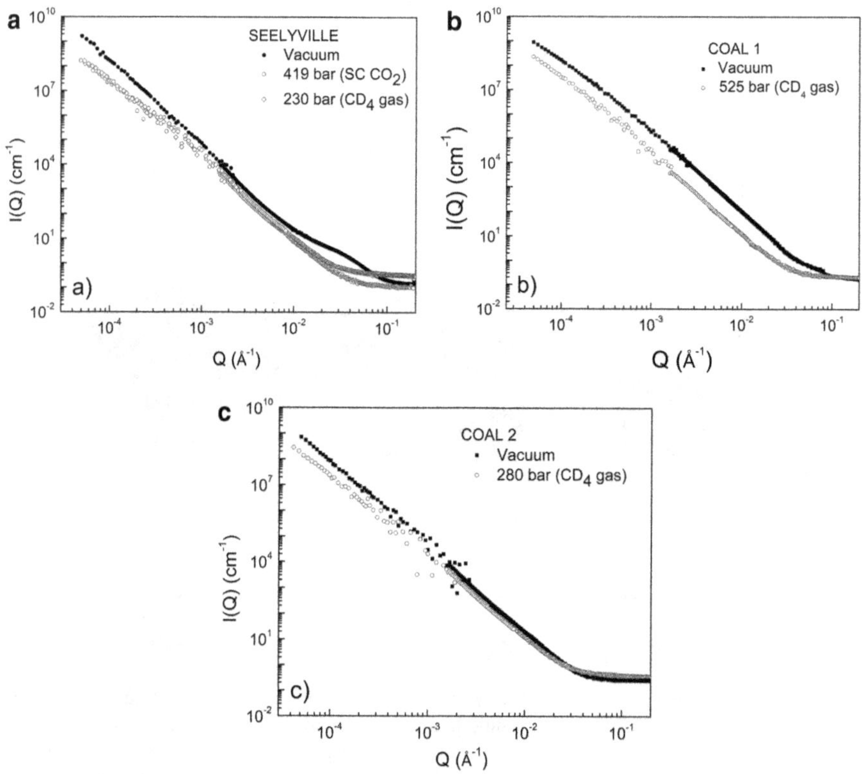

Fig. 10.36 Combined USANS and SANS curves acquired from coals in vacuum and at zero average contrast pressures (as indicated in the *insets*). Reprinted from [77]. Copyright (2012), with permission from Elsevier

Sample-detector distances were chosen to cover an overall range of scattering vectors $0.0016 < Q < 0.2$ Å^{-1}. The data were corrected for instrumental background as well as detector efficiency and put on absolute scale [cross section $I(Q)$ in units of cm^{-1}] by means of pre-calibrated secondary standards. USANS experiments were performed at NIST, using the BT5 perfect crystal SANS instrument ($\lambda = 2.4$ Å, Q-range $5 \times 10^{-5} < Q < 0.003$ Å^{-1}). Application of these instruments allowed probing a broad range of pore sizes, from approximately 10,000 to 12 Å. Figure 10.36 shows the combined USANS and SANS patterns from the coals, measured in vacuum and at $P \approx P_{ZAC}$. The relatively strong residual scattering from contrast-matched coals indicates the presence of significant number of pores, inaccessible to neither supercritical CO_2 nor methane. The scattering patterns at $P = P_{ZAC}$ show distinctive deviations from $I(Q)$ in vacuum that vary with each sample. Whereas all samples reveal a substantial decrease in $I(Q)$ in the low-Q range (indicating that most of the large pores are accessible to the fluid), the reduction of intensity in the intermediate Q-range is less accentuated and is Q-dependent.

As demonstrated in [77] (see also Appendix to this Chapter), the ratio of $I(Q_i, P_{ZAC})$ and $I(Q_i, VAC)$ can be used to calculate the volume fraction of accessible pores $c_{AC}(Q_i)$ *at any arbitrary value of the scattering vector* Q_i (or, equivalently, at any pore size, $R_i \approx 2.5/Q_i$):

$$\frac{I(Q_i, P_{ZAC})}{I(Q_i, VAC)} = 1 - c_{AC}(Q_i), \tag{10.23}$$

where $c_{AC}(Q_i)$ is defined as the ratio of the volume of accessible pores to the total pore volume at a given pore size. Consequently, a negligible change in the scattering intensity measured at zero average contrast pressure $I(Q_i, P_{ZAC})$ relative to the intensity measured in vacuum $I(Q_i, VAC)$ in some region Q_i indicates a low accessibility of gases into pores of sizes $R_i \sim 2.5/Q_i$. Conversely, if gases can effectively penetrate into pores of that size, a relatively large change in that ratio may be anticipated. Figure 10.37 shows the variation of the volume fraction of accessible pores in the studied coals as a function of Q_i and R_i calculated using (10.23). As may be seen in Fig. 10.37, the variation of $c_{AC}(R)$ varies from coal to coal. For the Seelyville coal, macropores larger than ~1000 Å are equally accessible to both CO_2 and d-methane molecules, and the value of c_{AC} for both fluids gradually decreases from ~0.9 to ~0.55. At the same time, pores inside the size range $1000 > R > 100$ Å appear to be more accessible to CO_2 molecules (by ~10 %). This result may be tentatively attributed to a smaller size of CO_2 molecules, which makes it easier to penetrate narrow capillary constrictions joining the pores. For Coal 1, the ratio of $I(Q_i, VAC)$ to $I(Q_i, P_{ZAC})$ and thus c_{AC} is approximately constant over a large range of Q_i, and the volume fraction of accessible pores is generally much larger than for the other coals. About 80–85 % of both macro- and mesopores within the size range from 100 to 25,000 Å are accessible to d-methane. For Coal 2, the variation of c_{AC} with pore size is qualitatively similar to that of the Seelyville except of the upturn for pore sizes $R < 700$ Å.

The described results provided first experimental evidence of the existence of closed pores in coal that are inaccessible to the molecules of supercritical CO_2 and d-methane on the time scale of performed experiments. It was demonstrated that each coal had its own "fingerprint" distribution of c_{AC} as a function of pore size in the meso- and macroporous regions. The fraction of pores accessible to CO_2 and methane appeared to be relatively large in highly porous inertinite-rich Coal 1. It was much lower in vitrinite-rich, low-porosity Seelyville and Coal 2 coals, both of

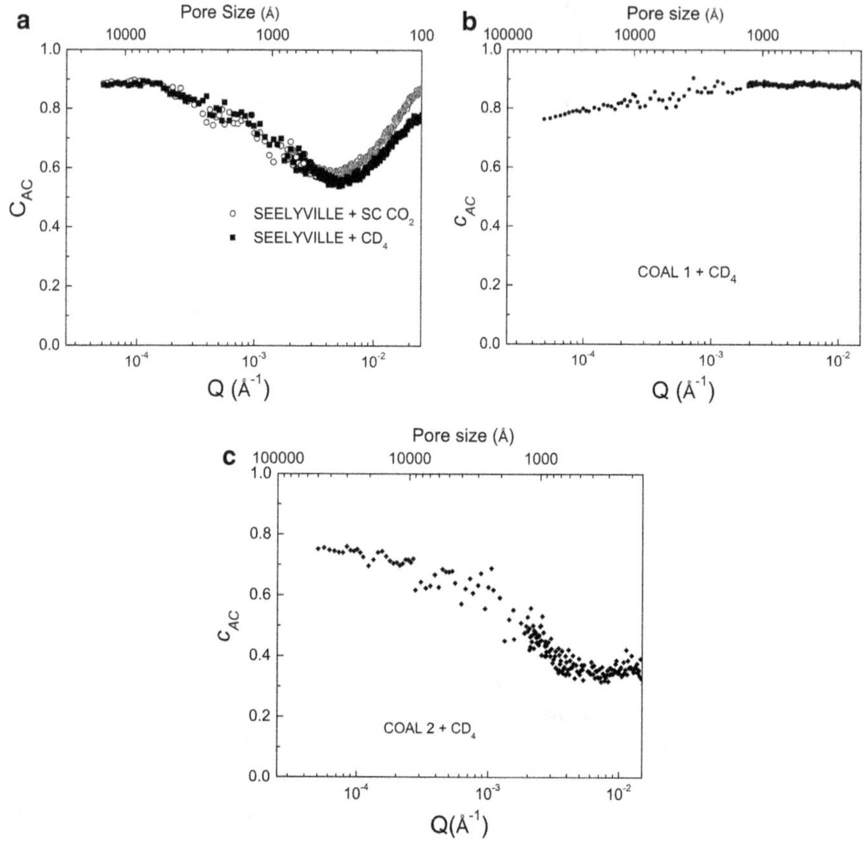

Fig. 10.37 The variation of the volume fraction of pores accessible to greenhouse gases as a function of the pore size for samples of coal specified in the insets. Reprinted from [77]. Copyright (2012), with permission from Elsevier

which demonstrated qualitatively similar but yet not identical variation of $c_{AC}(R)$. The amount of accessible pores in coal may be directly related to the total porosity as probability of the formation of the interconnected network of pores accessible to fluids should be facilitated in higher porosity coals. The influence of the maceral (or hydrogen) composition on the pore accessibility cannot be eliminated [78].

Scattering curves from porous materials measured in vacuum and at contrast-matched condition can also be used for evaluation of the *total* and *accessible* porosity in the sample. For this purpose, combined SANS/USANS data shown in Fig. 10.36 were analyzed in [79] using Porod invariant and PDSP model approaches, which provide information on the total (samples in vacuum) and accessible (sample at contrast matched condition) porosity in each sample. The results obtained by these two methods are shown in Tables 10.2 and 10.3. The porosities obtained using both methods agree reasonably well. Coal 1 exhibits the highest total porosity (13 %) and the highest fraction of pores accessible to

Table 10.2 Results of the Porod invariant analysis [79]

Coal	Total porosity under vacuum (%)	Fraction of total porosity accessible to CD_4 (%)
Seelyville	10.4	82.8
Coal 1	13.0	87.5
Coal 2	6.7	63.6

Table 10.3 Results obtained from the PDSP model [79]

Coal	Total porosity under vacuum (%)	Fraction of total porosity C_{AC} accessible to CD_4 (%)
Seelyville	10.7	79.6
Coal 1	12.0	86
Coal 2	6.7	65.3

methane, whereas Coal 2 is the least porous sample (6.7 %) with the lowest fraction of the accessible porosity. The volume of accessible pores shows no evident correlation with rank or maceral composition. At the same time the inaccessible porosity decreases systematically as the total porosity increases.

The fact that 12–36 % of all pores may not be accessible to injected methane suggests that, if methane were present in closed pores originally, it would be difficult to release from these pores. Studies of the gas release from coal often find that "residual gas" (gas not adsorbed before the sample is pulverized) can be a considerable fraction of the total gas present. The residual gas requires extensive crushing of the coal to release. The results [79] suggest that crushing helps to release the gas trapped in the initially closed (i.e., inaccessible) pores.

The pore structure of organic-rich shale is even more complex than that of coal because shales combine uconventional coarse porosity from inorganic mineral matter such as quartz or carbonates and ultrafine porosity from both organic matter and clay materials [80]. Understanding the interrelation between the pore morphology, sorption characteristics, and potential gas storage capacities of organic-rich shales is important because of the discovered significant gas contents of shale gas reservoirs. The developed new technologies of hydrocarbon extraction from low-permeability, unconventional-type reservoirs utilizing hydrofracturing and horizontal drilling stimulated a significant increase in research in this area [81].

The pore structure of shale gas reservoirs is difficult to characterize because a major fraction of the total porosity may be distributed in ultrafine nanopores that are inaccessible to invasive fluids. Significant amount of closed porosity makes SAS techniques especially valuable for structural characterization of shales. Recently initiated SANS/USANS studies demonstrated high effectiveness of these techniques for characterizing the morphology of dry and wet shales, and providing information on their accessible and closed porosity as a function of pore size (Fig. 10.38, [81–83]).

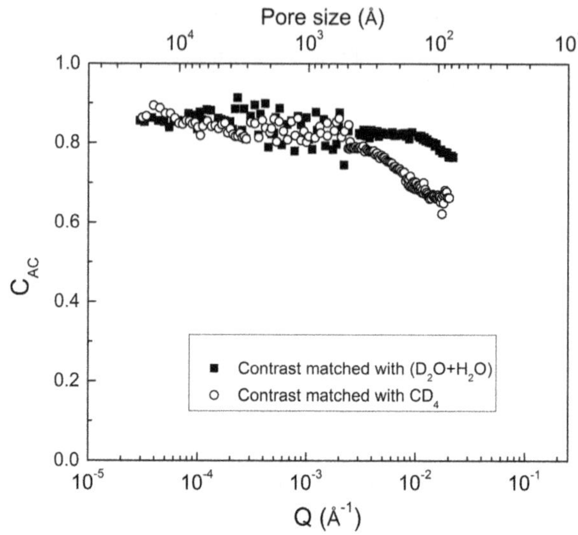

Fig. 10.38 Volume fraction of pores accessible to water and CD_4 in Barnett shale (sample 152) as a function of pore size. Reprinted with permission from [81]. Copyright 2013 American Chemical Society

10.8 Structural Stability of Porous Materials Under Pressure

Exposure of porous materials to high-pressure gases and supercritical fluids can modify the pore structure due to mechanical distortions. The occurrence of such distortions may affect interpretation of SANS data on the fluid adsorption in pores as the data analysis is usually performed under assumption of the porous structure being independent of the fluid pressure. Due to differences in mechanical properties of the solid matrix, various porous solids may respond differently to injection of pressurized fluids.

Due to the actively developing technologies of carbon sequestration in geological formations such as unminable coal seams, the behavior of coal upon exposure to high pressure of gases under supercritical conditions has become of great practical interest. To examine the effect of pressure alone on the pore size, a convenient gas to use is helium. Its low physical density even at high pressures (no significant condensation into a liquid phase, even in confinement) means that its SLD remains much lower than the SLD of most porous solids, which makes helium molecules effectively invisible to neutrons (Fig. 10.39). Thus, any possible changes seen in SANS and/or USANS patterns from helium-saturated coal (as well as any other porous solid) under pressure must be attributed to the effect of the mechanical pressure of helium on the pore size distribution.

Helium density determination is a standard method of estimating the volume occupied by materials. It is based on assumption that helium is not adsorbed by the material examined. If helium were sorbed, then the helium densities measured would be greater than the true density of the material. It is generally accepted

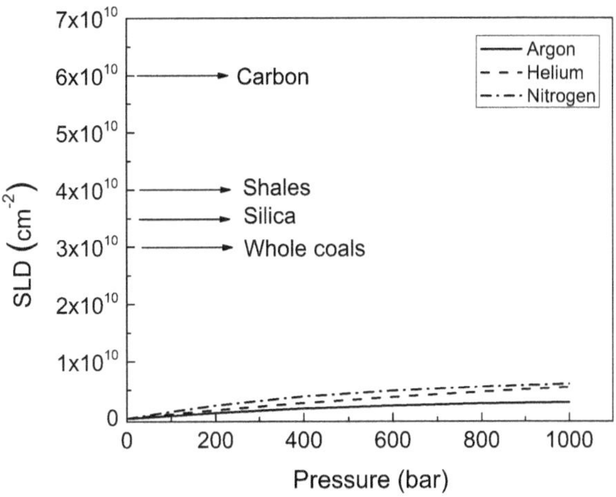

Fig. 10.39 SLD of helium, argon, and nitrogen as a function of pressure at T = 23 °C. SLDs of solid skeletons of several common natural and engineered porous materials are also shown for comparison

that helium penetrates coals completely. The helium densities obtained generally agree with theoretical expectations based on the molar volume of the constituents, which indicates that helium sorption in coals is negligible.

SANS and USANS measurements of the structure of two Australian bituminous coals before, during, and after exposure to 155 bar of helium were made to identify any possible effects of pressure alone on the pore size distribution and any irreversible effects upon exposure to high pressures of helium in [84]. Coal 1 (a highly porous, inertinite-rich coal) and coal 2 (a low porosity, vitrinite-rich coal) were selected because of their extremely different porosities. Coarse powder samples of the studied coals with the particle size between 1 and 0.5 mm were dried under vacuum at 60 °C and kept in a positive pressure oxygen- and moisture-free glove box prior to loading them in the SANS high-pressure cell.

SANS experiments were performed at the Oak Ridge National Laboratory using the reactor-based General-Purpose SANS instrument. The neutron wavelength was 4.8 Å and two sample-detector distances of 6 and 18 m were used for each pressure value, covering the range of scattering vectors Q from 2×10^{-3} to 0.2 $Å^{-1}$. The maximum Q value of the SANS spectra was determined by the cut-off acceptance angle of the pressure cell, which limited the minimum pore diameter that could be investigated in this system to about 30 Å. Scattering patterns of coal were measured using a 2D 1 m × 1 m detector, corrected for instrumental background as well as detector efficiency and put on absolute scale (neutron cross section per unit volume $I(Q)$ in units of cm^{-1}) using precalibrated secondary standards. USANS measurements were performed at the Institut Laue-Langevin using the double-crystal (Bonse-Hart geometry) USANS instrument S18 [85]. The range of scattering vectors investigated using this USANS instrumentation was 10^{-5} to 10^{-3} $Å^{-1}$. All measurements were conducted at room temperature.

Figure 10.40a demonstrates that the scattering pattern for coal 1 is unaffected by 155 bar pressure of helium either during or after the exposure. Within the

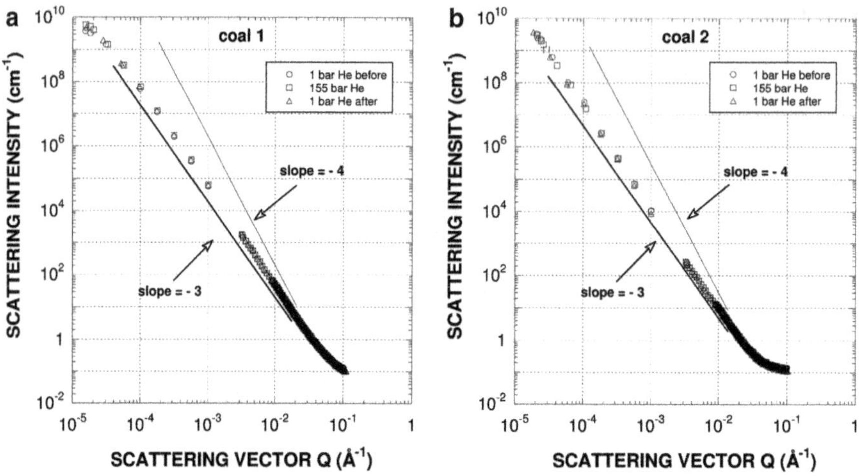

Fig. 10.40 A comparison of SANS/USANS data for coal 1 (**a**) and coal 2 (**b**) at three helium pressure values: 1 bar (before pressure cycling), 155 bar (maximum pressure) and 1 bar (after release of pressure). The (negative) slope for coal 1 is -3.1 ± 0.05 and -3.0 ± 0.05 for coal 2. Slope values between -3 and -4 indicate a surface fractal geometry of the coal matrix–pore interface. Reprinted with permission from [84]. Copyright 2009 American Chemical Society

experimental repeatability of the method (about ± 5 % of the absolute scattering intensity), in the pore size range 30 Å $<$ R $<$ 100 Å there is no deformation observed either during exposure to high pressure or as a consequence of the applied pressure.

Figure 10.40b shows the corresponding scattering pattern for coal 2. The pattern is not affected by exposure to pressure over the full Q range investigated, with an exception of the region around $Q = 10^{-4}$ Å^{-1}, where the scattering intensity decreases by about 15 % at 155 bar pressure and then returns to its original value after the pressure is released. This effect may be due to a partial contraction of closed macropores in the coal under pressurization. Similar to Coal 1, there was no appreciable difference in the scattering pattern detected before and after exposure to pressure.

The results demonstrate that the influence of 155 bar helium pressure on the microstructure of two Australian bituminous coals of very different porosities in the pore size range 30 Å to 10 μm appears negligible both in terms of the total porosity and pore size distribution. No irreversible structural effects on these coals as a consequence of pressurization were observed. The described "helium test" may be applied to other porous solids in order to check if changes seen in the SANS and/or USANS scattering patterns from the fluid saturated matrices should be interpreted as due to the formation of the dense adsorbed phase in pores and not the pressure itself.

Structural stability of the porous matrix of an ultramicroporous carbon (UMC) under applied pressure of inert gas argon and strongly adsorbing gas CO_2 was investigated using SANS in [86]. Understanding the adsorption-induced deformation of the porous solids remains one of the long-term problems in adsorption science [87]. The adsorption of fluid molecules may lead to development of stress (also known as solvation pressure [88]), which is responsible for modification of the nano-mechanical properties and shape of micropores. The origin of the stress is related to the tight packing of the fluid molecules in the micropores, and its magnitude may depend on the pressure, fluid–wall interactions and pore size distribution among other parameters. The extent of the pore deformation depends on the elastic modulus of the porous matrix. As the solvation pressure varies with the size of the pore, it may lead to a highly inhomogeneous distribution of the elastic strain for a sample with polydisperse pore size distribution [89].

In-situ SANS measurements have been carried out at the General Purpose SANS instrument of the High Flux Isotope Reactor, at ORNL with a neutron wavelength of 4.72 Å and wavelength resolution $\Delta\lambda/\lambda \sim 0.13$. The sample-to-detector distance was chosen to cover an extended range of scattering vectors $0.005 < Q < 1.0$ Å^{-1}. Powder carbon samples were outgassed for ~5 h at 300 °C and loaded into a thin-wall aluminum container with 1 mm internal thickness, and subsequently mounted inside the high-pressure cell.

SANS measurements were first conducted in vacuum to explore the texture and pore structure of the intact sample, and then argon or carbon dioxide was injected to monitor possible modification of the matrix structure due to the presence of the pressurized inert or strongly adsorbing fluid. As may be seen in Fig. 10.41a, the SANS profiles from UMC in argon remain unchanged within the experimental error in the range of $0 < P < 200$ bar, which indicates absence of the pore deformation due to hydrostatic pressure induced by inert, non-adsorbing fluid. At the same time, SANS patterns from UMC in CO_2 show complex Q- and pressure dependent variation of the scattering intensity as a function of pressure (Fig. 10.41b).

The measured scattering intensity was used to obtain size distribution as well as density of the adsorbed CO_2 in micropores at each pressure by fitting $I(Q)$ in the high Q domain to the polydisperse sphere model. The variation of the average size of micropores and density of adsorbed CO_2 as a function of pressure is shown in

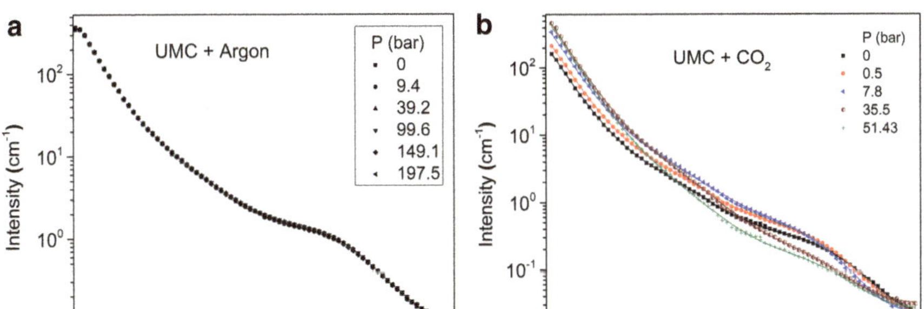

Fig. 10.41 SANS profiles of UMC porous carbon as a function of argon (**a**) and CO_2 pressure (**b**). *Solid lines* represent model fits to the data [86]

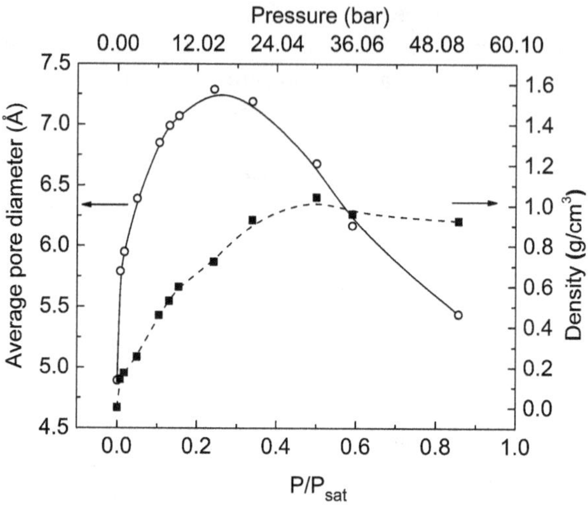

Fig. 10.42 Average diameter and density of bulk adsorbed CO_2 in micropores of UMC porous carbon as function of pressure [86]

Fig. 10.42. As may be seen, the average pore size initially increases by ~50 % and then starts to decrease at pressures >12 bar ($P/P_{sat} > 0.3$). The maximum pore size is reached when the initially increasing density of adsorbed CO_2 levels off at the value ~1 g/cm^3. This density value exceeds density of bulk CO_2 at the same temperature and pressure by a factor of ~20 and is close to the van der Waals density of CO_2.

The results obtained in [86] demonstrate that porous matrix of UMC does not reveal any measurable structural changes under high pressure of non-adsorbing inert gas argon. The observed micropore expansion in pressurized CO_2 can be attributed to the solvation stress, which develops due to strong adsorption and densification of CO_2 molecules in micropores. Similar experiments may be designed and conducted also using x-ray scattering. Thus, SAS and USAS methods can provide information of the pressure-induced deformation of pores in different solids as a function of pore sizes in the range between macro- and sub-micropores.

Appendix: Derivation of the Eq. (10.23) for Accessible Porosity

Porod Invariant

Following Porod [90], we define the static scattering structure factor *S(Q)* as:

$$S(Q) = \langle \Delta\rho(Q)\Delta\rho(-Q) \rangle = \left\langle |\Delta\rho(Q)|^2 \right\rangle, \tag{10.24}$$

where $\langle \ldots \rangle$ denotes the equilibrium statistical mechanical average while $\Delta\rho(\boldsymbol{Q})$ denotes the Fourier transformed fluctuation density. The experimentally measured data for $S(\boldsymbol{Q})$ can be used for restoration of the density-density correlator $\overset{\vee}{S}(\boldsymbol{r})$ in coordinate space. Indeed, we obtain:

$$\overset{\vee}{S}(\boldsymbol{r}) = \frac{1}{(\pi 2)^3}\int d\boldsymbol{Q}\exp(i\boldsymbol{Q}\cdot\boldsymbol{r})S(\boldsymbol{Q}). \tag{10.25}$$

Up to a constant, the Porod invariant can now be defined as $\overset{\vee}{S}(\boldsymbol{r}=0)$. At the same time, it is well known from thermodynamics [91] that $S(\boldsymbol{Q}=0)$ can be obtained with help of the thermodynamic sum rule as usual. In Porod notations (see e.g., page 28 in [90]) we write:

$$S(\boldsymbol{Q}=0) = V^2\left(\Delta\rho^2\right), \tag{10.26}$$

where V is the volume of the sample. To determine the volume, Porod defines the invariant:

$$\overset{\vee}{S}(\boldsymbol{r}=0) = 2\pi^2 V\left\langle\left(\Delta\rho^2\right)\right\rangle \tag{10.27}$$

known in the literature as Porod invariant (Sect. 6.2). Combining (10.26) and (10.27), the volume V can be determined. For the purposes of this work we would like to re-derive the Porod invariant and to explain why, indeed, it is an invariant. To do so, we derive the following chain of equalities:

$$\begin{aligned}\overset{\vee}{S}(\boldsymbol{r}=0) &= \frac{1}{(2\pi)^3}\int d\boldsymbol{Q}S(\boldsymbol{Q}) = \frac{1}{(2\pi)^3}\int d\boldsymbol{Q}\langle\Delta\rho(\boldsymbol{Q})\Delta\rho(-\boldsymbol{Q})\rangle \\ &= \frac{1}{(2\pi)^3}\int d\boldsymbol{Q}\int d\boldsymbol{r}\int d\boldsymbol{r}'\exp\left[-i\boldsymbol{Q}\cdot\left(\boldsymbol{r}-\boldsymbol{r}'\right)\right]\left\langle\Delta\rho(\boldsymbol{r})\Delta\rho\left(\boldsymbol{r}'\right)\right\rangle \\ &= \int d\boldsymbol{r}\left\langle(\Delta\rho(\boldsymbol{r}))^2\right\rangle = constV\left\langle(\Delta\rho(0))^2\right\rangle = 2\pi^2 V\left\langle\Delta\rho^2\right\rangle.\end{aligned} \tag{10.28}$$

The *const* was determined by the angular averaging, as usual. At the same time, the above can be also written as:

$$\int d\boldsymbol{Q}S(\boldsymbol{Q}) = \int d\boldsymbol{Q}\langle\Delta\rho(\boldsymbol{Q})\Delta\rho(-\boldsymbol{Q})\rangle = (2\pi)^3\int d\boldsymbol{r}\left\langle(\Delta\rho(\boldsymbol{r}))^2\right\rangle. \tag{10.29}$$

Equation (10.29) can be recognized as Parseval's formula used in the theory of Fourier transforms.

Porod Invariant and Parseval's Formula

To begin, we would like to remind our readers the basic facts about this formula. For this purpose, let *f(r)* be some arbitrary well behaved function whose Fourier transform is given by:

$$f(\boldsymbol{r}) = \frac{1}{(2\pi)^3}\int d\boldsymbol{Q}\,\exp(i\boldsymbol{Q}\cdot\boldsymbol{r})\cdot\ f(\boldsymbol{r}). \tag{10.30}$$

Accordingly, its inverse transform is given by:

$$f(\boldsymbol{Q}) = \int d\boldsymbol{r}\,\exp(-i\boldsymbol{Q}\cdot\boldsymbol{r})\cdot\ f(\boldsymbol{r}). \tag{10.31}$$

Using these definitions, consider the following chain of equalities:

$$\begin{aligned}\int d\boldsymbol{r}(f(\boldsymbol{r}))^2 &= \int d\boldsymbol{r} f(\boldsymbol{r})\frac{1}{(2\pi)^3}\int d\boldsymbol{Q} exp(i\boldsymbol{Q}\cdot\boldsymbol{r})f(\boldsymbol{Q})\\ &= \frac{1}{(2\pi)^3}\int d\boldsymbol{Q} f(\boldsymbol{Q})\int d\boldsymbol{r} f(\boldsymbol{r}) exp(i\boldsymbol{Q}\cdot\boldsymbol{r})\\ &= \frac{1}{(2\pi)^3}\int d\boldsymbol{Q} f(\boldsymbol{Q})f(-\boldsymbol{Q})\\ &\equiv \frac{1}{(2\pi)^3}\int d\boldsymbol{Q}|f(\boldsymbol{Q})|^2.\end{aligned} \tag{10.32}$$

In this expression $|f(\boldsymbol{Q})|^2 = f(\boldsymbol{Q})f(-\boldsymbol{Q}) \equiv f(\boldsymbol{Q})f(\boldsymbol{Q})^*$, where * denotes complex conjugation. The expression (10.32) is the Parseval identity. That is:

$$\int d\boldsymbol{r}(f(\boldsymbol{r}))^2 = \frac{1}{(2\pi)^3}\int d\boldsymbol{Q}|f(\boldsymbol{Q})|^2 \tag{10.33}$$

It is useful now to compare this result with (10.29). To do so, we have to make the following identifications:

$$(2\pi)^3\int d\boldsymbol{r}[f(\boldsymbol{r})]^2 \leftrightarrows (2\pi)^3\int d\boldsymbol{r}\left\langle \Delta\rho[(\boldsymbol{r})]^2\right\rangle$$

and

$$\int d\boldsymbol{Q}|f(\boldsymbol{Q})|^2 \leftrightarrows \int d\boldsymbol{Q} S(\boldsymbol{Q}).$$

Parseval's formula can be written in many different ways. For instance, instead of (10.31) we can write $f(\boldsymbol{r}) = \sum_i a_i \varphi_i(\boldsymbol{r})$, where we assume that $\int |\varphi_i(\boldsymbol{r})|^2 d\boldsymbol{r} = 1$ and $\int \varphi_i^*(\boldsymbol{r})\varphi_j(\boldsymbol{r})d\boldsymbol{r} = \delta_{ij}$. Using this result we obtain:

$$\int d\boldsymbol{r}[f(\boldsymbol{r})]^2 = \sum\nolimits_i |a_i|^2. \tag{10.34}$$

This result can be looked upon as follows. Replace integration by summation in (10.33), that is write:

$$\sum\nolimits_i (\Delta \boldsymbol{r})^3 [f(\boldsymbol{r}_i)]^2 = \sum\nolimits_i \frac{(\Delta \boldsymbol{Q})^3}{(2\pi)^3} |f(\boldsymbol{Q}_i)|^2. \tag{10.35}$$

By doing so, we effectively put our system onto some (say, cubic) lattice with effective size of the cell of order $\Delta \boldsymbol{r}$ in real space and $\Delta \boldsymbol{Q}$ in reciprocal space. By analogy with quantum mechanics, let now $\Delta \boldsymbol{r}\ \Delta \boldsymbol{Q} = \Theta$, where Θ is some constant. Such result makes sense in view of the wave nature of neutrons. Such Heisenberg-type relation was used successfully already by Radlinski [92] in his computer simulations of neutron scattering from coal. In his work, the constant Θ was estimated as 2.5. With account of such a relation, we can rewrite (10.35) as follows:

$$\sum\nolimits_i \left\langle [\Delta\rho(\boldsymbol{r}_i)]^2 \right\rangle = \frac{(\Delta \boldsymbol{Q})^6}{(2\pi)^3} \frac{1}{\Theta^3} \sum\nolimits_i S(\boldsymbol{Q}_i). \tag{10.36}$$

Equivalently, the above results are just the discretised form of (10.29). We shall use this form of Parseval's identity below.

From Porod Invariant to Eq. (10.23)

We would like now to generalize the obtained result, Eq. (10.36), by extending it to two-phase systems. For this purpose, in view of the fact that Eq. (10.36) is written for the cubic lattice, we can use known results from the scattering theory for solid alloys [93]. We begin by introducing random numbers c_i such that $c_i = 1$ (if the i_{th} site is occupied by phase 1) and $c_i = 0$ (if the ith site is occupied by phase 2). The average $\langle c_i \rangle$ can now be defined as:

$$\langle c_i \rangle = c = \frac{1}{N} \sum\nolimits_1^N c_i,$$

where summation takes place over all lattice sites. Also, in view of definition of c_i it follows that $c_i^2 = c_i$. Consider now $\Delta\rho(\boldsymbol{r}_i)$ for such two phase system. In the case of just one phase we define fluctuation of density as $\Delta\rho(\boldsymbol{r}_i) = \rho(\boldsymbol{r}_i)-\rho$, where ρ is homogeneous reference density. In the case of two phases we have $\rho \to c\rho_1 + (1-c)\rho_2$ and

$$\Delta\rho(\boldsymbol{r}_i) = c_i\rho_1 + (1-c_i)\rho_2 - [c\rho_1 + (1-c)\rho_2] = (c_i - c)\rho_1 - (c_i - c)\rho_2.$$

Consider now the average:

$$\begin{aligned}\left\langle \Delta\rho(\boldsymbol{r}_i)^2 \right\rangle &= \left\langle [(c_i - c)\rho_1]^2 \right\rangle + \left\langle [(c_i - c)\rho_2]^2 \right\rangle - 2\rho_1\rho_2\left\langle (c_i - c)^2 \right\rangle \\ &= (\rho_1 - \rho_2)^2\left\langle (c_i - c)^2 \right\rangle = (\rho_1 - \rho_2)^2 c(1-c).\end{aligned}$$

Using this result we obtain:

$$\sum\nolimits_i S(\boldsymbol{Q}_i) = 2\pi^2 V(\rho_1 - \rho_2)^2 c(1-c), \tag{10.37}$$

where averaging over angles was made. Explicitly, we took into account that $\frac{(2\pi)^3}{4\pi} = 2\pi^2$ and that $(\Delta\boldsymbol{r})^3\sum_i = V$. Here and below we shall assume that $(\Delta\boldsymbol{r})^3$ is of order of Θ^3 so that $V = N\Theta^3$. The constant factor $(\Delta\boldsymbol{Q})^3$ was absorbed into definition of $S(\boldsymbol{Q})$ since it is not essential (see below). Thus, (10.37) is the standard result by Porod 91.

This result should now be looked upon as follows. Following book by Krivoglaz [93], especially taking into account his Eq. (1.15) on page 8, we can think not only about the averages of the type $\frac{1}{N}\sum_{1=1}^{N}(c_i - c)^2 = c(1-c)$ in the *direct* lattice but also about analogous averages in the *dual* lattice, which in the present case is cubic also. In such a case, we can use Eqs. (10.24, 10.35, 10.36) in order to write:

$$S(\boldsymbol{Q}_i) = \left\langle |\Delta\rho(\boldsymbol{Q}_i)|^2 \right\rangle. \tag{10.38}$$

In order to perform averaging over random variables $\Delta\rho(\boldsymbol{Q}_i)$ we have to take into account that:

$$\Delta\rho(\boldsymbol{Q}_i) = \frac{1}{V}\sum\nolimits_j \Delta\rho(\boldsymbol{r}_j)\exp(i\boldsymbol{Q}_i \cdot \boldsymbol{r}_j) \tag{10.39}$$

and $V = N\Theta^3$ as before. To evaluate $\langle \Delta\rho[(\boldsymbol{Q}_i)]^2 \rangle$ using (10.39), following [35], we introduce density-related variable $c(\boldsymbol{Q}_i)$ in such a way that

$$\frac{1}{N}\sum\nolimits_{i=1}^{n} c(\boldsymbol{Q}_i)[(1 - c(\boldsymbol{Q}_i))] = c(1-c). \tag{10.40}$$

In view of Eqs. (10.36) and (10.37):

$$S(\boldsymbol{Q}_i) = \omega c(\boldsymbol{Q}_i)(1 - c(\boldsymbol{Q}_i)). \tag{10.41}$$

The constant ω is known in principle but unimportant since it will be subsequently eliminated. To get rid of this constant, we determine the ratios of the type:

$$\frac{S(\boldsymbol{Q}_i, P_{ZAC})}{S(\boldsymbol{Q}_i, VAC)} = \frac{c_{IN}(\boldsymbol{Q}_i)[1 - c_{IN}(\boldsymbol{Q}_i)]}{c(\boldsymbol{Q}_i)[(1 - c_i(\boldsymbol{Q}_i)]}, \tag{10.42}$$

where $S(\boldsymbol{Q}_i, P_{ZAC})$ and $S(\boldsymbol{Q}_i, VAC)$ is the structure factor from fluid saturated coal at zero average contrast pressure P_{ZAC} and the structure factor of the coal under vacuum, respectively. Furthermore, $c(\boldsymbol{Q}_i)$ is the volume fraction of all pores and $c_{IN}(\boldsymbol{Q}_i)$ is the volume fraction of *inaccessible* pores defined as the ratio of the volume of *inaccessible pores* to the *total pore volume*. For low porosity samples such as coal both c and $c_{IN} \ll 1$, and for any arbitrary value of the scattering vector $\boldsymbol{Q}_i$ (or equivalently at corresponding pore size $R_i = 2.5/Q_i$) we have:

$$\frac{c_{IN}(\boldsymbol{Q}_i)(1 - c_{IN}(\boldsymbol{Q}_i))]}{c(\boldsymbol{Q}_i)(1 - c(\boldsymbol{Q}_i))} \cong \frac{c_{IN}(\boldsymbol{Q}_i)}{c(\boldsymbol{Q}_i)} = 1 - c_{AC}(\boldsymbol{Q}_i) \tag{10.43}$$

or

$$\frac{S(\boldsymbol{Q}_i, P_{ZAC})}{S(\boldsymbol{Q}_i, VAC)} \cong 1 - c_{AC}(\boldsymbol{Q}_i), \tag{10.44}$$

where $c_{AC}(\boldsymbol{Q}_i)$ is the volume fraction of *accessible pores* at $\boldsymbol{Q}_i$ is defined as the ratio of the volume of *accessible pores to the total pore volume*. Equation (10.44) is Eq. (10.23) given in Sect. 10.7. It can be used for evaluating the volume fraction of accessible pores as a function of $\boldsymbol{Q}_i$ (or $\boldsymbol{R}_i$) by measuring SANS/USANS patterns from the "dry" and contrast matched samples and finding the ratio of the scattering intensities at each $\boldsymbol{Q}_i$ [77].

References

1. Anisimov, M.A., Sengers, J.V.: Equations of state for fluids and fluid mixtures. In: Sengers, J. V., Kayser, R.F., Peters, C.J., White Jr., H.J. (eds.) Critical Region. Elsevier, Amsterdam (2000)
2. Kiran, E., Levelt-Sengers, J.-M.-H. (eds.): Supercritical Fluids: Fundamentals for Applications. Kluwer, Dordrecht (1993)
3. Sircar, S.: Gibbsian surface excess for gas-adsorption – revisited. Ind. Eng. Chem. Res. **38**, 3670 (1999)
4. Do, D.D., Do, H.D.: Adsorption of supercritical fluids in non-porous and porous carbons: analysis of adsorbed phase volume and density. Carbon **41**, 1777 (2003)

5. Findenegg, G.H., Thommes, M.: High pressure physisorption of gases on planar surfaces and in porous materials. In: Fraissard, J. (ed.) Physical Adsorption: Experiment, Theory and Application. Kluwer, Dordrecht (1997)
6. Egorov, S.A.: Adsorption of supercritical fluids and fluid mixtures: inhomogeneous integral equation study. J. Phys. Chem. B **105**, 6583 (2001)
7. Monsalvo, M.A., Shapiro, A.A.: Study of high-pressure adsorption from supercritical fluids by the potential theory. Fluid Phase Equilib. **283**, 56 (2009)
8. Dundar, E., Zacharia, R., Chahine, R., Bernard, P.: Modified potential theory for modeling supercritical gas adsorption. Int. J. Hydrogen Energy **37**, 9137 (2012)
9. Tan, Z., Gubbins, K.: Adsorption in carbon micropores at supercritical temperatures. J. Phys. Chem. **94**, 6061 (1990)
10. Cracknell, R., Gordon, P., Gubbins, K.: Influence of pore geometry on the design of microporous materials for methane storage. J. Phys. Chem. **97**, 494 (1993)
11. Fischer, J., Bohn, M., Korner, B., Findenegg, G.H.: Supercritical gas adsorption in porous materials. II. Prediction of adsorption isotherms. Germ. Chem. Eng. **6**, 84 (1983)
12. Brochard, F., de Gennes, P.G.: Phase transitions of binary mixtures in porous media. J. Phys. Lett. (Paris) **44**(785) (1983)
13. Wong, A.P.Y., Chan, M.H.W.: Liquid-vapor critical point of He-4 in aerogel. Phys. Rev. Lett. **65**, 2567 (1990)
14. Tan, H.W., Beamish, J.R.: Liquid-vapor critical behavior of neon confined in aerogel. Physica B **284**, 389 (2000)
15. Chan, M.H.W., Blum, K.I., Murphy, S.Q., Wong, G.K.S., Reppy, J.D.: Disorder and the superfluid transition in liquid He-4. Phys. Rev. Lett. **61**, 1950 (1988)
16. Mulders, N., Chan, M.H.W.: Heat capacity measurements of He-3–He-4 mixtures in aerogel. Phys. Rev. Lett. **75**(3705) (1995)
17. Herman, T., Beamish, J.R.: Low frequency acoustic resonance studies of the liquid-vapor transition in silica aerogel. J. Low Temp. Phys. **141**, 193 (2005)
18. Melnichenko, Y.B., Wignall, G.D., Cole, D.R., Frielinghaus, H.: Density fluctuations near the liquid-gas critical point of a confined fluid. Phys. Rev. E **69**, 057102 (2004)
19. Melnichenko, Y.B., Wignall, G.D., Cole, D.R., Frielinghaus, H., Bulavin, L.A.: Liquid-gas critical phenomena under confinement: small-angle neutron scattering studies of CO_2 in aerogel. J. Mol. Liq. **120**, 7 (2005)
20. Anisimov, M.A., Kiselev, S.B., Sengers, J.V., Tang, S.: Crossover approach to global critical phenomena in fluids. Physica A **188**, 487 (1992)
21. Sinha, S.K.: Small-angle scattering from porous materials. In: Wong, P.-Z. (ed.) Methods in the Physics of Porous Media. Academic, London (1999)
22. Vacher, R., Woignier, T., Pelous, J., Courtens, E.: Structure and self-similarity of silica aerogels. Phys. Rev. B **37**, 6500 (1988)
23. Stanley, H.E.: Introducton to Phase Transitions and Critical Phenomena. Oxford University Press, New York, NY (1971)
24. Frisken, B.J., Cannell, D.S., Lin, M.Y., Sinha, S.K.: Neutron scattering studies of binary mixtures in silica gels. Phys. Rev. E **51**, 5866 (1995)
25. Tremblay, L., Socol, S.M., Lacelle, S.: Simple views on critical binary liquid mixtures in porous glass. Phys. Rev. E **61**, 656 (2000)
26. Wallace, D.J., Zia, R.K.P.: Eucledean group as a dynamical symmetry of surface fluctuations – planar interface and critical behavior. Phys. Rev. Lett. **43**, 808 (1979)
27. Liu, A.J., Durian, D.J., Herbolzheimer, E., Safran, S.A.: Wetting transitions in a cylindrical pore. Phys. Rev. Lett. **65**, 1897 (1990)
28. Strunz, P., Saroun, J., Keiderling, U., Wiedenmann, A., Przenioslo, R.: General formula for determination of cross-section from measured SANS intensities. J. Appl. Crystallogr. **33**, 829 (2000)
29. Cambedouzou, J., Diat, O.: Quantative small-angle scattering on mesoporous silica powders: from morphological features to specific surface estimation. J. Appl. Crystallogr. **45**, 662 (2012)

30. Bostain, D.A., Brenizer Jr., J.S., Norris, P.M.: Neutron radioscopic measurement of water adsorption coefficients in aerogels. Res. Nondestruct. Eval. **14**, 47 (2002)
31. Melnichenko, Y.B., Wignall, G.D., Cole, D.R., Frielinghaus, H.: Adsorption of supercritical CO_2 in aerogels as studied by small-angle neutron scattering and neutron transmission techniques. J. Chem. Phys. **124**, 204711 (2006)
32. Melnichenko, Y.B., Wignall, G.D.: Density and volume fraction of supercritical CO_2 in pores of native and oxidized aerogels. Int. J. Thermophys. **30**, 1578 (2009)
33. Melnichenko, Y.B., Wignall, G.D.: Small-angle neutron scattering in materials science: recent practical applications. J. Appl. Phys. Rev. **102**, 021101 (2007)
34. Tajiri, K., Igarashi, K., Nishio, T.: Effects of supercritical drying media on structure and properties of silica aerogel. J. Non-Cryst. Solids **186**, 83 (1995)
35. Humayun, R., Tomasko, D.L.: High-resolution adsorption isotherms of supercritical carbon dioxide on activated carbon. AIChE J. **46**, 2065 (2000)
36. Gao, W., Butler, D., Tomasko, D.L.: High-pressure adsorption of CO_2 on NaY zeolite and model prediction of adsorption isotherms. Langmuir **20**, 8083 (2004)
37. Schneider, M.S., Grunwaldt, J.D., Baiker, A.: Near-critical CO_2 in mesoporous silica studied by in situ FTIR spectroscopy. Langmuir **20**, 2890 (2004)
38. Stefanopoulos, K.L., Steriotis, T.A., Katsaros, F.K., Kanellopoulos, N.K., Hannon, A.C., Ramsay, J.D.F.: Structural characterization of subcritical carbon dioxide confined in nanoporous carbon by in situ neutron diffraction. J. Phys. Conf. Ser. **340**, 012049 (2012)
39. Steriotis, T.A., Stefanopoulos, K.L., Mitropoulos, A.C., Kanellopoulos, N.K., Hoser, A., Hofmann, M.: Structural studies of supercritical carbon dioxide in confined space. Appl. Phys. **A74**, S1333 (2002)
40. Steriotis, A., Stefanopoulos, K.L., Kanellopoulos, N.K., Mitropoulos, A.C., Hoser, A.: The structure of adsorbed CO_2 in carbon nanopores: a neutron diffraction study. Colloids Surf. A **241**, 239 (2004)
41. Rother, G., Melnichenko, Y.B., Cole, D.R., Frielinghaus, H., Wignall, G.D.: Microstructural characterization of adsorption and depletion regimes of supercritical fluids in nanopores. J. Phys. Chem. **111**, 15736 (2007)
42. http://www.nist.gov/srd/nist23.htm
43. Rother, G., Vlcek, L., Gruszkiewicz, M.S., Chialvo, A.A., Anowitz, L.M., Banuelos, J.L., Wallacher, D., Grimm, N., Cole, D.R.: Sorption phase of supercritical CO_2 in silica aerogel: experiments and mesoscale computer simulations. J. Phys. Chem. C **118**, 15525 (2014)
44. Ciccariello, S., Melnichenko, Y.B., He, L.: Supercritical carbon dioxide behavior in porous silica aerogel. J. Appl. Crystallogr. **44**, 43 (2011)
45. Ciccariello, S., Melnichenko, Y.B., He, L.: Phase behavior of carbon dioxide confined in silica aerogel in the vicinity of the bulk critical point. J. Phys. Chem. C **115**, 22336 (2011)
46. Melnichenko, Y.B., Ciccariello, S.: Small-angle neutron scattering study of deuterated propane adsorption in silica aerogel. J. Phys. Chem. C **116**, 24661 (2012)
47. Wilmer, C.E., Farha, O.K., Bae, Y.S., Hupp, J.T., Snurr, R.Q.: Structure-property relationships of porous materials for carbon dioxide separation and capture. Energ. Environ. Sci. **2**, 9849 (2012)
48. Melnichenko, Y.B., Mayama, H., Cheng, G., Blach, T.: Monitoring phase behavior of sub- and supercritical CO_2 confined in porous fractal silica with 85% porosity. Langmuir **26**, 6374 (2010)
49. Mayama, H., Tsujii, K.: Menger sponge-like fractal body created by a novel template method. J. Chem. Phys. B **125**, 124706 (2006)
50. Wignall, G.D., Littrell, K.C., Heller, W.T., Melnichenko, Y.B., Bailey, K.C., Lynn, G.W., Myles, D.A., Urban, V.S., Selby, D.L.: The 40 m general purpose small-angle neutron scattering instrument at Oak Ridge National Laboratory. J. Appl. Crystallogr. **45**, 990 (2012)
51. Pone, J.D.N., Halleck, P.M., Mathews, J.P.: Sorption capacity and sorption kinetic measurements of CO_2 and CH_4 in confined and unconfined bituminous coal. Energy Fuel **23**, 4688 (2009)

52. Bee, M.: Quasielastic Neutron Scattering. Adam Hilger, Bristol (1988)
53. Chathoth, S.M., Mamontov, E., Melnichenko, Y.B., Zamponi, M.: Diffusion and adsorption of methane in nano-porous carbon aerogel: a combined quasi-elastic and small-angle neutron scattering study. Micropor. Mesopor. Mater. **132**, 148 (2010)
54. Melnichenko, Y.B., Radlinski, A.P., Mastalerz, M., Cheng, G., Rupp, J.: Characterization of the CO_2 fluid adsorption in coal as a function of pressure using neutron scattering techniques (SANS and USANS). Int. J. Coal Geol. **77**, 69 (2009)
55. Mamontov, E., Herwig, K.W.: A time-of-flight backscattering spectrometer at the Spallation Neutron Source, BASIS. Rev. Sci. Instrum. **82**, 85109 (2011). doi:10.1063/1.3626214
56. Woessner, D.E., Snowden, B.S., George, R.A., Melrose, J.C.: Dense gas diffusion coefficients for methane-propane system. Ind. Eng. Chem. Fundam. **8**, 779 (1969)
57. Ohba, T., Omori, T., Kanoh, H., Kaneko, K.: Cluster structures of supercritical CH_4 confined in carbon nanospaces with in situ high-pressure small-angle X-ray scattering and grand canonical Monte Carlo simulation. J. Phys. Chem. B **108**, 27 (2004)
58. Losano-Castello, D., Cazorla-Amoros, D., Linares-Solano, A., Hall, P.J., Gascon, D., Galan, C.: In situ small angle neutron scattering study of CD_4 adsorption under pressure in activated carbons. Carbon **39**, 1343 (2001)
59. Gallego, N.C., He, L., Saha, D., Contescu, C.I., Melnichenko, Y.B.: Hydrogen confinement in carbon nanopores: extreme densification at ambient temperature. J. Am. Chem. Soc. **133**, 13794 (2011)
60. Debye, P., Anderson, H.R., Brumberger, H.: Scattering by an inhomogeneous solid. 2. The correlation function and its application. J. Appl. Phys. **28**, 679 (1957)
61. He, L., Melnichenko, Y.B., Gallego, N.C., Contescu, C.I., Guo, J., Bahadur, J.: Investigation of morphology and hydrogen adsorption capacity of disordered carbons. Carbon **80**, 82 (2014)
62. http://webbook.nist.gov/chemistry/fluid
63. Ihm, Y., Cooper, V.P., Peng, L., Morris, J.R.: The influence of dispersion interactions on the hydrogen adsorption properties of expanded graphite. J. Phys. Condens. Matter **24**, 424205 (2012)
64. Morris, J.R., Contescu, C.I., Chisholm, M.F., Cooper, V.R., Guo, J., He, L., Ihm, Y., Mamontov, E., Melnichenko, Y.B., Olsen, R.J., Pennycook, S.J., Stone, M.B., Zhang, H., Gallego, N.C.: Modern approaches to studying gas adsorption in nanoporous carbons. J. Mater. Chem. A **1**, 9341 (2013)
65. Bastos-Neto, M., Patzschke, C., Lange, M., Mollmer, J., Moller, A., Fichtner, S., Schrage, C., Lassig, D., Lincke, J., Staudt, R., Krautscheid, H., Glaser, R.: Assessment of hydrogen storage by physisorption in porous materials. Energ. Environ. Sci. **5**, 8294 (2012)
66. Zoback, M.D.: Reservoir Geomechanics. Cambridge University Press, Cambridge (2007)
67. White, C.M., Smith, D.H., Jones, K.L., Goodman, A.L., Jikich, S.A., LaCount, R.B., DuBose, S.B., Ozdemir, E., Morsi, B.I., Schroeder, K.T.: Sequestration of carbon dioxide in coal with enhanced coal bed methane recovery – a review. Energy Fuel **19**, 659 (2005)
68. Melnichenko, Y.B., Radlinski, A.P., Mastalerz, M., Cheng, G., Rupp, J.: Characterization of the CO_2 fluid adsorption in coal as a function of pressure using neutron scattering techniques (SANS and USANS). Int. J. Coal Geol. **77**, 6979 (2009)
69. Mahajan, O.P.: CO_2 surface area of coals – the 25-year paradox. Carbon **29**, 735 (1991)
70. Bond, R.L.: Capillary structure of coal. Nature **178**, 104 (1956)
71. Larsen, J.W., Wernett, P.: Pore structure of Illinois #6 coal. Energy Fuel **2**, 719 (1988)
72. Sakurovs, R., Day, S., Weir, S.: Relationships between the critical properties of gases and their high pressure sorption behavior on coals. Energy Fuel **24**, 1781 (2010)
73. Gethner, J.S.: The determination of the void structure of macroporous coals by small-angle neutron scattering: void geometry and structure in Illinois No-6 bituminous coal. J. Appl. Phys. **59**, 1068 (1986)
74. Hall, P.J., Antxustegi, M., Ruiz, W.: Contrast-matching small-angle neutron scattering evidence for the absence of a connected pore system in Pittsburgh No. 8 coal. Fuel **77**, 1663 (1998)

75. Antxustegi, M.M., Hall, P.J., Calo, J.: Development of porosity in Pittsburgh No. 8 coal char as investigated by contrast-matching small-angle neutron scattering and gas adsorption techniques. Energy Fuel **12**, 542 (1998)
76. Hall, P.J., Brown, S.D., Calo, J.M.: The pore structure of the Argonne coals as interpreted from contrast matching small angle neutron scattering. Fuel **79**, 1327 (2000)
77. Melnichenko, Y.B., He, L., Sakurovs, R., Kholodenko, A.L., Blach, T., Mastalerz, M., Radliński, A.P., Cheng, G., Mildner, D.F.R.: Accessibility of pores in coal to methane and carbon dioxide. Fuel **91**, 200 (2012)
78. Sakurovs, R., He, L., Melnichenko, Y.B., Radlinski, A.P., Blach, T., Lemmel, H., Mildner, D. F.R.: Pore size distribution and accessible pore size distribution in bituminous coals. Int. J. Coal Geol. **100**, 51 (2012)
79. He, L., Melnichenko, Y.B., Mastalerz, M., Sakurovs, R., Radlinski, A.P., Blach, T.: Pore accessibility by methane and carbon dioxide in coal as determined by neutron scattering. Energy Fuel **26**, 1975 (2012)
80. Mastalerz, M., He, L., Melnichenko, Y.B., Rupp, J.A.: Porosity of coal and shale: insights from gas adsorption and SANS/USANS techniques. Energy Fuel **26**, 5109–5120 (2012)
81. Ruppert, L.F., Sakurovs, R., Blach, T.P., He, L., Melnichenko, Y.B., Mildner, D.R.F., Alcantar-Lopez, L.: A USANS/SANS study of the accessibility of pores in the Barnett scale to methane and water. Energy Fuel **27**, 772 (2013)
82. Clarkson, C.R., Solano, N., Bustin, R.M., Bustin, A.M.M., Chalmers, G.R.L., He, L., Melnichennko, Y.B., Radlinski, A.P., Blach, T.P.: Pore structure characterization of North American chale gas reservoirs using SANS/USANS, gas adsorption, and mercury intrusion. Fuel **103**, 606 (2013)
83. Bahadur, J., Melnichenko, Y.B., Mastalerz, M., Furmann, A., Clarkson, C.R.: Hierarchical pore morphology of Cretaceoius chale: a small-angle neutron scattering and ultrasmall-angle neutron scattering study. Energy Fuel **28**, 6336 (2014)
84. Sakurovs, R., Ralinski, A.P., Melnichenko, Y.B., Blach, T., Cheng, G., Lemmel, H., Rauch, H.: Stability of the bituminous coal microstructure upon exposure to high pressures of helium. Energy Fuel **23**, 5022 (2009)
85. Hainbuchner, M., Villa, M., Kroupa, G., Bruckner, G., Baron, M., Amenitsch, H., Seidl, E., Rauch, H.: The new high resolution ultra small-angle neutron scattering instrument at the High Flux Reactor in Grenoble. J. Appl. Crystallogr. **33**, 851 (2000)
86. Melnichenko, Y.B., et al., unpublished
87. Tvardovskiy, A.V.: Sorbent Deformation. Elsevier, Oxford (2007)
88. Yang, K., Lu, X., Lin, Y., Neimark, A.V.: Deformation of coal induced by methane adsorption at geological conditions. Energy Fuel **24**, 5955 (2010)
89. Striolo, A., Chialvo, A.A., Cummings, P.T., Gubbins, K.E.: Water adsorption in carbon-slit nanopores. Langmuir **19**, 8583 (2003)
90. Porod, G.: General theory. In: Glatter, O., Kratky, O. (eds.) Small-Angle X-Ray Scattering. Academic, New York, NY (1982)
91. Higgins, J., Benoit, H.: Polymers and Neutron Scattering. Clarendon, Oxford (1994)
92. Radlinski, A.P., Boreham, C.J., Lindner, P., Randl, O., Wignall, G.D., Hinde, A., Hope, J.M.: Small angle neutron scattering signature of oil generation in artificially and naturally matured hydrocarbon source rock. Org. Geochem. **31**, 1 (2000)
93. Krivoglaz, M.: Theory of X-Ray and Thermal Neutron Scattering by Real Crystals. Plenum, New York, NY (1969)

Index

A

Absolute scale, 76, 83–85, 211, 268, 277, 280, 292, 297
Absorption, 11–12, 16
Accessible pores, 141, 189, 244, 289–291, 293–296, 300–305
Activated carbon, 94, 155–158, 210, 263, 275–282, 289
Adsorbate, 174, 181, 194, 198, 253
Aerogels, 148–149, 162–165, 209, 212–221, 226
Alloys, 166
Aluminum, 9
Amplitude, 8, 80, 96, 98, 109, 154, 181, 182, 198, 225, 239, 257
Angle, 17, 18, 223
Angstroms, 7, 35, 51, 75, 194, 222
Aperture, 40, 71, 73–75, 83, 84
Argon, 189, 297–300
Autocorrelations, 110, 167, 278

B

Background
  scattering, 41, 65, 66, 79, 81, 82, 155, 190, 217, 244
  subtraction, 5, 52, 69, 76, 93, 116, 156, 190, 229
Beam
  incident, 4, 5, 11, 40, 41, 43, 71, 83, 85, 109, 200, 226
  scattered, 4, 6, 63, 207
Beamtrap, 72, 74–76, 78
Beryllium, 22, 23, 35, 64
Bragg law, 7, 134

C

Calcination, 150, 152
Calibration, 83–85
Capillary condensation, 174, 175
Carbon, 9, 142, 158–165, 225, 236
Carbonaceous material, 140, 143, 189–194
Carbon aerogels, 162, 225
Cavitation, 217–232
Ceramics, 166
Chord, 113, 119, 177
CO_2, 8, 16, 52, 54, 89, 95–98, 100, 101, 103, 142, 146, 167, 194–197, 205, 208–225, 228, 230, 232, 236–245, 247, 252–254
Coals, 142, 195–197, 225, 236–249, 251, 252
Coherent scattering, 152
Collimation, 39, 40, 45, 52, 76
Composite material, 130, 166
Compressibility, 14, 254, 256
Computer simulations, 239, 253, 267, 303
Concentrations, 14, 79, 93, 118, 142, 162, 199, 200, 207, 233, 236, 271, 282
Configuration, 71–75
Contrast, 7–11, 92–98, 101, 130–131, 155, 178, 180–187, 189, 233
Contrast matching, 93, 94, 96, 101, 155, 189, 233
Contrast variation, 10, 93, 96, 98, 178
Controlled porous, 143
Core shell model, 83, 225, 227
Correlation
  function, 109–112, 114, 115, 125, 130, 132, 167, 232
  length, 115, 116, 120, 125, 147, 155, 177, 180, 208, 210–212, 232–234, 245, 246
CO_2 sequestration, 236–242

Y.B. Melnichenko, *Small-Angle Scattering from Confined and Interfacial Fluids*, DOI 10.1007/978-3-319-01104-2

Critical temperature
liquid-gas, 148, 174, 205, 209, 210, 214, 220
liquid-liquid, 211, 232
Crossover, 8, 149, 209, 256
Cross section, 65

D
Dark current, 76, 83–84
Debye-Bueche function, 110, 115
Debye formula, 278
Deformation, 63, 154, 183–184, 252–254, 298–300
Densities, 7–11, 100–104, 208–212, 236
Density profile, 131, 240, 241
Desmearing, 49
Detectors, 11, 76–77
Detector sensitivity, 76–77
Deuterated methane, 227
Deuteration, 9, 94, 178, 234, 273, 290, 291
Deuterium labeling, 3
Diamonds, 9, 16, 65, 66, 227
Differential scattering cross section, 6–8, 83, 110
Diffraction, 17, 18, 99, 224
Diffuse interface, 121
Diffuse scattering, 65, 244
Diffusion coefficient, 194, 271
Disc scattering, 136
Disorder, 240–245
Distance, 7, 40, 74, 111, 123, 134, 234, 241, 253, 255, 260, 268, 277, 279, 284, 292, 297, 299
Distribution, 14, 37, 39, 49, 85, 87, 88, 111, 112, 126, 135, 220, 229, 254–255, 261, 262, 268, 271, 289, 290, 293, 295–299
D_2O, 9, 15, 19, 27, 45, 46, 78, 86, 87, 93, 96, 101, 183, 189–193, 206–209, 211, 212, 223–227, 234, 238–240, 244
Dynamic range, 72

E
Elastic neutron scattering, 3
Electrode, 192, 193, 205–210
Electrolytes, 205–213
Electron density, 10, 93, 110, 176, 187, 188, 220
Electron density distribution, 126
Energy
of neutron, 3, 5, 12, 15, 19, 23, 37
of x-ray photons, 3, 11, 195
Energy storage, 205
Equation of state, 264
Excess adsorption, 215

F
Fluctuations, 208–212
Flux, 15, 16, 19, 24, 28, 49, 233, 253
Fourier transformation, 110, 124, 301
Fractal
dimension, 89, 122–124, 126, 148, 150, 167, 181, 210

G
Gas, 208–212
Glass, 143, 147–148, 217
Gravimetry, 131
Guinier approximation, 114, 117, 118, 164
Guinier plots, 86, 118, 119, 157, 159

H
Helium, 250
Heterogenuity, 177, 187
Hexagonal, 96, 140, 144, 152, 165, 182, 211, 229
H_2O, 9, 15, 19, 27, 78, 85–87, 93, 96, 101, 183, 192, 193, 206, 208, 209, 211, 212, 223–225, 227, 238–240, 244
Hydrocarbons, 213–216
Hydrogen (H_2), 27, 158, 206, 207, 230–236
Hysteresis, 174, 180, 187, 254

I
Inaccessible pores, 290, 293, 295, 305
Incoherent scattering, 84
In-situ measurements, 59, 183, 208, 209, 279, 299
Intensity
at zero angle, 96, 208, 210, 247
Ion, 206–210
Ionic liquids, 210
Isotherm, 232
Isotopic substitution, 92, 94

K

Kelvin equation, 173, 175

L

Lattice, 152, 183, 211, 212, 257, 258
Layer thickness, 38, 89, 128
Light source, 31, 65, 66, 200, 220, 229
Log-normal distribution, 88, 160

M

Macropores, 165, 173, 191, 196, 197, 253, 263, 293, 298
Maxwell distribution, 37, 49
Mass
  pore, 90, 91, 123, 125
  surface, 125, 149, 150
MCM-41, 140, 144, 152, 154, 155, 165, 181–183, 198–199, 211, 238, 239, 241
Membranes, 165, 199–201
Mercury porosimetry, 146, 166, 289–290
Mesopores, 232–237
Methane, 225–229
Mobility, 271, 273, 275
Moderator, 27
Momentum, 4, 255, 260
Monochromatization, 38
Monodisperse, 89, 94, 112, 116, 134, 135, 174, 268
Multiple scattering, 79

N

Nanoparticles (NP), 14
Neutron guide, 38, 40, 44, 45
Neutron scattering, 2, 3, 9, 15, 46, 181, 182, 193
Nitrogen, 184
Number-average, 119

O

Oil, 213–216

P

Parseval's formula, 301–302
Phase diagram, 232, 241, 243, 254
Photon, 29–31, 195, 217
Polydisperse spherical model, 122–127
Polydispersity, 13, 87, 89, 135, 184
Polymer monolith, 146, 165–166
Pores, 98, 122–127, 147, 155, 174, 243–249
Porod law, 13, 114, 116, 119, 121, 125, 126, 133, 148, 155, 192, 209, 210, 219, 235
Porosimetry, 173–176
Porosity, 142, 254–259
Position sensitive detector, 41, 82
Power law, 120, 122, 190, 191, 213, 214, 239
Pressure, 59–63, 63–67, 228, 250–254
  relative, 59, 173–176, 180, 182, 183, 185–188, 193
Pressure cell, 54, 59–67, 229, 255, 268, 272, 277, 279–280, 284, 297, 299
Pycnometer, 90

Q

Quartz, 9, 102, 103
Quasi elastic neutron scattering (QENS), 3, 194, 271–273, 275

R

Radiation, 16, 30, 235
Radius of gyration, 75, 117–119, 124, 156
Random phase approximation, 112
Reactors, 15, 19–23, 28, 29, 42, 49, 233, 253
Reciprocal space, 134, 303
Resolution, 39, 47, 48, 85–89
Rocks, 167, 213–216
Rod scattering, 122

S

Sample
  effective thickness, 69, 89–92
Sandstone, 101, 141
Sapphire window, 245, 59–62
SBA-15, 96, 140, 144, 150, 152, 153, 165, 182–184, 187, 188, 211
Scattered intensity
  desmeared, 49, 52
  smeared, 49, 52, 79, 88
  at zero angle, 14, 49, 93, 150, 216
Scattering, 1–7, 7–11, 14–18, 28, 32, 45, 49, 75, 77–79, 82, 100–104, 113–115, 118–127, 122–131, 134, 159, 173–176, 181, 194, 209, 214, 219, 223, 231–233, 235, 238, 246, 248, 251
Scattering function, 93, 120–121, 127–129, 162, 227, 273
Scattering length, 2, 9, 246
Scattering length densities, 2, 246

Scattering vector, 4–7, 43, 60, 83, 134, 147, 181, 234, 268, 274, 277–280, 284, 285, 287, 288, 292, 293, 297, 299, 305
Schultz distribution, 88
Shales, 103, 142, 243–249
Shape, 117
Silica, 143, 147–155, 148–149, 149–150, 150–155, 180–189, 212–224, 212–221, 221–224
Solvent, 10, 11, 79, 92, 93, 101, 148, 166, 175, 176, 180, 217, 251, 290
Sorption, 194–201, 195–197
Source
  light, 66, 200, 220, 229
  neutron, 19, 20, 23, 25, 28, 38, 73
  synchrotron, 19, 30, 31, 54, 67, 70, 93, 194, 220, 229
  x-ray, 15, 19, 30, 31
Spallation source, 20, 28, 43
Sphere scattering, 112, 126, 135
Structural stability, 252
Structures, 12–14, 17, 18, 51, 167, 176–180, 222
Structure factor, 12–14, 112, 123, 147, 300, 305
Subcritical fluid, 252, 253
Supercooled water, 238–240
Supercritical fluid, 95, 140, 189, 251–305
Surface roughness, 148, 185, 192
Surface
  specific, 100, 115, 116, 119, 127, 140, 152, 162–164, 167, 183, 187, 211
Synchrotron, 16, 31–33, 67, 70, 187–189, 195, 217, 220, 229, 235

T

Target, 27
Temperature, 209, 234, 238, 243
Thickness, 77–79, 89–92, 221
Transition
  order-disorder, 240–245
Transmission, 11–12, 39, 77–79, 215
Two-phase approximation, 98, 227, 290

U

Undulators, 30
USANS, 7, 16, 24, 27, 35, 46–50, 50–51, 57, 60, 73, 96, 104, 126, 148, 150, 166, 213, 216, 238, 239, 242, 244, 246, 248–252, 259
USAXS, 7, 16, 35, 46, 47, 51–54, 59, 104

V

Vanadium, 23, 84–86
Van der Waals, 263, 282, 300
Vapor, 180–189, 193, 198–199, 199–201
Velocity selectors, 38
Voids, 90, 91, 177, 181
Volume fractions, 134, 220, 250
Vycor porous glass, 96, 97, 134, 233

W

Water, 9, 58, 199–201, 238–240
Wavelength
  distribution, 37, 85, 87

X

X-ray scattering, 1, 3, 7, 14–16, 70, 93, 173, 218, 221, 300

Z

Zeolites, 143

www.ingramcontent.com/pod-product-compliance
Lightning Source LLC
Chambersburg PA
CBHW071425230926
51846CB00001B/128
* 9 7 8 3 3 1 9 3 4 6 4 6 5 *